Glimpses of CYANOBACTERIA

Glimpses of CYANOBACTERIA

– Editors –

Dr. Rajan Kumar Gupta
Reader at the Department of Botany
Govt. P.G. College
Rishikesh - 249 201, Dehradun (Uttarakhand)

Dr. Mukesh Kumar
Reader at the Department of Botany
Sahu Jain P.G. College,
Najibabad - 246 763, Bijnor (U.P.)

Professor G.S. Paliwal
Senior Consultant,
Regional Centre of the
National Afforestation and Eco-development Board,
Ministry of Environment and Forests,
Government of India, New Delhi

2018

Daya Publishing House®

A Division of

Astral International Pvt. Ltd.

New Delhi - 110 002

First Published, 2006
Reprinted, 2018

ISBN: 978-81-7035-435-2 (Hardbound)

Publisher's note:

Every possible effort has been made to ensure that the information contained in this book is accurate at the time of going to press, and the publisher and author cannot accept responsibility for any errors or omissions, however caused. No responsibility for loss or damage occasioned to any person acting, or refraining from action, as a result of the material in this publication can be accepted by the editor, the publisher or the author. The Publisher is not associated with any product or vendor mentioned in the book. The contents of this work are intended to further general scientific research, understanding and discussion only. Readers should consult with a specialist where appropriate.

Every effort has been made to trace the owners of copyright material used in this book, if any. The author and the publisher will be grateful for any omission brought to their notice for acknowledgement in the future editions of the book.

Published by : **Daya Publishing House®**
A Division of
Astral International Pvt. Ltd.
– ISO 9001:2008 Certified Company –
4736/23, Ansari Road, Darya Ganj
New Delhi-110 002
Ph. 011-43549197, 23278134
E-mail: info@astralint.com
Website: www.astralint.com

Preface

Algae are the architects of earth atmosphere, they are responsible for the atmosphere we are enjoying. Cyanobacteria, a large and diverse group of photosynthetic prokaryotes, have been the subject of both basic and applied research over the years. Their peculiar features and potential applications in agriculture, aquaculture, bioremediation, bioenergy, human nutrition and pharmaceutical industries have attracted the attention of workers from diverse fields. Cyanobacterial biomass production has assumed industrial dimension. In recent years, we are witnessing interdisciplinary research in cyanobacteriology.

Different aspects of cyanobacteria and their potential biotechnological applications are being revealed by workers world over. India is richly endowed with the cyanobacterial flora, has long tradition of cyanobacterial research. Cyanobacteria form a significant part of the subjects like Botany, Microbiology, Agriculture and Biotechnology being taught in the Indian Universities.

The rapid advances and developments in cyanobacteriology prompted us to assemble the up-to-date information in the form of present book targeted to both post graduate students and researchers. This book consists of 20 Chapters which encompasses topics on cyanobacterial ecology, toxicology, taxonomy, morphology, physiology, stress responses, biotechnology and molecular biology. Various information incorporated in the book by authors who are internationally acknowledged experts in the field of phycology. The book with the intention of providing a sufficient depth of the subject to satisfy the needs at a level which will be comprehensive and interesting. We hope that this book would be informative to the students, teachers, scientists and researchers in the field of basic and applied cyanobacteriology.

The authors wish to thank and give appreciations to the scientists whose contributions have enriched this volume. We wish to express our sincere thanks to Dr. V.D. Pandey for the help extended by him, we also express our deep sense of gratitude to our parents whose blessings have always prompted us to pursue academic activities deeply. Last but not least we are also thankful to Mr. Anil Mittal, Publisher of Daya Publishing House, New Delhi for publishing this book. Finally we will be always remain debtor to all our well wishers for their blessings without which this book would not have come to light.

Contents

List of Contributors

A. Khare
Department of Botany, Sahu Jain P.G. College, Nazibabad, Bijnore – 246 763, U.P.

A.K. Kashyap
Centre of Advanced Study in Botany, Banaras Hindu University, Varanasi – 221 005, U.P.

A.K. Mishra
Centre of Advanced Study in Botany Banaras Hindu University, Varanasi – 221 005, U.P.

Anuradha Tiwari
Algal Biotechnology Laboratory, Department of Biological Sciences, Rani Durgavati University, Jabalpur – 482 001, M.P.

Archita Ganguly
Algal Biotechnology Laboratory, Department of Biological Sciences, Rani Durgavati University, Jabalpur – 482 001, M.P.

B.V. Singh
Centre for Conservation and Utilization of Blue-Green Algae, Indian Agricultural Research Institute, New Delhi – 110 012

D. Vyas
Department of Botany, Dr. H.S. Gaur University, Sagar, M.P.

G.S. Paliwal
Regional Centre of the National Afforestation and Eco-development Board, Ministry of Environment and Forests, Government of India, New Delhi

H.D. Kumar
Centre of Advanced Study in Botany, Banaras Hindu University, Varanasi – 221 005, U.P.

I. Habib
Department of Botany, Govt. P.G. College, Sitapur – 261 125

J. Tirkey
P.G. Department of Botany and Biotechnology, Utkal University, Bhubaneshwar – 751 004, Orissa

K.D. Pandey
Centre of Advanced Study in Botany, Banaras Hindu University, Varanasi – 221 005, U.P.

M. Kumar
Department of Botany, Sahu Jain P.G. College, Nazibabad, Bijnore – 246 763, U.P.

M. Masroor A. Khan
Centre for Advanced Study in Botany, Aligarh Muslim University, Aligarh – 202 002, U.P.

M.K. Chaudhary
Department of Botany, Sahu Jain P.G. College, Nazibabad, Bijnore – 246 763, U.P.

Mihir K. Das
P.G. Department of Botany, G.M. College, Sambalpur – 768 004, Orissa

P. Jaiswal
Centre for Conservation and Utilization of Blue-Green Algae, Indian Agricultural Research Institute, New Delhi – 110012

R.C. Gupta
Department of Botany, Kumaun University, SSJ Campus, Almora

R.K Gupta
Department of Botany, Govt. P.G. College, Rishikesh – 249 201, U.A.

R.K. Gupta
Centre for Conservation and Utilization of Blue-Green Algae, Indian Agricultural Research Institute, New Delhi – 110 012

S. Roy
Centre of Advanced Study in Botany, Banaras Hindu University, Varanasi – 221005, U.P.

S. Saxena
Centre for Conservation and Utilization of Blue-Green Algae, Indian Agricultural Research Institute, New Delhi – 110012

S.K. Singh
Centre of Advanced Study in Botany Banaras Hindu University, Varanasi – 221 005, U.P.

S.P. Adhikary
P.G. Department of Botany and Biotechnology, Utkal University, Bhubaneshwar – 751 004, Orissa

S.S. Singh
Centre of Advanced Study in Botany, Banaras Hindu University, Varanasi – 221 005, U.P.

Shalini
Centre for Conservation and Utilization of Blue-Green Algae, Indian Agricultural Research Institute, New Delhi – 110 012

Surendra Singh
Algal Biotechnology Laboratory, Department of Biological Sciences, Rani Durgavati University, Jabalpur – 482 001, M.P.

U. Pandey
Department of Botany, Guru Nanak Girls P.G. College, Udaipur – 313 001, Rajasthan

U.K. Chaturvedi
Department of Botany, Bareilly College, Bareilly – 243 005, U.P.

V. Prasad
Department of Botany, Thakur Ram Multiple Campus, Tribhuvan University, Birgunj, Nepal

V.D Pandey
Department of Botany, Govt. P.G. College, Rishikesh – 249 201, U.A.

Y. Singh
Department of Botany, Govt. Degree College, Bhadohi, Sant Ravidasnagar, U.P.

Y.V. Singh
Centre of Conservation and Utilization of Blue-Green Algae, Indian Agricultural Research Institute, New Delhi – 110 012

Chapter 1

Cyanobacteria in the Soil Crusts and their Significance

J. Tirkey and S.P. Adhikary

Contents

Glimpses of Cyanobacteria, 2006
Daya Publishing House, New Delhi

ABSTRACT

Biological Soil crust is a complex mosaic of living organisms formed by an intimate association between soil particles and cyanobacteria, microalgae, lichens and bryophytes. These are common worldwide in arid and semi-arid shrublands, grasslands and woodlands. They prevent soil erosion, increase water holding capacity of soil, reduce evaporative water loss, and increase carbon and nitrogen levels of the soil. Due to their effective adaptive and tolerance mechanisms against various abiotic stresses, cyanobacteria constitute the dominant organisms in many biological crusts.

1. Introduction

Ever walked across the rangeland and heard a little crunch under your feet? If so, you have just discovered the biological soil crust. It is a complex mosaic of living organisms *i.e.* algae, Cyanobacteria (blue-green algae), bacteria, lichens, liverworts, mosses and fungi, that grow on or just below the soil surface. Biological soil crusts are common worldwide in arid and semi-arid shrublands, grasslands and woodlands.

Biological crust on soil result from an intimate association between soil particles and Cyanobacteria, algae, microfungi, lichens and bryophytes (in different proportions), which live within or immediately on top of the uppermost millimeters of soil. Binding of these organisms with soil particles is because of their by-products (extracellular carbohydrate matrix) forming a surface layer is a common phenomenon in these crusts and such layers are sometimes called "films" or "biofilms"(*e.g.* Büdel, 1999; Castenholz and Garcia-Pichel, 2000). Biological soil crusts are highly variable in appearance. Which organisms dominate the crusts is determined by several factors including chemical and physical characteristics and weather patterns. Biological soil crusts are distinguishable from bare soil by a bumpy appearance, forming sort of a mini-landscape on the soil surface. They tend to be dark in colour especially when dry. However, add a little moisture and it become vivid greens, yellows or reddish-brown as the mosses come out of dormancy and spring to life.

Biological soil crusts are also known as cryptogramic, macrobiotic, cryptobiotic and microphytic crusts, which leads to some confusion. The names are all meant to indicate common features of the organisms that compose the crusts as this distinguish them from physical crusts while not limiting crust components to plants. Whatever be name used, there remains an important distinction between these formations and physical or chemical crusts.

2. Biological Crusts as Part of Arid Ecosystems

These are assemblages of tiny, often microscopic organisms, such as Cyanobacteria, green algae, fungi, lichens, and mosses, living on or just beneath the soil surface. Although inconspicuous, biological crusts are critical to the productivity of many arid land ecosystems and in some places account for 70 per cent of the living plant cover on soils. Unfortunately, the important role of biological crusts has been unnoticed or ignored by many people, including most range managers and livestock-grazing proponents. Traditionally, only the impact of livestock grazing on vascular plants has been a concern in evaluation of rangeland health. Yet recent research suggests that even if vascular plant communities are not affected in any detectable way by livestock, there can be significant differences between grazed and ungrazed sites in the proportion of ground covered by biological crust. And over time, livestock damage to biological crusts can lead to the declining health of the entire ecological system from increased soil erosion, diminished water-holding capacity of the soil, and less favorable nutrient flow, to greater vulnerability to invasion by exotic plants.

Biological crusts are native elements of most Western public lands. As a group they are amazingly diverse and often account for a far greater number of species than the vascular plants with which they are associated. For example, in Southern Idaho, 16 vascular plant species and 39 biological soils crust species in 140 plots placed throughout the rangeland plant community has been found (Rosenstreter, 2001). Biological crusts help to hold the soil surface together and thus reduce soil erosion from wind and water. They play an important role in reducing the impact of raindrops on unprotected soils (lacking biological crusts). Heavy rain breaks up soil aggregates, which leads to the clogging of soil pores and reduces water infiltration rates. The crusts create small-scale roughness or depressions in the surface of the soil that catch water, allowing it to infiltrate, thus reducing soil erosion. Some biological crusts have microfilaments that weave soil particles together, again anchoring the soil against erosion. Biological soil crusts also act as mulch, reducing evaporative water losses. Some biological crusts capture and fix atmospheric nitrogen and all of them can contribute to carbon fixation, providing an important source of carbon for microbial soil populations. Since nitrogen and carbon are both limiting factors in arid environment, maintaining normal nitrogen cycles and carbon deposition is critical to soil fertility and prevention of desertification. Vascular plants growing in soils with intact biological crusts have been found to have a higher concentration of nitrogen than plants growing in soils lacking such crusts. By occupying the spaces between perennial plants, biological crusts also prevent the establishment and spread of exotic weeds. Most native perennials found in North American deserts tend to have seeds with self-burial mechanisms or rodents–ensuring that they will be covered by soil or plant litter and will be able to germinate. However, the seeds of most exotic species, such as cheat grass, do not use these strategies; rather they germinate on the soil surface. Where biological crusts are intact, seeds of exotics generally fail to germinate successfully. Indeed, the loss of crusts in the bunchgrass communities of the Intermountain West may be largely responsible for the widespread establishment of cheat grass and other exotic annuals (Johansen, 1993).

Most biological crusts are dark and can raise temperatures as much as 23 degrees Fahrenheit above that of adjacent surfaces. Heightening soil temperature can increase nutrient uptake and speed seed germination, photosynthetic rates, and nitrogenase activity for associated vascular plants. Higher temperatures may be critical in many desert environments since soil moisture is typically higher during the cooler fall, winter and spring months, and biological activity may be dependent on favorable soil temperature and moisture (Johansen and Rushforth, 1986). When the dark-colored biological soil crusts are eliminated, the result can be lowered biological activity, with green-up pushed back to later in the spring and early summer. This can negatively affect vascular plants, since they are usually limited by soil moisture, and soils generally dry out as the season progresses into the warmer months.

Biological crusts also play a role in moderating fire frequency and intensity. Native plants in the most arid parts of the West are naturally widely spaced, and fires usually do not carry far because of the discontinuous and patchy distribution of fuels. Biological crusts occupy the open spaces between the larger plants–impeding the establishment of exotics, such as cheat grass, which allow fires to carry farther and also increase fire frequency. So long as the crust help maintain these mini firebreaks, fires are slowed, and their intensity is decreased. Furthermore, under low-intensity blazes, soil crusts remain intact, limiting potential erosion that may occur in the aftermath of a fire (Belnap, 1993).

Biological soil crusts are formed by living organisms activities and their byproducts, coating a crust of soil particles bound together by organic materials. Whereas chemical and physical crusts are inorganic features such as salt crusts or platy surface crusts, often formed by trampling. Living organisms and their by-products, creating a surface crust of soil particles bound together by organic materials, form crusts. Above ground crust thickness can reach up to 10 cm. The general appearance of the crusts in terms of color, surface topography, and surficial coverage varies widely. The color is due in part to the density of the organisms and to the often-dark color of Cyanobacteria, lichens, and mosses. Crusts generally cover all soil spaces not occupied by vascular plants, which may be 70 per cent or more of the living cover crusts are characterized by their marked increase in surface topography, often referred to as pinnacles or pedicles. The process of creating surface topography, or pinnacling, is due largely to the presence of filamentous Cyanobacteria and green algae. These organisms swell when wet, migrating out of their sheaths. After each migration new sheath material is exuded, thus extending sheath length. Repeated swelling leaves a complex network of empty sheath material that maintains soil structure after the organisms have dehydrated and decreased in size. Frost heaving, subsequent uneven erosion, and lack of surface plant roots results in high pedicles.

3. Crust Types and Distribution

Biological soil crusts can be classified into several broad types based on habitat conditions, taxonomic composition, physical appearance and functions. The species composition of soil crusts is still not adequate for a large-scale flouristic, phyto-sociological grouping of soil crusts communities. There have been several suggestions for crusts classification using the growth forms of dominant species, soil substrate and development process as the criterion (West, 1990). Soil algal growth types were also described by Komaromy (1979). Following Johansen (1993), Evan and Johansen (1999) typify crust formation encountered in the western United States. For the rangelands of Australia, Eldrige and Greene (1994) identified three forms of biological crusts: hypermorphs (above ground), perimorph (at ground) and cryptomorph (hidden ground). Based on these earlier definitions, Belnap (2001) proposed a topographic classification of crust types that facilities a worldwide comparison of crusts morphology. Four crusts categories were used: smooth, rugose, pinnacled and rolling. Smooth crusts are found in hot deserts where soil do not freeze and are almost exclusively containing endephic cyanobacteria, algae and fungi. The other three crusts categories generally have epedaphic colonizer (lichen and mosses) in addition to endephic autotrophs and fungi. Rugose crusts have low surface roughness (<2cm), generally from scattered lichen and/or moss clumps. Under moisture conditions in temperate regions, an epedic layer of filamentous algae can also dominate rugose crusts. Pinnacled and rolling crusts occur only where soils freeze in winter. Pinnacled crusts are dominated by Cyanobacteria, but locally can have up to 40 per cent moss-lichen cover. Pinnacled range up to 15cm high. Rolling crusts occur in regions where high precipitation results in heavy lichen-moss cover generally about 5 cm high. This classification is generalized and intermediate crust types connect the four categories. The precise definition of the crusts types associated with soil has been presented in Table 1.1.

Biological crust characteristics are quite variable. Above ground crust thickness can be up to about 10 cm (about 4 inches), but is generally much less. Color, surface topography and surface coverage vary. The dark color is due in part to the density of the organisms and to the often-dark color of the cyanobacteria, lichens, and mosses. The presence or absence of a crust is partly determined by soil texture and conductivity, pH, moisture and possibly temperature. Crust soil surface coverage varies from less than 10 per cent to nearly 100 per cent . Some crusts increase surface topography by forming pinnacles or pedicles. Other crusts are merely rough or even smooth and flat. The process of creating surface topography, or pinnacling, is due largely to the presence of filamentous Cyanobacteria and green algae. They swell when wet and migrate out of their sheaths. After each migration, new sheath material is exuded, thus extending sheath length. Repeated swelling leaves a complex network of empty sheath

Table 1.1: Definitions of Crusts Types Associated with Soil (Edephic)

Author	*Term*	*Descriptions*
Friedmann *et al.* (1967)	Epedaphic	Cyanobacteria, algae, lichens; surface of the soil
	Endephic	Cyanobacteria, algae; beneath the surface of the soil
Komaromy (1976)	Dispers	Mainly green algae, but also Cyanobacteria, inside the soil
	Stratose	Filamentous green algae like *Zygogonium* or *Klebsormidium*; soil surface
	Microcoleus-type	Filamentous non-branched Cyanobacteria, *e.g. Microcoleus* spp. Inside the soil
	Ramose	Filamentous branched Cyanobacteria; inside the soil surface
	Mucose	Unicellular and filamentous Cyanobacteria and algae with strong production of mucilage; soil surface.
	Glutinose	Mainly mucilaginous diatoms; soil surface
Eldrige and Green (1994)	Hypermorph	Above ground
	Perimorph	At gound level
	Crytomorph	Hidden below ground
Belnap (2001)	Smooth	Cyanobacteria, algae; inside the soil; surface roughness, 0- 1cm
	Rugose	Lichens and mosses above the soil surface (in addition to organisms inside the soil); roughness 1-3 cm; no frost heave.
	Rolling	Lichen and mosses above the soil surface (in addition to organisms inside the soil); roughness 1-5 cm, frost heave.
	Pinnacled	Cyanobacteria, algae, etc. inside the soil with or without lichens and mosses above ground; roughness 3-10 cm; frost heave

material that maintains soil structure after the organisms have dehydrated and decreased in size. Frost heaving can leave mounds bound by crust organisms. Crust cover is inversely related to vascular plant cover.

In other words, crusts do not compete well with higher plants, so they are more common where there are fewer plants. When all crust types are considered, cover is greatest at lower elevation, inland sites (<3000 feet) compared to mid-elevation sites (3000 to 7500 feet). However, relative lichen and moss cover increases with elevation and effective precipitation until vascular plant cover precludes their growth.

Shallow soils often support a wide variety of Cyanobacteria, lichens, and mosses regardless of soil texture. Soil texture does influence species composition. More stable fine-textured soils (such as gypsum and silty loams) support greater cover and more varied populations of Cyanobacteria, lichens, and mosses, than less stable, coarse-textured soils. Severely, newly, or frequently disturbed soils are generally dominated by low successional, large filamentous Cyanobacteria. When disturbance is less severe, less frequent, or some time has elapsed since disturbance, crusts are generally in a mid-successional state with some mosses or lichens present. Predominant crusts for the Great Basin are mosses and lichens. Predominant crusts for the Columbia Basin are tall mosses and green algae.

Biological soil crusts occurs in different substrata *e.g.* soil surface, desert soils, bark of trees, inselbergs etc. Several authors in different regions of globe has described the worldwide distribution of soil crust communities *viz.*, Rosenstreter and Belnap (2001) for North America, Büdel (2001a,b, c) for South America, European and Mediterranean regions and same parts of Asia, Hansen (2001) for Artic Greenland, Turk and Gartner (2001) for the Alps, Galun and Garty (2001) for the Middle East, Ullmannand and Büdel (2001a) for Africa, Eldrige (2001) for Australia and Green and Broady (2001) for Antarctica. Recent papers dealing with the structure and composition of soil crust are that of Reynaud and Lumpkin (1988) for the Lanzhou region in China, Dor and Danin (1996) for the Dead Sea Valley in Israel, Malam Issa (1998), Malam Issa *et al.* (1999), Hahn and Kusserow (1998) for the Sahel region in Africa, Eldrige and Koen (1998) and De Winder (1990) for coastal dunes in northern Europe.

The work so far carried out on the types of the biological crusts on various substratum, feature of the crusts and the salient findings of this work have been summarized in the Table 1.2.

4. Functions of Biological Crust

Crusts contribute to a number of environmental functions, primarily at the soil surface. These include soil stability and erosion, atmospheric nitrogen fixation, nutrient contributions to plants, soil-plant-water relations, infiltration, seedling germination and plant growth. Very little research has been conducted on biological soil crusts. Most of the research that has been conducted on crusts has focussed on identification and relationships with vascular plants. Not much of the research has focused on crust influences on the soil hydrologic or soil stability processes.

4.1 Soil Stability

Cyanobacteria and green algae have filamentous growth forms that bind soil particles. These filaments exude sticky polysaccharide sheath around their cells that aid in soil aggregation by cementing particles together. Fungi contribute to soil stability by binding soil particles with hyphae (thread like filaments). Lichens and mosses bind particles with rhizines/rhizoids (filaments that act as roots for lichens and mosses but lack the complex vascular system found in true roots or rhizomes). The aggregate stability from these binding actions, and the increased surface topography of some crusts, improves resistance to wind and water erosion (Williams *et al.*, 1999).

4.2 Nutrient Contributions

Cyanobacteria can increase available nitrogen as well as other nutrients in the soil. Crusts can be the dominant source of fixed nitrogen in some semi-arid ecosystems. The nitrogen appears to be available to higher plants (Belnap *et al.*, 1993).

4.3 Water Relations

The water holding capacity of crusts benefit the surrounding vegetation by slowing evaporation. This ability to hold water may be so strong as to prevent vegetation from accessing it, thereby decreasing available water. So far, a conclusion has not been reached on this issue (Budel, 1999).

4.4 Infiltration

Crusts can alter infiltration. However infiltration has found increased in the presence of crusts, usually attributed to increased aggregate stability. Other studies showed that crusts decrease infiltration or had no effect. Differences in findings seemed to be site specific and were often related to soil texture and chemical properties of the soil (Rosenstreter and Belnap, 2001).

4.5 Effects on Plant Germination and Growth

Results have varied among studies investigating the role of crusts in plant germination. Increased surface relief is presumed to provide safe sites for seeds while darker surface color increases soil temperatures to those required for germination earlier in the season, coinciding with spring water availability. The above conditions should favour seed germination, but not all studies have supported this conclusion (Williams *et al.*, 1999).

Several studies have shown that survival and/or nutrient content in crust increased in covered environments as opposed to bare soil. Nutrients shown to increase in plant tissues grown in the presence of crusts are nitrogen, phosphorus, potassium, iron, calcium, magnesium and manganese.

5. Species Compostion of Biological Soil Crust/Mat

Composition of the biological soil crusts varies according to different environmental conditions and different nature of substratum. There are reports that environmental condition can change the structure and the behavior of an organism. Biological crusts from Utah's Colorado platue when developed on different substrata like gypsum, sand silt and shale, the filamentous Cyanobacterial species *Phormidium* grew were phylogenetically distant from the Cyanobacterium from other environment (Garcia-Pichel *et al.* 2001). Another Cyanobacterial species *Merismopedia* is a dominant organism in most of the microbial crusts samples. The *Merismopedia* species collected from the field of Mellum after prolonged maintenance under laboratory condition lost the capacity of forming platelet as a result of changed environmental condition

Table 1.2: Summary of the Findings of the Research Work on Biological Soil Crusts/Mats on Various Substratums and their Ecophysiological Significance

Crusts Collected from the Habitat/ Country	*Feature of Soil Crusts*	*Findings*	*Reference*
China	Algal mat on soil of hilly arid regions	The algal mat contain *Nostoc commune*, *Nostoc flagilleforme* and *Nostoc* sp. Respiration started after 30 mins of rewetting whereas photosynthesis started within 6-8 hours which was inhibited in darkness. Recovery of nitrogen fixation was dependent upon newly differentiated heterocysts.	Scherer *et al.*, 1984
Wuhan, Habei Province, China	Dried colony in crusts	Dried colonies of *Nostoc commune* showed regular pattern of synthesis and degradation of glucose.	Earnst *et al.*1987
The Hague, Netherlands	Dune soil crusts	*Oscillatoria* species was the first colonizer of soil crusts that helps the green alga *Klebsormidium flaccidum* to colonize after progressing improvement of the water retention properties. After rehydration *Oscillatoria* sp. showed photosynthesis instantaneously and revealed 100 per cent viability, whereas the green alga showed only 35 per cent viability and recovery of photosynthesis.	Winder *et al.*, 1989
Western Negev desert, Israel	Biological soil crusts in sand dunes	The soil crusts proved to be a complex system maintained by at least one moss, four blue-green algae, and two green algae species and by a variety of undefined bacteria. *Microcoleus sociatus* was present in all of the samples studied. The blue-green alga was sometimes the only microphyte in the crusts samples, either being the first colonizer or being able to grow under the more extreme condition.	Lange *et al.*, 1992

Contd.

Table 1.2–Contd...

Crusts Collected from the Habitat/ Country	*Feature of Soil Crusts*	*Findings*	*Reference*
Soil surface	Cells incubated to slit clay, calcareous clay and slit loam	Cyanobacterial strains inoculated in slit clay, calcareous clay and silt loam showed 50 per cent survival, after 1 year *Anabaena* sp., *Nostoc* sp., *Calothrix* sp., *Cylindrospermum* sp., *Fischerella* sp.and *Scytonema* sp. were the best survivors in all kinds of soils. *Fischerella* sp. showed significant survival capability whereas *Anabaena* sp., *Cylindrospermum* sp. and *Calothrix* sp. showed very low survivability. Soil inoculated with *Scytonema* sp and *Fischerella* sp. had a higher total count.	Tomas Elli and Giovannetti 1993
In tropics (Europe) Africa and South America) rock mountain sand rock outcrops and bare soil	Dried crusts with Cyanobacteria and cryptograms	*Oscillatoria sancta*, *Stigonema mamillosum*, *S. lyngbioides*, *Gloeocapsa sanguienea*, *Gloeocapsa alphina* were the dominant Cyanobacteria adopted to a very wide range of light <20mmol $m^2 s^{-1}$ to full sunlight, descication tolerant, assimilate atmospheric dinitrogen and independent of N_2 input from the substratum	Luttge *et al.*, 1995
Solar Lake, Sinai, Egypt	Microbial mat, dark blue green in colour	The microbial mat dominated by photoautotrophic, cyanobacterium *Microcoleus chthonoplastes.* Upon Exposure to UV-B (i) the colour of the mat changed from Dark blue green to lighter blue green (ii) decrease in the pore water oxygen concentration in the surface layer, (iii) a shift downward of the depth at which the maximum pore water Oxygen concentration occurred.	Bebout and Garcia-Pichel 1995

Contd...

Table 1.2–Contd...

Crusts Collected from the Habitat/ Country	*Feature of Soil Crusts*	*Findings*	*Reference*
Dead sea valley, Israel	Desert crusts	Cyanobacteria are profusely accompanied by lichen and one green algae. At the lowest, youngest terraces only two Cyanobacterial species *Microcoleus vaginatus* and *Schizothrix friessi* were found. Chlorophyll-a and polysaccharides concentrations were lowest at the close of wide bottomland and increases with degree of elevation.	Dor and Danin 1996
Nizzana field of Arid ecosystem, Jerusalem	Microbial crusts on soil surface	*Microcoleus* sp. and *Nostoc* sp. excrete polysaccharides to the sand. The microbial EPS increases the water holding capacity of the sand surface.	Mazor *et al.*, 1996
Norderney, Island	Giant blooms of Unicellular benthic Cyanobacteria observed within the microbial mat	Five *Merismopedia* like Cyanobacterial strains isolated from microbial mat after prolonged maintence under laboratory conditions. The platelet like colonies formation stopped, whereas cell sizes, production of extracellular polymeric substances, division patterns were stably maintained. From SDS-PAGE of the total cell protein analysis *Merismopedia* strains were found to be actually one species, *i.e.Mersmopedia punctata.*	Palinska *et al.*, 1996
South–eastern Australlia	Rangelands, hills and foot slopes of shallow loams sand-planes of calcareous earth or red earths and floodplains	Some Lichens and Bryophytes were consistently associated with preliminarily dominated Cyanobacteria, which were dominant. Disturbance influencing soil stability and soil surface condition lead to change in the composition of soil crusts communities.	Eldridge *et al.*, 1998

Contd...

Table 1.2–Contd...

Crusts Collected from the Habitat/ Country	*Feature of Soil Crusts*	*Findings*	*Reference*
Antarctic desert habitats	Dried cyano-bacterial mat, sandy soil crusts	*Nostoc* sp., *Scytonema* sp. and *Lyngbya aestuarii* possess the pigments scytonemin, carotenoid, anthroquinones and Mycosporine-like amino acids to protect them against enhanced UV-B damage.	Wynn-Williams *et al.*, 1999
Soil surface of fallow rice field in summer season	Brownish wooly crusts	*Aulosira* sp., the corresponding organism in the rice field soil crusts absorbed water rapidly and became saturated within 10 mins of wetting. The organisms synthesized scytonemin pigment only in desiccated state and when exposed to bright sunlight, scytonemin was lost within 48 hours of wetting. Soon after wetting the photosynthesis and respiration rate increased up to 6h and then stabilized. Nitrogenase activity was faster after 24 h of wetting.	Adhikary and Sahu, 2000
Inselbergs, dry thorn buss Savanna, humid Savanna	Microbiotic crusts: black, green or to red in colour	Algae, fungi and lichen were the dominant organisms on the surfaces: *Chlorococcidiopsis* grew mostly inside the rock (endolithic in 1-4mm depth), *Gloeocapsa*, *Schizothrix*, *Scytonema* and *Stigonema* grew in epilithic habitat.	Budel *et al.*, 2000
Arches National park in the Colorado Plateau, Utah	Crusts from sandy soil alluvial soil, Manco shales and gypsiferous outcrops	*Microcoleus vaginatus* dominate in most of samples. A new phylogenetic cluster named "Xeronema" grouped a series of filamentous *Phomidium* like Cyanobacteria. Gypsum crusts were most deviant from the rest while sandy, slit and shale crusts were relatively more similar among themselves.	Garcia-Pichel *et al.*, 2001

Contd...

Table 1.2–Contd...

Crusts Collected from the Habitat/ Country	*Feature of Soil Crusts*	*Findings*	*Reference*
Asia including Donau and Volga region	Crusts, soil as substratum	The most prominent Cyanobacteria in these soil crusts is *Micrococus vaginatus* followed by *Scytonema ocellatum* and *Nostoc commune*	Budel, 2001a
European temperate and Mediterranean regions, North Africa	Lichen dominated soil, black crusts crusts in saline soil	Cyanobacteria occurred in large numbers in the soil crusts communities composed of algae, lichen and mosses along with Cyanobacteria.	Budel, 2001b
Trachypogon Savanna, Sahara Desert of South America	Crusts in older, flat rock pools of inselbergs and endephic soil crusts	Cyanobacterial genera dominated are *Lyngbya, Oscillatoria, Porphyrosiphon, Anabaena, Nodularia, Plectonema, Scytonema* and *Stigonema*.	Budel, 2001c
Namib desert, South-Western Africa	Rocky terranean inselbergs, rocky ridges, laterite soil ridges, laterite soil	Crusts dominated by Cyanobacteria genera is *Schizothrix* with co-occurring *Aphannizonnenon, Nostoc, Scytonema*, coccal Cyanobacteria and green algae.	Ullmann *et al.* 2001
World distribution	Soil crust	Lichen, cyanoabacteria, algae, moss and liverworts genera are component of soil crusts of different geographical regions. Cyanobacterial genera occurred pre-dominantly	Budel, 2001d
On soil surfaces	Blackish-brown crusts	The cyanobacterium, *Microcoleus* provide most of the cohesive quality of the biological soil crusts in arid and semiarid regions. Soil crust cyanobacteria contribute carbon and nitrogen to soils, lichens penetrate uppermost soil layer and bryophytes attached to soil by rhizoid form aggregates with soil particle.	Belnap *et al.* 2001

(Lange, 2001). It is also expected that several isolates of coccoid Cyanobacteria identified or maintained as *Synechococcus* or *Synechocystis* are actually *Merismopedium* strains that lost the capacity of forming platelets (Palinksa and Krumbein, 1994). Palinksa *et al.* (1996) also found that in five *Merismopedia* strains–like Cyanobacterial strains collected from microbial mat at Norderny island after prolonged maintence under laboratory condition, the formation of platelet like colonies stopped, whereas cell size, production of extracellular polymeric substances and division pattern were stably maintained. By cell protein assay all analysed strains were combined into one species *i.e. Merismopedia puncta*. In diverse environments, usually Cyanobacteria were the primary colonizers. The reason for the success of Cyanobacteria as pioneers on bare substrata of different kinds of is that generally they produced shade and sun forms adapting to a very wide range of sunlight, are desiccation tolerant and assimilate atmospheric dinitrogen, thus are independent of nitrogen input of the substratum (Luttge *et al.*, 1995). Generally large filamentous Cyanobacteria colonies appear first. They possess thick sheath or slime layer around the filament that improves the water retention properties in the organisms and gives a suitable environment for growth of green algae, fungi and lichen (Winder *et al.*, 1989; Büdel, 1999; Büdel *et al.*, 2000). They can grow with very low nutrient concentration absorbing volatiles and aromates from the atmosphere.

After prolonged succession and formation of a crust, the water retention properties in the top layer of the soil became improved through colonization of Cyanobacteria, which helps the green algae *Klebsormidium flaccidum* to survive (Winder *et al.*, 1989). Usually Cyanobacteria species can survive under drought condition for several years. Survival depends on the capability of overcome adverse condition. *Scytonema* sp., *Fischerella* sp. and *Nostoc* sp. showed significant survival in different soil types (silt clay, calcareous soil and silt loam) and had a higher N content than the non-inoculated soils (Tomaselli and Giovannetti, 1983). Drought resistant Cyanobacteria *Nostoc flagilliforme, Nostoc commune* and *Nostoc* sp. showed rapid absorption of water upon rewetting (0.6, 3.3 and 15.5 min) respectively even after 2 years dry period (Scherer *et al.*, 1984).

6. Response to UV-Radiation

For the photosynthetic microbial mats developed in phototropic environments, solar UV (wavelength shorter than 400 nm) is a significant physiological stress. Under such conditions they possess efficient mechanism to prevent or to counteract the harmful effects of ionizing radiations in the ultraviolet regions. Ultraviolet radiation region in the spectrum are UV-A (315-400 nm) and UV-B (280-315 nm). UV-A, UV-B, UV-C radiation inhibit several physiological and morphological processes. Among possible mechanisms to prevent UV photodamage are avoidance response to light (Richardson and Castenholz, 1987) and synthesis of UV sunscreen compounds (Garçia-Pichel and Castenholz, 1991). Scytonemin a yellowish-brown pigment occurred in the outermost part or top layer of Cyanobacteria mats, crusts or colonies that are exposed to intense solar radiation. The term scytonemin was coined by Nageli; 1849. It is lipid soluble dimeric pigment with a molecular mass of 544 Da and has a prominent absorption maximum in the near Ultraviolet region of the spectrum (384 nm, 370nm *in vivo)* with a long tail extending to the infrared region (Garcia-Pichel and Castenenholz, 1991; Ehling-Schulz and Scherer, 1999)

In addition to scytonemin, many Cyanobacterial species possess mycosporine amino acid–like substances (MAAs) absorbing prominently in the UV-B region of the spectrum (Bohm *et al.*, 1995). MAAs protects Cyanobacteria and other lower groups of organisms by absorbing harmful UV radiation (Scherer *et al.*, 1988; Karentz *et al.*, 1991; Karsten and Garcia–Pichel, 1996; Ehling-Schulz *et al.*, 1997; Roy *et al.*, 1997; Xiong *et al.*, 1997). Mycosporine amino acid are water soluble, substituted cyclohexanes which are linked to amino acids and amino alcohols and have absorption maxima between 310 nm to 360nm (Nakamura *et al.*, 1982; Scherer *et al.*, 1988; Ehling-Schulz and Scherer, 1999) and an average molecular weight around 300.The relative protection against UV-B damage provided by MAAs depend on the species and the location of the pigment. Significant protection has been reported for various Cyanobacteria with MAAs located in the cytoplasm.

Mat forming Cyanobacteria could escape high solar radiation by downward migration into mat communities (Quesada and Vincent, 1997). Daily vertical migration to avoid periods of incident high solar irradiation has been reported for *Oscillatoria* sp. and *Spirulina* sp. (Garcia-Pichel *et al.*, 1994). The vertical migration of *Microcoleus chthonoplastes* has been shown to be UV and PAR inducible. Since UV-B by far the most effective wave band-promoting migration, vertical migration may be an effective strategy to minimize UV–induced damage. Migration led to an overall decrease of productivity of the crust ecosystem (Bebout and Garcia- Pichel, 1995).

7. Stress Response

There are reports that biological soil crusts and the corresponding organism are often exposed to various stresses like drought, water deficit, UV–radiation etc. (Scherer and Potts, 1989; Hill *et al.*, 1994; Tripathy and Srivastava, 2001). The desiccation tolerant cyanobacterium *Nostoc commune* in one such crusts accumulate novel group of acidic water stress proteins (Wsp) (Hill *et al.*, 1994; Scherer and Potts, 1989). Water stress proteins present in the cell subjected to drying and desiccation but not in cells grown without water stress. Wsp fraction withstands prolonged desiccation more than other soluble proteins.

Mat forming desiccation tolerant *Lyngbya* sp. inhibiting in the bark surface of *Magnifera indica* is also exposed to various stresses (Sinha *et al.*, 1999; Tripathy and Srivastava, 2001). The light harvesting pigments, phycocyanin and phycoerythrin operate as a defence against photodamage in this organisms and help to survive in hot, dry and highly irradiated environments. Dry Cyanobacteria bears high ability to maintain stability of its enzymes under extremes of environmental conditions. Further, Cyanobacteria did not show complete stability of the enzymes on storage for longer duration.

8. Response to Disturbance

Biological soil crusts are well adapted to severe growing conditions, but they do not respond well to compression. Trampling by large animals; by tourist activities such as hiking, biking, or ORVs; and by military activities (*e.g.* track-vehicle maneuvers) can take a toll on crust integrity. Trampling breaks up the sheaths and filaments holding the soil together and reduces the capability of the soil organisms to function, particularly in nitrogen fixation and soil holding capability.

Other disturbances that impact soil crusts are (possibly) allelopathic effects, burial, and fire. Sagebrush, saltbush and winterfat may have allelopathic effects on the nitrogen fixing capabilities of crusts. Crusts can be buried by eroding soils. They can tolerate shallow burial by extending sheaths to the surface to begin photosynthesis again. Deeper burial can kill them. Fire is a common component of many regions where crusts grow. Fire can cause severe damage to crusts, but recovery is possible. Intensity of fire can make a difference. Low intensity fires do not remove all the structure of crusts allowing for re-growth without significant soil loss (Lange, 2001).

9. Conclusion

Crusts contribute to a number of environmental functions, primarily at the soil surface. These include soil stability and erosion, atmospheric nitrogen fixation, nutrient contributions to plants, soil-plant-water relations, infiltration, seedling germination and plant growth. After prolonged succession and formation of a crust, the water retention properties in the top layer of the soil became improved through colonization of Cyanobacteria, which help the green algae to survive (De Winder *et al.*, 1989). Usually Cyanobacteria species can survive under drought condition for several years. Survival depends on the capability of overcome adverse condition. Resistance to desiccation has been attributed to various characteristics; ultra-structural, biochemical as well as physiological are the sites of research work yet to be emphasized.

A general attempt can be taken regarding the cellular response or mechanisms of adaptation of desiccant-tolerant Cyanobacteria in the drought and direct exposure to high solar radiation. Very little research has been conducted on biological soil crusts. Most of the research that has been conducted on crusts has focussed on identification and relationships with vascular plants. Not much of the research has focused on crust influences on the soil hydrologic or soil stability processes.

As a concluding comment, soil crusts have not received much attention all over the world. Evidently, experimental research is needed to determine the influence of crust on nutrient cycling, vascular plant regeneration and vigor, soil hydrologic and stability properties, plant community development and structure.

10. References

Adhikary S P and Sahu J K 2000 Survival strategies of Cyanobacteria occurring as crust in the rice fields under drought conditions. *Ind J Microbiol* 40: 53-56.

Bebout B M and Garcia–Pichel F 1995 UV-B induced vertical migrations of Cyanobacteria in a microbial mat. *Appl Environ Microbiol* 61: 4215-4222.

Belnap J 2001 Comparative structure of physical and biological soil crusts In: Lange O L and Belnap J (eds) *Biological Soil Crusts: Ecological Studies.* Springer Verlag, Berlin, Heidelberg and New York.

Belnap J, Büdel B and Lange O L 2001 *Biological Soil Crusts: Ecological Studies*, Springer Verlag, Berlin, Heidelberg and New York, 150: 1-30.

Belnap J, Büdel B and Lange O L 2001 "Biological Soil Crust: Characteristics and Distribution" In: Lange O L and Belnap J (eds) *Biological Soil Crusts: Ecological Studies.* Springer Verlag, Berlin, Heidelberg and New York.

Belnap J, Harper K T and Waren S D 1994 Surface disturbance of cryptobiotic soil crusts nitrogenase activity, chlorophyll content and chlorophyll degradation. *Arid Soil Res Rehabil* 8:1-8.

Belnap J 1993 Recovery rates of cryptobiotic soil crust: inoculant use and assessment methods. *Great Basin Nat* 53: 89-95.

Belnap J, Sanfort R A and Lungu L 1996 Biological soil crusts: ecological roles and response in Miombo Woodlands of Zimbawe. Trans Zimbawe, *Sci Assoc* 70: 14-20.

Böhm G A, Pfleiderer W, Böger P and Scherer S 1995 Structure of a novel oligosaccharide mycosporine amino acid ultraviolet A/B sunscreen pigment from the terestrial cyanobacterium *Nostoc commune*. *J Biol Chem* 270: 8536-8539.

Büdel B 2001a "Biological soil crust of South America" In: Lange O L and Belnap J (eds) *Biological Soil Crusts: Ecological Studies*. Springer Verlag, Berlin, Heidelberg and New York, 150: 51-55.

Büdel B 2001b "Biological soil crust of European temperate and mediterranian regions. In: Lange O L and Belnap J (eds) *Biological Soil Crusts: Ecological Studies*. Springer Verlag, Berlin, Heidelberg and New York.

Büdel B 2001c "Biological soil crust of Asia including the Don and Volga region. In: Lange O L and Belnap J (eds) *Biological Soil Crusts: Ecological Studies.* Springer Verlag, Berlin, Heidelberg and New York.

Büdel B 1999 Ecology and diversity of rock-inhabitating Cyanobacteria in tropical regions. *Eur J Phycol* 34: 361-370.

Büdel B 1999 Ecology and diversity of rock-inhabitating blue-green algal/Cyanobacteria from hot arid regions. *Arch Hydrobiol Algological Studs* 64: 384-398.

Büdel B, Becker U, Follmann G and Sterflinger K 2000 Algae, Fungi and lichens on inselbergs. *Ecologica Stud* 146: 69-90.

Castenholz R W and Garcia-Pichel F 2000 Cyanobacterial response to UV-response. In: Whitton B A and Potts W (eds) *The Ecology of Cyanobacteria.* Kluwer, Dordrecht.

De Winder B D 1990 *Ecophysiological Strategies of Drought Tolerant Phototrophic Microorganisms in Dune Soils.* Academisch Proeschrift, University of Amsterdam, Amsterdam.

Dor I and Danin A 1996 Cyanobacterial desert crusts in the Dead Sea Valley, Israel. *Algol stud* 83: 197-206.

Ehling-Schulz M and Schere S 1999 UV protection in Cyanobacteria. *Eur J Phycol* 34: 329-338.

Ehling-Schulz M, Bilger W and Scherer S 1997 UV-B induced synthesis of photo protective pigments and extracellular polysaccharides in the terrestrial Cyanobacterium *Nostoc commune. J Bacteriol* 179: 1940-1945.

Eldridge D J 2001 Biological soil crusts of Australia. In: Lange O L and Belnap J (eds) *Biological Soil Crusts. Ecological Studies.* Springer Verlag. Berlin, Heidelberg and New York.

Eldridge D J and Green R S B 1994 Macrobiotic soil: review of their roles in soil and ecological processes in the rangelands of Australia. *Aust J soil* 32: 389-415.

Eldridge D J and Koen T B 1998 Cover and floristic of microphysics soil crusts in relation to indices of heath. *Plant Eco* 137: 101-114.

Ernst A, Chen T W and Boger P 1987 Carbohydrate formation in rewetted terrestrial Cyanobacteria. *Oecologia* 72: 574-576.

Evans R D and Johansen J R 1999 Microbiotic crusts and ecosystem process. *Cric Rev Plant Sci* 18: 183-225.

Friedmann E I, Lipkin Y and Ocampo-Paus P 1967 Desert algae of the Negev, Israel. *Phycologia* 6:185-200.

Galun M and Garty J 2001 Biological soil crusts of the middle East In: Lang O L, Belnap J (eds) *Biological Soil Crusts. Ecological Studies* Springer Verlag, Berlin, Heidelberg and New York.

Garcia-Pichel F and Castenholz R W 1991 Characterization and biological implications of scytonemin, a Cyanobacterial sheath pigment. *J Phycol* 27: 395-409.

Garcia-Pichel F, Looez-Cortes A and Nubel U 2001 Phylogenetic and morphological diversity of Cyanobacteria in soil desert crusts from the Colorado Plateau. *Appl Environ Microbiol* 67: 1902-1910.

Garcia-Pichel F, Mechling M and Castenholz R W 1994 Dial migrations of microorganisms with in a benthic, hypersaline mat community, *Appl Environ Microbiol* 60: 1500-1511.

Green T G A and Broady P 2001 Biological soil crusts of Antarctica. In: Lange O L and Belnap J (eds) *Biological Soil Crusts. Ecological Studies.* Springer Verlag, Berlin, Heidelberg and New York.

Hahn A and Kusserow H 1998 Spatial and temporal distribution of algae in soil crusts in the Sahel of W. Africa: preliminary results. *Wildenowia* 28: 227-238.

Hansen E S 2001 Lichen-rich soil crusts of Artic Greenland.In: Lange O L and Belnap J (eds) *Biological Soil Crusts. Ecological Studies.* Springer Verlag. Berlin, Heidelberg and New York.

Hill D R, Hladun S L, Scherer S and Potts M 1994 Water stress proteins of *Nostoc commune* (Cyanobacteria) are secreted with UV-A/B absorbing pigments and associate with 1,4-β-D-xylanohydrolase activity. *J Biol Chem* 269: 7726-7734.

Johansen J R 1993 Cryptogamic crusts of semi arid and arid lands of North America. *J Phycol* 29: 140-147.

Johansen J R and Rushforth S R 1986 Cryptogamic soil crusts: seasonal variation *Great Basin Nat* 46: 632-640.

Karentz D, Mceuch F S, Land M C and Dunalp W C 1991 Survey of mycosporine-like amino acid compounds in Antarctic marine organisms: potential protection from ultraviolet exposure. *Mar Biol* 108: 157-166.

Karsten U and Garcia-Pichel P F 1996 Carotenoids and mycosporine like amino acid compounds in members of the genus *Microcoleus* (Cyanobacteria): a chemosystematic study. *Syst Appl Microbiol* 19: 285-294.

Kindron G 1995 The Impact of Microbial Crust upon Rainfall Rundruff Sediment Yield Relationships on Longitudinal Dune Slopes. Nizzana, Western Negev Desert, Israel. Ph.D. Thesis. The Hebrew University of Jerusalem (English Summary).

Komaromy Z P 1976 Soil algal growth types as edaphic adaption in Hungarian forest and grass steppe ecosystems. *Acta Bot Acad Sci Hung* 22: 73-379.

Lange O L 2001 Photosynthesis of soil crust biota as dependent on environmental factors. In: Lange O L and Belnap J (eds) *Biological Soil Crust: Ecological Studies*. Springer Verlag, Berlin, Heidelberg and New York.

Lange O L, Kidron G J, Büdel B, Meyer A, Killan E and Abeliovich A 1992 Taxonomic composition and photosynthetic characteristics of the "Biological soil crusts" covering sand dunes in the Western Negev Desert. *Funct Ecol* 6: 519-527.

Lüttge U, Büdel B, Ball E, Strube F and Weber P 1995 Photosynthesis of terrestrial Cyanobacteria under light and desiccation stress as expressed by chlorophyll fluorescence and gas exchange. *J exp Bot* 46: 309-319.

Malam Issa O, Trichet J, Defargee C, Coute A and Valentin C 1999 Morphological and micro-structure of microbiotic soil crusts on a tiger bush sequence (Niger, Sahel). *Catena* 37: 175-196.

Malam Issa O 1998 *Role of microbiotic soil crust in two Sahelian ecosystems (fallow lands* and *tiger bush)* of Niger. Micromorphology, phycal and biogeochemical properties. Diss. CNRS-Universite d' Orleans.

Mazor G, Kindron G J, Vonvashak A and Albeliovich A 1996 The role of Cyanobacterial exo polysachharides in structuring desert microbial crust. *FEMS Microbiol Ecol* 21:121-130.

Nakamura H, Kobayashi J and Hirta Y 1982 Separation of mycosporine-like amino acids in marine organisms using reverse-phase high performance liquid chromatography. *J Chromatogr* 250: 115-118.

Palinska K A, Liesack W, Rhiel E and Krumbein W E 1996 Phenotype variability of identical genotypes: the need for a combined approach in Cyanobacterial taxonomy demonstrated on *Merismopedia*–like isolates. *Arch Microbiol* 166: 224-233.

Palinska K A and Krumbein W E 1994 Ecotype-Phenotype.An approach to the *Synechococcus-Synechocystis-Merismopedia-* Eucapsis complex. *Arch. Hydrobiol. Algological Studs* 75: 213-227.

Quesada A and Vincent W F 1997 Strategies of adaption by Antarctic Cyanobacteria to ultraviolet radiation. *Eur J Phycol* 32: 315-323.

Reynaud P A and Lumpkin T A 1988 Mikroalgae of the Lanzhou (China) cryptogamic crust. *Arid Soil Res Rehab* 2: 145-155.

Richardson L L and Castenholz R W 1987 Diel vertical movements of cyanobacterium *Oscillatoria terebriformis* in a sulfide-rich hot spring microbial mat. *Appl Environ Microbiol* 53: 2142-2150.

Rosentreter R and Belnap J 2001 *Biological soil crusts. Ecological studies.* Springer Verlag, Berlin, Heidelberg and New York, pp 31-50 - Stigonematales. *Arch Hydrobiol Algolog stud.* 59: 1-73.

Roy A, Tripathy P and Adhikary S P 1997 Epilithic blue-green algae/ Cyanobacteria from temples of India and Nepal: presence of UV sunscreen pigments. *Arch Hydrobiol Algolog Studs* 86: 147-161.

Scherer S and Potts M 1989 Novel water stress protein from a desiccation tolerant cyanobacterium. *J Biol Chem* 264: 12546-12553

Scherer S, Chen T W and Böger P 1988 A new UV-A/B protecting pigment in the terrestrial cyanobacterium *Nostoc commune*. *Plant Physiol* 88: 1055-1057.

Scherer S, Ernst A, Chen T W and Böger P 1984 Rewetting of drought resistant blue-green algae; time source of water uptake and reappearance of respiration, photosynthesis and nitrogen fixation. *Oecologia* 62: 418-423.

Sinha R P, Klishch M and Häder D P 1999 Induction of a mycosporine like amino acid (MAAs) in the rice field cyanobacterium *Anabaena* sp. by UV-irradiation. *J Photochem Photobiol* 52: 59-64.

Tomaselli L and Giovannetti L1993 Survival of Algae in a desiccated soil: A 35-year study. *Phycologia* 34: 113-116.

Tripathy S N and Srivastava P 2001 Presence of stable active oxygen scavenging enzymes superoxide dismutease, ascorbate peroxidase and catalase in a desiccation-tolerant cyanobacterium, *Lyngbya arboricola* under dry state. *Curr Sci* 81: 197-200.

Turk R and Gartner G 2001 Biological soil crust of the Sub alpine and naval areas in the Alps. In: Lange O L and Belnap J (eds) *Biological Soil Crust, Ecological Studies*. Springer Verlag, Berlin, Heidelberg and New York.

Ullmann I and Büdel B 2001a Biological soil crusts of Africa. *Ecological Studies* 150: 107-118.

Ullmann I and Büdel B 2001b Ecological determinants of species composition of biological soil crusts on landscape scale. In: Lange O L and Belnap J (eds) *Biological Soil Crusts: Ecological Studies*. Springer Verlag, Berlin, Heidelberg and New York.

West N F 1990 Structure and function of soil microphytic crusts in wild land ecosystems of arid and semi-arid regions. *Adv Ecol Res* 20: 179-223.

Winder B D, Matthijis H C P and Mur L R 1989 The role of water retaining substrate on the photosynthetic response of three drought tolerant phototropic microorganisms isolated from a terrestrial habitat. *Arch Microbiol* 152: 458-462.

Wynn-Williams D D, Edward H G M and Garcia-Pichel F 1999 Functional biomolecules of Antarctic stomatolitic and endolithic Cyanobacterial communities. *Euro J Phycol* 34: 381-391.

Xiong F, Komenda J, Kopecky J and Nedbal L 1997 Strategies of ultraviolet-B protection in microscopic algae. *Physiol Plant* 100: 378-388.

Chapter 2
Regulation of Heterocyst and Nitrogen Metabolism

A.K. Mishra, S.K. Singh, S.S. Singh and V.D. Pandey

Contents

Glimpses of Cyanobacteria, 2006
Daya Publishing House, New Delhi

26–47

ABSTRACT

Upon nitrogen starvation, filaments of *Anabaena/Nostoc* species start a programme of differentiation that leads to the formation of about 5-10 per cent heterocysts (site of nitrogen fixation) at semi-regular interval along the filaments, generating spacing pattern. The presence of fixed nitrogen in the medium repressed the development of such cells of *Anabaena/Nostoc* species and produced undifferentiated filaments composed of only vegetative cells. The spacing pattern depend upon the interaction between diffusible substances (probably glutamine) that originate from the heterocyst, and move along the filament and other factors, presumably proteins, that interact with diffusible substances and directly or indirectly control the gene expression. Recently, numerous genes have been identified which are involved in heterocyst differentiation and nitrogen fixation. The most exciting discovery is that developing heterocysts of *Anabaena* 7120 is accompanied by the rearrangement of nitrogen fixing genes (*nifHDK*). The *NifD* gene is interupted by 11kb, *fdxN* gene is interrupted by 55 kb, which are excised from the main genome by the action of site specific- recombinase enzyme. Similarly, *hupL* gene is also interrupted by 10.5 kb which encodes uptake hydrogenase that can recycle hydrogen produced during nitrogen fixation. The earliest documented gene that is activated by nitrogen deprivation is *ntcA* (regulatory gene), which encodes DNA-binding protein not only essential for *glnA* (encoding GS) but also for *hetR* and *hetC* (*i.e.* expressed early in the course of nitrogen starvation, and require their own functions for full expression). Two genes *hetF* and *patA*, appear to be involve in heterocyst and pattern formation are activated by HetR protein. Other heterocyst-specific genes whose products are required for heterocyst differentiation and nitrogen fixation are *hetP, hglb, hglCD, hglE, hglK, devABC, hepA, hepB, hepC* and *hepK, Xis* and *nif* genes. Recently, *patS* gene has been discovered, whose products control the intercellular inhibition of heterocyst development Mutants of filamentous, N_2-fixing cyanobacteria selectively impaired in specific enzymatic pathways have been found useful tools towards deciphering the regulation of heterocyst and nitrogen metabolism.

1. Introduction

In many filamentous cyanobacteria such as *Anabaena/Nostoc* species, vegetative cells differentiates into specialized cells called heterocysts, at regular interval when deprived of combined nitrogen. These cells are the site of synthesis and activity of the nitrogen-fixing enzyme nitrogenase following *nif* gene rearrangement while the vegetative cells continue to carry out oxygenic photosynthesis. The nitrogen fixed in the heterocysts moves to the vegetative cells in the form of glutamine and CO_2 fixed in vegetative cells is transported to the heterocyst through plasmodesmata. The intercellular exchange of metabolites between these cells is not only required for diazotrophic growth but also for generating spacing pattern. Heterocyst formation requires substantial changes in gene expression which alter the physiology, biochemistry and morphology of the differentiating cells *e.g.* the genes responsible for nitrogen fixation my be switched on while genes necessary for CO_2 fixation and O_2–evolving reactions of photosynthesis are switched off. Recently, numerous genes have been identified which are assigned for heterocyst differentiation, pattern formation and nitrogen fixation (*http://www.kazusa.or.jp/cyanobase*). A main objective of this review is to consider the factors regulating heterocyst differentiation and nitrogen metabolism.

2. Cyanobacteria: General Characteristics and Evolution

Cyanobacteria (also called blue-green algae) are gram-negative photosynthetic procaryotes that are capable of fixing carbon dioxide with water as the reductant and oxygen as a byproduct. They are a highly diverse, widely distributed and ecologically important oldest group of microorganisms that are capable of splitting water during the photosynthetic process. They have been on earth for over three billion years. During the course of evolution, they have provided the major source of oxygen and acquired several morphological specializations, both structural and functional, in order to survive best in diverse ecological conditions. These prokaryotes lack the conventional mode of sexual reproduction but show great capacity to differentiate various cells, facilitating the asexual reproductive function or dispersal of species. Among the multicellular Cyanobacteria belonging to the order Nostocales and stigonematales, species of *Nostoc* and *Anabaena*, capable of producing nitrogen fixing cells (heterocysts) at evenly space intervals in filaments (Wolk, 1982), provide a procaryotic systems with great potential for the elucidation of problem of wide interest in developmental biology as they show a high degree of morphological and developmental complexity. Therefore, heterocytous cyanobacteria are of great scientific interest for a number of reasons.

3. Heterocysts and N_2-Fixation

Heterocystous Cyanobacteria are able to fix dinitrogen (N_2) despite the presence of photosynthetically derived and ambient level of oxygen. This is possible because aerobic fixation of dinitrogen is confined to heterocysts while photosynthesis is confined to vegetative cells. Fixation of dinitrogen is an energetically expensive process and known to be highly sensitive to very low level of oxygen. This important biological process (N_2-fixation) is restricted only to procaryotic organisms. Nitrogen-fixing bacteria have devised complicated mechanisms to protect the nitrogen fixation apparatus from oxygen (Fay, 1992). In heterocystous cyanobacteria, it is achieved through the differentiation of a 5-10 per cent of vegetative cells into heterocysts under the limitation of fixed or combined nitrogen (Fogg, 1949). This process of differentiation is repressed by external source of fixed nitrogen (nitrate, ammonia or glutamine). So, the primary function of heterocysts is to provide anaerobic conditions necessary for the production and proper functioning of nitrogenase (N_2-reducing enzyme complex) and other proteins related to N_2-fixation. Since this process requires reductants (fixed carbon), the vegetative cells that serve as the sites of oxygenic-photosynthesis, export fixed carbon to heterocysts (N_2-fixing cells) and in return, receive fixed nitrogen from heterocysts. This separation of function is necessary since nitrogenase is extremely sensitive to oxygen. The heterocyst a near anoxic state mainly by three means: Dismantle of photosystem II responsible for photolysis of water and evolution of oxygen during heterocyst differentiation, second heterocyst invested with specialized envelope that limits the influx of gases. Two layers within the envelope have been implicated in O_2 protection: an inner layer composed of hydroxylated glycolipid layer and an outer polysaccharide layer. Neither layer is found in vegetative cells. Third, an outer much of the oxygen that overcome these barriers is consumed by the high oxidase activity (heme protein) localized mainly at the polar nodules (protoplasmic bridge) of heterocysts (Haselkorn, 1978; Wolk, 1982; Wolk *et al.*, 1994).

3.1 Intercellular Communications

The nitrogen fixed in the heterocysts moves to vegetative cells in the form of glutamine (Thomas *et al.*, 1977) and carbon dioxide fixed in vegetative cells is transported to heterocysts through plasmodesmata. Therefore, the burden of providing the filament carbon and nitrogen is thus divided between these cell types, and under conditions of azotrophic (N_2-fixing) growth the filament might be considered simple multicellular organism, with each cell type dependent on the other for growth. The intercellular exchange of metabolites between these cells is required not only for the diazotrophic growth but also for interheterocyst spacing pattern in filaments (Tiwari, 1977; Adams, 1992). Any factors that interfere with the process of such cellular intercourse would affect the cell from heterocyst differentiation or pattern formation. The control of heterocyst differentiation and pattern formed in the filament is poorly understood.

3.2 Control of Heterocyst Formation

It is well-established fact that the presence of fix nitrogen (such as ammonia or nitrate) in the medium causes repression of heterocyst formation in most *Anabaena / Nostoc* strains. These organisms in the absence of fixed nitrogen produce heterocysts, which are formed midway between two existing heterocysts. So a nitrogen-fixing product of heterocyst metabolism while diffusing out may be inhibiting nearby cells from differentiating into heterocysts (Wolk, 1967; Fay *et al.*, 1968; Singh *et al.*, 1972). It was first thought that ammonia, the product of dinitrogen fixation, might be the repressor of heterocyst differentiation (Fogg, 1949), however, ammonia does not always repress heterocysts as reported in an isolate of *Anabaena* sp. strain CA (Bottomley *et al.*, 1979). Singh and Tiwari (1970) reported the frequent germination of heterocysts in *Gloeotrichia ghosei* in the presence of ammoniacal nitrogen, clarifying the role of fixed nitrogen in regulating a functional role of heterocysts. Further, it was reported that methionine sulfoximine (MSX), a known inhibitor of glutamine synthetase activity, causes heterocysts to form in the presence of ammonia (Stewart and Rowell, 1975; Ownby, 1977; Singh *et al.*, 1977). Similarly, rifampicin, an inhibitor of procaryotic transcription, blocks the inhibitory function of heterocysts causing development of strings of heterocysts in the filaments (Wolk and Quine, 1975). The mechanism(s) by which heterocyst differentiation is repressed during the growth on a source of fixed nitrogen is not clearly known. However, it has been suggested that glutamine or a metabolic product of glutamine, rather than ammonia, is the regulatory molecule (Thomas *et al.*, 1977; Haselkorn, 1978; Wolk, 1982, 2000; Mishra, 1997,2003b). Several observations provide evidences to support the hypothesis that glutamine repress heterocyst development. These are: (i) MSX inhibits synthesis of glutamine as reported by Stewart and Rowell (1975) and Meeks *et al.* (1978) in *Anabaena* cylindrica, but the addition of glutamine to cultures containing both MSX and ammonia results in the suppression of heterocysts (Ownby, 1977), (ii) glutamine is the first organic product of nitrogen fixation and this product ($^{13}N_2$ derived) is transported to vegetative cells (Wolk *et al.*, 1976; Meeks *et al.*, 1978; Thomas *et al.*, 1977; Wolk, 2000; Mishra, 2003b), (iii) *Anabaena azollae, Azolla cyanobiont*, is defective in GS and produces more heterocysts (upto 24 per cent) than the free living heterocystous cyanobacteria (Ray *et. al.*, 1978). These points support the idea that it is glutamine or derivative of glutamine, which could be critical inhibitory substance produced by heterocysts. The induction of heterocysts by certain tryptophan analogues may be attributed to their inhibitory effect on glutamine-dependent synthesis of tryptophan, rather than synthesis of glutamine itself (Wolk, 1982).

3.3 Pattern Formation

The filamentous cyanobacteria *Anabaena/Nostoc* species maintain a normal pattern of spaced heterocysts during growth in the presence of atmospheric nitrogen. Division of vegetative cells widens the space between heterocysts, but the spacing pattern is maintained by the formation of new heterocysts midway between two existing heterocysts (Wolk and Quine, 1975). Apparently, most of the cells in the filament have potential to differentiate into heterocysts. In contrast, in *Gloeotrichia* spp., *Cylindrospermum* spp., *Anabaenopsis* spp. *Rivularia* spp. heterocysts are formed at the ends of a filament (Wolk, 2000). It has been suggested that establishment and maintenance of the heterocyst spacing pattern depends on the interaction between diffusible substances that originate in the heterocysts and move along the filament and other factors, presumably proteins, that interact with the diffusible substances and control directly or indirectly, the gene expression (Haselkorn, 1978; Wolk, 1989). Mutants altered in pattern formation of heterocysts have been found useful tools in resolving the nature of regulatory proteins, encoded by a gene, which is required for the pattern formation. Mutants of *Nostoc* and *Anabaena* sp. have been reported with the formation of strings of heterocysts or with only terminal heterocysts, which grow poorly on N_2 (Singh and Tiwari, 1969; Wilcox *et al.*, 1975; Buikema and Haselkorn, 1991). Subsequently, Liang *et al.* (1992) isolated a mutant and characterized *PatA* gene, which controls the development of intercalary heterocysts. Such mutants producing only terminal heterocysts do contain a gene, called *hetR*, which is required for heterocyst differentiation. However, inactivation of *hetR* gene by mutation results in the loss of both terminal and intercalary heterocyst formation (Buikema and Haselkom, 1991). This indicates that both *hetR* and *PatA* genes are the components of the same regulatory circuit for controlling the heterocyst development. The genus *Cylindrospermum*, having similar pattern of heterocyst formation (*i.e.* short trichomes producing only terminal heterocysts) as found in *PatA* mutant of *Anabaena* 7120, grows very well on atmospheric nitrogen, while *PatA* mutants are unable to grow properly under similar conditions. It is not clearly known why N_2-growth is normal in *Cylindrospermum* but not in *PatA* mutant. This means *patA* gene product is required for the development of intercalary heterocysts. This also suggests that *PatA* is required to interpret the environment only in intercalary cells. Recently, Black and Wolk (1994) reported *het:N* gene that controls heterocyst formation in N10 mutant of *Anabaena* 7120 as reported by Ernst *et al.* (1992). Simultaneously, Wei *et al.* (1994) came up with another novel mutant of the same cyanobacterial strain, which requires ammoniacal nitrogen for growth. The characteristic of this mutant is that it is unable to grow on nitrate and lacks heterocyst formation. These workers have identified the mutated gene *ntcA*, which is required for growth on nitrate and heterocyst formation. Failure of this mutant to rearrange *nifD* element may be attributed to the early block in heterocyst differentiation and aerobic nitrogen fixation. They have provided further evidence to demonstrate that *ntcA* required for transcription of a *glnA* nif like promoter and is also required for heterocyst differentiation. The gene was originally identified and characterized as sequence of specific DNA-binding protein which interacts with *Anabaena* 7120 *xisA* gene, which encodes a developmentally regulated site-specific recombinase (Chastain *et al.*, 1990). It is however, not known why heterocyst differentiation is blocked in *ntcA* mutant.

4. Nitrogen Starvation and Associated Processes

Adaptive response of heterocystous cyanobacteria to nitrogen starvation does not lead changes only in cell morphology (heterocyst development) but also in reorganisation of gene expression and gene rearrangement associated with the cellular process (Stock *et al.*, 1989). The most recent and exciting discovery is that a developing heterocyst of *Anabaena* 7120 is accompanied by the rearrangement of nitrogen fixing genes (operon nifHDK). The *nifD* gene is interrupted by 11 kb DNA element that is excised from the main genome involving site-specific recombinasé enzyme (Golden *et al.*, 1985; Haselkorn, 1992). This important process requires an anaerobic condition, which is provided by the differentiating heterocysts during late stage. Simultaneous excision of 55 kb element also occurs which blocks the transcription of *nifS* gene of *fdxN* operon. In the absence of excision of these elements it is impossible for the interrupted genes to transcribe and produce active nif gene proteins. But once the nif genes (nifHDK) are reconstructed, the three active polypeptide components of the nitrogenase complex are encoded by three linked genes, *nifHDK*, and transcribed in that order (Haselkorn, 1992). *hupL* encodes the large subunits of a membrane-bound NiFe uptake hydrogenase that can reassimilate hydrogen produced during N_2-fixation. Within SalD fragment of the chromosome of *Anabaena* sp. strain PCC7120 (Matveyev *et al.*, 1994) but not in *Anabaena* variabilities ATCC 294133 (Bohme, 1998) or *Nostoc* sp. strain PCC 73102 (Oxelfelt *et al.*, 1998), *hupL* is interrupted by a third excisable element 10.5 kb in length. Most heterocyst-forming cyanobacteria tested possess the 11 kb element, whereas 55 kb element is less common (Carrasco and Golden, 1995; Golden 1998; Meeks *et al.*, 1994).

5. Nitrogen Regulating Proteins

Many genes are expressed in two types of cells of heterocystous cyanobacteria, but the factors controlling differential transcription are not known. Two sigma factors are produced in nitrogen-starved filaments but the inactivation of genes encoding these factors does not affect the process of heterocyst differentiation (Brahamsha and Haselkorn, 1992). In recent years, several genes have been identified whose function is required for nitrogen deprivation to evoke heterocyst differentiation and pattern formation. One of these, ntcA/ NtcA, a global nitrogen regulatory protein in the cyclic AMP receptor protein family act as nitrogen sensor and also as transcriptional activators, is required for the utilization of nitrate and for heterocyst differentiation (Frias *et al.*, 1994; Meeks and Elhai, 2002). NtcA binds to a putative consensus sequence that is found upstream of the promoter of a number of Cyanobacterial genes and is presumed to exert its activity by activating expression. It is required for the induction of *hetR* to the moderate level of expression only in a fraction of cells, those that are in a critical stage of the cell cycle, mediated through HetF. Expression of HetR indeces the synthesis and release of an inhibitory signals, presumably PatS (HetN-related products). A relatively high intracellular level of HetR would, through positive feedback, tend to reduce expression of hetR, committing the cell to differentiation. Other heterocyst-specific genes whose products are required for heterocyst differentiation and nitrogen fixation are *hetP, hglb, hglCD, hglE, hglK, devABC, hepA, hepB, hepC and hepK, Xis* and nif genes (Wolk 2000; Kaneko *et al.* 2001; Meeks and Elhai 2002). Conversely, a high level of PatS transported from adjacent initiate cells, would tend to decrease the level of hetR expression, leading to regression (Yoon and Golden, 1998).

5.1 N_2-Fixation

Cyanobacterial nitrogen fixation, the conversion of dinitrogen to ammonia, is a reaction that is catalyzed by the enzyme complex nitrogenase. This is made up of two components. Component Ist (MO-Fe protein/nitrogenase) and component IInd (Fe protein/ nitrogen reductase). Substrate binding and reduction takes place on component Ist and the role of component IInd is to donate electrons, one at a time to component Ist. Both components are irreversibly inactivated by exposure to O_2 (Haselkorn, 1978). The considerable amount of ATP required for nitrogen fixation is derived from cyclic photophosphorylation and from respiration (Houchins, 1985; Tel-Or and Stewart, 1977). The presence of contiguous *nifHDK* and *fdXN* operons reconstructed during heterocyst differentiation in *Anabaena* PCC 7120 encodes the active structural components of the enzyme complex. The biological nitrogen fixation activity associated with heterocystous cyanobacteria makes the organisms to grow in the absence of fixed nitrogen.

6. Regulation of Nitrogen Metabolism

Cyanobacteria, in addition to their N_2-fixing ability, can utilize a wide range of nitrogenous compounds for their growth although the preferred source of nitrogen is ammonia. The assimilation of either exogenous ammonia or ammonia derived from N_2-fixation or NO_3^- metabolism by cyanobacteria is accompanied by the ATP-dependent reaction catalyzed by glutamine synthetase. When cells are grown under conditions in which ammonia is limiting, there is a dramatic increase in the transcription of a number of genes. This aspect in respect of nitrogen metabolism has been worked out in clear details in three members of enterobacteria (Stock *et al.*, 1989) and the problem of preference of one nitrogen source like ammonia over another nitrogen source like NO_3^-, N_2, and amino acids has been worked out at molecular level. The enterobacterial system generates repressor signal for assimilation of these nitrogen sources when supplied with exogenous ammonia. In the absence of ammonia the condition of nitrogen limitation sets in, resulting in the production of signal essential for activation of gene transcription for N_2-fixation or nitrate assimilation. The *Ntr* regulon constitutes a subset of these nitrogen-regulated genes that are regulated by common regulators (Magasanik, 1982; Magasanik and Neidhardt, 1987; Reitzer and Magasanik, 1987; Stock *et al.*, 1989). This means that Ntr-dependent adaptive response to nitrogen starvation involves an increase in the level of glutamine synthetase to enhance the ability of cells to use low concentrations of ammonia and then the *Ntr* operon product such as histidase and nitrogenase to provide ammonia from sources of organic nitrogen or from fixation of dinitrogen.

Although starvation for a source of fixed nitrogen is known to be the primary signal for differentiation, the biochemical mechanisms by which N-starvation mediates control of differentiation are not fully known. One of the earliest detectable events upon nitrogen-step-down is degradation of proteins (Wood and Haselkorn, 1978; 1980), and Fleming and Haselkorn (1974) suggested that the product of proteolysis might regulate heterocyst differentiation. Two proteins are known to degrade proteins during nitrogen starvation: (a) protease, a soluble, monospecific, Ca_2^+ requiring protein, that acts early (Wood and Haselkorn, 1978) and cyanophycinase, a specific protein to degrade cyanophycin reserve, is preferentially active in heterocysts (Gupta and Carr, 1981). The role of protein turnover may be attributed to change the gln: a-KG ratio in cells, a condition required for a global nitrogen regulatory protein to switch on genes required for specific nitrogen assimilation associated with heterocyst development. The presence of ammonia in MSX-treated cells suggests that ammonia limitation is not the signal for heterocyst differentiation. The identification of chemical molecule(s) may lead to identification of molecule(s) that regulate differentiation of heterocysts (Thiel, 1990).

The nitrogen status of the cell is measured in terms of gln: α-KG ratio. Its high and low ratio determines whether the cells are sufficient or deficient in nitrogen. In nitrogen deficient cells the glnD product like uridyl transferase (UTase) function as a sensor of nitrogen status and catalyzes uridylation of gln B gene product. The PII and PII-UMP formed interact with the product of glnE. The complex of PII-UMP plus glnE protein functions like deadenylylating enzyme (adenyl-removing enzyme) on adenylated inactive GS. The result is the production of deadenylylated GS fully active in ammonia assimilation within nitrogen starved cells. The same PII-UMP also interacts in other direction involving activation of genes for nitrogen-regulated operons. The PII-UMP also interacts with NtrB protein malting it to function like kinase on Ntr protein. Converting the later into phosphorylated form *i.e.* NtrC-P, which is real activator of nif gene. Infact, NtrC is essential for activation of LA operon of nif gene complex and the product of nifA gene activates, in turn, the genes of other nif operons leading to production of functional nitrogenase and fixation of dinitrogen. When N_2-fixing cells are supplied with ammonia as nitrogen source the ratio of GS: α-KG become high. This results in activation of uridyl-removing activity on PIT-UMP. The consequence is production of PII, which in turn, promotes adenylylating activity of glnE product on GS causing generation of inactive GS-UMP complex. It also complexes with NtrB product making it function like phosphatase on NtrC-P. This phosphorylated NtrC, thus formed, cannot function in activation of nif LA operon, thus leading to switching off the phenomenon of nitrogen fixation. The given mechanism of nitrogen signal transduction is the mechanism of why a bacterial cell does not fix dinitrogen when supplied with ammonia or vice-versa.

6.1 Regulation of Cyanobacterial Glutamine Synthetase

The glutamine synthetase in cyanobacteria is similar to classical procaryotic GS type I that has been widely studied in enterobacteria and whose regulation is well known as described above. In contrast to the enterobacterial enzyme, cyanobacterial GS is not regulated by adenylylation in response to nitrogen source (Fisher *et al.*, 1981; Merida *et al.*, 1991). However, in *Synechocystis* sp. short-term inactivation of G31 promoted by ammonia has been reported. This inactivation seems to involve a phosphorylated compound as has been reported by Merida *et al.* (1991). This also provides evidence for the absence of an enterobacterial-type of nitrogen-regulated (Ntr) system and thus of PII, protein known to play a central role in the modulation of both activity and synthesis of GS in enteric bacteria. It has been recently shown that a glnA mutant of *Agmenellum guadruplicatum* is able to grow in the absence of glutamine, indicating that *glnA* is non-essential for ammonia assimilation. This also suggests that another enzyme is responsive for the glutamine synthesis (Wagner *et al.*, 1993). A new type of GS in the cyanobacterium *Synechocystis* sp. has been reported which is encoded by gin N gene (Reyes and Florencio, 1994). However, this enzyme is not found in N_2-fixing filamentous species. In heterocystous cyanobacteria, GS is found both in heterocysts and vegetative cells (Stewart *et al.*, 1985). The amount of enzyme is two fold more in heterocysts than in vegetative cells (Stewart *et al.*, 1980). The regulation of GS synthesis in *Anabaena* sp. 7120 has been characterized: several promoters have been found for the structural GS gene, *glnA* (Tumer *et al.*, 1983). In N_2-fixing cells, the promoter resembles with *Anabaena nif* genes. Selective transcription from promoter determines that in NH_4^+, grown cells, the level of GS protein is reduced to 50 per cent than that in N_2 grown cells. The divalent cation Mg_2^+ and Mn_2^+ are required for stability of the GS enzyme which has high affinity for ammonium.

Regulation of GS which operates both at the level of enzyme synthesis and enzyme activity, involves several individual or complementary mechanisms like (a) repression and derepression of its biosynthesis by alterations in the nitrogen supply, (b) end product inhibition by glutamine, (c) cumulative feedback, inhibition by multiple end products of glutamine metabolism, (d) NH_4^+ modulation, (e) light-dark modulation, (f) activation by phosphorylated compounds and (g) possible existence of PII system in its regulation. The regulation of most gram-positive bacteria appears to involve a single feedback mechanism in which GS is strongly inhibited by glutamine (Hubbard and Stadtman, 1967). GS deficient, glutamine auxotrophic mutants of *Nostoc muscorum* have been reported (Verma *et al.*, 1990; Srivastava and Amla, 1993). These mutants are incapable of growth on N_2 or NH_4^+ but capable of growth with glutamine as nitrogen source, suggesting that GS is primarily involved in production of glutamine. A preliminary report has been made by Verma *et al.* (1990) in which they have claimed lo transfer this character to the wild strain of *Nostoc muscorum*. Studies with enzyme inhibition with MSX suggest for its involvement, directly or indirectly, in regulation of nitrogenase and heterocyst differentiation in cyanobacteria. The MSX-inhibition of GS activity in N_2-fixing cyanobacteria leads to extracellular release of ammonia (Stewart *et al.*, 1975). Since the chemical inhibitor is very expensive, an alternative way for ammonia excretion is the use of mutant strains defective in GS activity (Spiller *et al.*, 1986; Hien *et al.*, 1988; Tiwari *et al.*, 1991). The immobilized cells of N_2-fixing cyanobacteria have been also found suitable for continuous photoproduction of ammonia (Kerby *et al.*, 1986). The work on auxotrophic mutants like those of GS-defective mutants of cyanobacteria (glutamine requiring mutants for growth) is rather limited. Singh *et al.* (1983) reported GS defective, MSX-resistant mutants of *Nostoc muscorum* capable of producing extracellular ammonia in N_2-fixing growth conditions. Similar mutants have been reported in varied diazotrophic cyanobacteria by other workers (Thomas *et al.*, 1990; Tiwari *et al.*, 1991; Srivastava and Amla, 1993; Mahasneh *et al.*, 1994). Most of these mutants have been found to be resistant to MSX or ED (Ethylenediamine) and are capable of releasing ammonia produced by fixation of dinitrogen.

7. Role of Mutants Impaired in Heterocyst or Metabolism

Mutants of N_2-fixing cyanobacteria that are detective in heterocyst differentiation and nitrogen metabolism serve as premising tools in deciphering the structure and function of genes as reported in N_2-fixing cyanobacteria (Elhai and Wolk, 1990; Buikema and Haselkorn, 1993). An important break through was made by Wolk *et al.* (1984) who developed a conjugative transfer system for *Anabaena* sp. PCC 7120. The system requires the presence of a suitable conjugative plasmid such as RP-4, which mediates the transfer of cargo plasmids from *E. coli* to *Anabaena* cells in the presence of helper plasmid. The shuttle vectors contained a portion of PBR 322 required for replication and mobilization, with deletion of the sites for *AvaI* and *AvaII* (the restriction enzyme present in Anabaena cells). The cargo plasmid carrying transposons (Tn5-1063) is based on PRL 1058 (Wolk *et al.*, 1991). Transposon mutagenesis, using conjugation with a plasmid, has been successful (Borthakur and Haselkorn, 1989; Ernst *et al.*, 1992; Mahasneh *et al.*, 1994; Mishra, 2003a and b). Transposons are known to disrupt the genes and cause production of various mutant phenotypes. Such transposon-tagged gene mutations are advantageous in the sense that these genes can be marked, cloned and exploited in complementation studies. Also, the useful drug markers in transposons, in addition to transposition genes, are quite helpful in positive selection of transformants/exconjugants. The conventional positive selection markers for bacterial crosses are genes conferring resistance to antibiotics. These are described in details by Elhai and Wolk (1988). For *Anabaena* 7120, useful selection can be achieved with neomycin, streptomycin, spectinomycin or chloramphenicol. The possibilities of selection with metal ions such as Hg_2^+ have not been fully explored.

Mutants of diazotrophic cyanobacteria defective in nitrogen assimilating process and heterocyst differentiation have been found useful and important tools in understanding the above closely related processes (Tiwari, 1977; Haury and Wolk, 1978). Several workers described a large number of developmentally impaired mutants of *Anabaena* 7120 and they have been divided into various classes based on morphology and nitrogen fixation ability. These classes are: Het^-(for absence of visible heterocyst development); Fix^- (for absence of nitrogen fixation under any conditions; Fox^-(for no or limited nitrogen fixation in the presence of O_2); Hen^-(for the absence of heterocyst envelopes); Hgl^-(for absence of heterocyst glycolipids); Hep^-(for defects in heterocyst envelop polysaccharide); and Dab^- (for reduced diaminobenzidine oxidation in heterocysts). Many mutants displayed more than a single class of defects, such as Fox^- Hen^-. Among these mutants, Het^-class of mutants are important because they are likely to have defects in early genes possibly involved in the sensing of fixed nitrogen and carbon levels and the commitment to heterocyst development. The interesting point associated with Het^-class of mutants is that these Het^-mutants are incapable of even anaerobic N_2-fixation, with or without dithionite, suggesting a direct linkage of nif gene transcription to heterocyst development (Ernst *et al.*, 1992; Buikerma and Haselkorn, 1993; Mishra 2003a).

8. Acknowledgements

We are thankful to the Head and Programme Co-ordinator (CAS) of the Centre of Advanced Study, Banaras Hindu University, Varanasi, for providing the necessary facilities. Dr. S.K. Singh is also thankful to the CSIR for providing financial support in the form of a Research Associate fellowship.

9. References

Adams D G 1992 Multicellularity in Cyanobacteria. *Symp Soc Gen Microbiol* 47: 341-384.

Black T A and Wolk C P 1994 Analysis of Het⁻mutation in *Anabaena* sp. strain PCC7120 implicates a secondary metabolite in the regulation of heterocyst spacing. *J Bacteriol* 176: 2282–2292.

Bohme H 1998 Regulation of nitrogen fixation in heteroqyst forming Cyanobacteria. *Trends Plant Sci* 3: 346-351.

Borthakur D and Haselkorn R 1989 Tn5 mutagenesis of *Anabaena* sp. strain PCC 7120: Isolation of a new mutant unable to grow without combined nitrogen. *J Bacteriol* 171: 5759-5761.

Bottomley P J, Grillo J F, Van Baalen C and Tabita F R 1979 Synthesis of nitrogenase and heterocysts by *Anabaena* sp. Ca in the presence of high level of ammonia. *J bacterial* 140: 938-943.

Brahamsha B and Haselkorn R 1992 Identification of multiple RNA polymerase sigma factor homologs in the cyanobacterium *Anabaena* sp. PCC 7120: cloning, expression and inactivation of the sigB and SigC genes. *J Bacteriol* 174: 7273-7282.

Buikema W J and Haselkorn R 1993 Molecular genetics of Cyanobacterial development. *Ann Rev Plant Physiol Mole Biol* 44: 33-52.

Buikema W J and Haselkorn R 1991 Characterization of a gene controlling heterocyst differentiation in the cyanobacterium *Anabaena* 7120. *Genes Dev* 5: 321–330.

Carrasco C D and Golden J W 1995 Two heterocyst-specific DNA rerrangement of *Nif* operons in *Anabaena cylindrica* and *Nostoc* sp. strain *Mac Microbiol* 141: 2479-2487.

Chastain C J, Brusca J S, Ramasubranian T W, Wei T F and Golden J W 1990 A sequence specific DNA- binding factor (VFI) from *Anabaena* sp. strain PCC 7120 vegetative cells binds to three adjacent sites in the *XISA* upstream region. *J Bacteriol* 172: 5044-5051.

Elhai J and Wolk C P 1988 Conjugal transfer of DNA to cyanobacteria. *Methods in Enzymol* 167: 747–754.

Elhai J and Wolk C P 1990 Developmental regulation and spatial pattern of expression of the structural genes for nitrogenase in the cyanobacterium *Anabaena*. *EMBO J* 9: 3379–3388.

Ernst A, Black T, Cai Y, Panoff J M, Tiwari D N and Wolk C P 1992 Synthesis of nitrogenase in mutants of the cyanobacterium *Anabaena* sp. strain PCC 7120 affected in heterocyst development or metabolism. *J Bacteriol* 174: 6025-6032.

Fay P 1992 Oxygen relations of nitrogen fixation in cyanobacteria. *Microbiol Rev* 56:340–373.

Fay P, Stewart W D P, Walsby A E and Fogg G E 1968 Is the heterocyst the site of nitrogen fixation in blue green algae. *Nature* 220: 810–812.

Fisher R, Tuli R and Haselkorn R 1981 A cloned cyanobacterial gene for glutamine synthetase function in *Escherichia coli*, but the enzyme is not adenylylated. *Proc Natl Acad Sci USA* 78: 3393-3397.

Fleming, H. and Haselkorn, R. 1974. The program of protein synthesis during heterocyst differentiation in nitrogen fixing blue green algae. *Cell* 3: 159-170.

Fogg G E 1949 Growth and heterocyst production in *Anabaena* cylindrical Lemm. In relation to carbon and nitrogen metabolism. *Amm Bot* 13: 241-259.

Frias J E, Flores E and Herrero A 1994 Requirement of the regulatory protein NtcA for the expression of nitrogen assimilation and heterocyst development genes in the Cyanobacterium *Anabaena* sp. PCC 7120. *Mol Microbiol* 14: 823–832.

Golden J W 1998 Programmed DNA rearrangement in cyanobacteria. In: de Bruiijn FJ, Lupski JR and Weinstock GM (eds) Bacterial Genome, Physical Structure and Analysis. Chapman and Hall New York.

Golden J W, Robinson S J and Haselkorn R 1985 Rearrangement of nitrogen fixation genes during heterocyst differentiation in the Cyanobacterium *Anabaena*. *Nature* 314: 419-423.

Gupta M and Carr N J 1981 Enzymology of arginine metabolism in heterocyst forming Cyanobacteria. FEMS *Microbiol Letts* 12: 179-181.

Haselkorn R 1978 Heterocysts. *Ann Rev Plant Physiol* 29: 319-344.

Haselkorn R 1992 Developmentally regulated gene rearrangements in prokaryotes. *Ann Rev Gene* 26: 113-130.

Haury J F and Wolk C P 1978 Classes of *Anabaena variabilis* mutants with oxygen-sensitive nitrogenase activity. *J Bacteriol* 136: 688-692.

Hien N T, Kerby N W, Machray G C, Rowell P and Stewart W D P 1988 Expression of glutamine synthetase in mutant strains of the cyanobacterium *Anabaena variabilis* which liberate ammonia. FEMS *Microbiol Lett* 56: 337-342.

Houchins J P 1985 Electron transfer chains of cyanobacterial heterocysts. In Ludden P W, Burris J E (eds.) Nitrogen Fixation and CO_2 Metabolism. Elsevier New York.

Hubbard J S and Stadtman E R 1967 Regulation of glutamine synthetase. II. Patterns of feedback inhibition in microorganisms. *J Bacteriol* 93: 1045-1055.

Kaneko T, Nakamura Y, Wolk C P, Kuritz T, Sasamoto S, Watanabe A, Iriguchi M, Ishikawa A, Kawashima K, Kimura T, Kishida Y, Kohara M, Matsumoto M, Matsuno A, Muraki A, Nakazaki N, Shimp O, Sugimoto M, Takazawa M, Yamada M, Yasuda M and Tabata S 2001 Complete genomic sequence of the filamentous nitrogen-fixing cyanobacterium *Anabaena* sp. Strain PCC 7120. *DNA Res* 8: 205–213.

Kerby N W, Musgrave S C, Rowell P, Shestakov S V and Stewart W D P 1986 Photoproduction of ammonium by immobilized mutant strains of *Anabaena variabilis*. *App Microbiol Biotechnol* 24: 42–46.

Liang J, Scappino L and Haselkorn R 1992 The *patA* gene product, which contains a region similar to CheY of *Escherichia coli*, controls heterocyst pattern formation in the cyanobacterium *Anabaena* 7120. *Proc Natl Acad Sci* USA 89: 5655–5659.

Magasanik B 1982 Genetic control of nitrogen assimilation in bacteria. *Ann. Rev Genet* 16: 135-168.

Magasanik B and Neidhardt F C 1987 The regulation of carbon and nitrogen utilization. In: Neidhardt F C, Ingraham J L, Low K B, Magasanik B, Schaechter M and Umbarger H E (eds) *Escherichia coli* and *Salmonella typhimurium:* Cellular and Molecular Biology. American society for microbiology, Washington, DC. pp 1328-1325.

Magasanik B, Schaechter M and Umbarger H E (eds) *Escherichia coli* and *Salmonella typhimurium:* Cellular and Molecular biology. American society for microbiology, Washington, DC. pp, 302-320.

Mahasneh I A, Mishra A K and Tiwari D N 1994 Transposon–induced mutants of the Cyanobacterium *Anabaena* sp. PCC 7120 capable of ammonium liberation. *Biotechnol Letts* 16: 765-770.

Matveyev A V, Rutgers E, Soderback E and Bergman B 1994 A novel genome rearrangement involved in heterocyst differentiation of the cyanobacterium *Anabaena* sp. PCC 7120. FEMS *Microbiol Letts* 116: 201-208.

Meeks J C, Campbell E L and Bisen P S 1994 Elements interrupting nitrogen fixation genes in cyanobacteria: presence and absence of a *nif* D element in clones of *Nostoc* sp. Strain. *Mac Microbiol* 140: 3225-3232.

Meeks J C, Wolk C P, Lackau W, Schilling N, Shaffer P W and Chein W S 1978 Pathways of assimilation of (^{13}N) N_2 and $^{13}NH_4^+$ by cyanobacterium with and without heterocysts. *J Bacteriol* 134: 125-130.

Meeks J C and Elhai J 2002 Regulation of cellular differentiation in filamentous cyanobacteria in free-living and plant associated symbiotic growth states. *Microbiol Molecu Biol Rev* 66: 94–121.

Merida A, Candau P and Florencio F J 1991 *In vitro* reactivation of *in vivo* ammonium–inactivated glutamine synthetase from *Synechocystis* sp. PCC 6803. *Biochem Biophys Res Commun* 181: 780-786.

Mishra A K 1997 Regulation of cellular constituents, heterocyst development, photosynthetic O2-evolution and enzyme activities of Anabaena sp. PCC 7120 by nitrogen sources. *Cytobios* 89: 173–182.

Mishra A K 2003a. Fox⁻and Fix⁻mutants of *Anabaena* 7120 defective in heterocyst development and nitrogen fixation. *Algological Studies* 108: 75-85.

Mishra A K 2003b MSX-resistant mutants of *Anabaena* 7120 with derepressed heterocyst development and nitrogen fixation. *World J Microbiol Biotechnol* 19: 675-680.

Ownby J D 1977 Effect of amino acids on methionine sulfoximine induced heterocyst formation in *Anabaena*. *Planta (Berl)* 136: 277–289.

Oxelfelt F, Tamagnini P and Lindblad P 1998 Hydrogen uptake in *Nostoc* sp. strain PCC 73102. Cloning and characterization of *hup*SL homologue. *Archiev of Microbial* 169: 267-274.

Ray T B, Peters G A, Toia R E and Mayne B C 1978 *Azolla- Anabaena* relatioship VII. Distribution of ammonium assimilating enzymes, protein and chlorophyll between whost and symbiont. *Plant Physiol* 62: 463- 467.

Reitzer L J and Magasanik B 1987 Ammonia assimilation and the biosynthesis of glutamine, glutamate, aspartate, asparagines, L-alanine and D-alanine. In: Neidhardt F C, Ingraham J L, Low K B, Magasanik B, Schaechter M and Umbarger H E (Eds) *Escherichia coli* and *Salmonella typhimurium:* Cellular and Molecular Biology. American society for microbiology, Washington, DC. pp, 302-320.

Reyes J C and Florencio F J 1994 A new type of glutamine synthetase in cyanobacteria: the protein encoded by the *glnN* gene supports nitrogen assimilation in *Synecocystis* sp. strain PCC 6803. *J Bacteriol* 176: 1260-1267.

Singh H N Ladha J K and Kumar H D 1977 Genetic control of heterocyst formation in the blue green algae *Nostoc muscorum* and *Nostoc linckia*. *Archiev Microbiol* 114: 155-159.

Singh R N and Tiwari D N 1969 Induction of ultraviolet irradiation of mutation in the blue green alga *Nostoc linckia* (Roth.) Born et Flah. *Nature* 221: 62-64.

Singh R N and Tiwari D N 1970 Frequent heterocyst germination in the blue green alga *Gloeotrichia ghosei* Singh. *J Phycol* 6: 172-176.

Singh R N, Tiwari D N and Singh V P 1972 Genetic basis of cellular differentiation in blue green algae. In: Desikachary TV (Ed.) Taxonomy and Biology of Blue- green Algae. Madras. pp, 27-37.

Singh H N, Rai U N, Rao U V and Bagchi S N 1983 Evidence for ammonia as an inhibitor of heterocyst and nitrogenase formation in the cyanobacterium *Anabaena cycadae*. *Biochem Biophys Res Commun* 111: 180–187.

Spiller H, Latorre C, Hassan M E and Shanmugam K T 1986 Isolation and characterization of nitrogenase-derepressed mutant strains of cyanobacterium *Anabaena variabilis*. *J Bacteriol* 165: 412–419.

Srivastava R and Amla D V 1993 Physiological and biochemical analysis of the glutamine synthetase impaired mutants of the nitrogen fixing cyanobacterium *Nostoc muscorum*. *Curr Microbiol* 26: 205-215.

Stewart W D P and Rowell P 1975 Effect of L-Methionine-DL-Sulfoximine on the assimilation of newly fixed NH_3, acetylene reduction and heterocyst production in *Anabaena cylindrica*. *Biocem Biophys Res Commun* 65: 846-856.

Stewart W D P, Rowell P, Cossak J D and Kerby N W 1985 Physiological studies on N_2-fixing cyanobacteria. In: Ludden PW and Burris J (eds). Nitrogen Fixation and CO_2 Metabolism. Elsevier Science Publishing, Amsterdam. Pp, 269-279.

Stock J B, Ninfa A J and Stock A M 1989 Protein phosphorylation and regulation of adaptive responses in bacteria. *Microbiol Review* 53: 450-490.

Tel-Or E and Stewart W D P 1977 Photosynthetic components and activities of nitrogen fixing, isolated heterocysts of *Anabaena cylindrica*. *Proc R Soc Lond* Ser. B 198: 61–86.

Thiel T 1990 Protein turnover and heterocyst differentiation in the Cyanobacterium *Anabaena variabilis*. *J Phycol* 26: 50–54.

Thomas J, Meeks J C, Wolk C P, Shaffer P W, Austin S M and Chien W B 1977 Formation of glutamine from (^{13}N) ammonia (^{13}N) dinitrogen and (^{14}C) glutamate by heterocysts isolated from *Anabaena cylindrical*. *J Bacteriol* 141: 1375-1385.

Tiwari D N 1977 Pattern of heterocyst differentiation as a function of nitrogen–fixing ability in the blue green alga *Nostoc linckia*. *Beitr Biol Pflanzen* 53: 177-186.

Tiwari D N, Ashok Kumar and Mishra A K 1991 Use of Cyanobacterial diazotrophic technology in rice agriculture. *Appl Biochem Biotech* 28/29: 387-396.

Tumer N E, Robinson S J and Haselkorn R 1983 Different promoters for the *Anabaena* glutamine synthetase gene during growth using molecular or fixed nitrogen. *Nature* 306: 337-342.

Verma S K, Singh A K, Katiyar S and Singh H N 1990 Genetic transformation of glutamine auxotrophy to prototrophy in the cyanobacterium *Nostoc muscorum*. *Archiev Microbiol* 154: 414-416.

Wagner S J, Thomas S P, Kaufman R T, Nixon B T and Stevens S E J 1993 The *glnA* gene of the cyanobacterium *Agmenellum* quadruplicatum PR-6 is non-essential for ammonium assimilation. *J Bacteriol* 175: 604-612.

Wei T F, Ramasubramanian T S and Golden J W 1994 *Anabaena* sp. strain PCC 7120 *ntcA* gene required for growth on nitrate and heterocyst development. *J Bacteriol* 176: 4473–4482.

Wilcox M, Mitchison G J and Smith R J 1975 Mutants of *Anabaena cylindrical* altered in heterocyst spacing. *Archiev Microbiol.*, 130: 219-223.

Wolk C P 1967 Physiological basis of the pattern of vegetative growth of a blue–green alga. Proc *Natl Acad Sci* USA 57: 1246-1251.

Wolk C P 1982 Heterocysts. In: Carr NG and Whitton BA (eds) The Biology of Cyanobacteria. Blackwell Scientific Publications, Oxford. Pp, 359-386.

Wolk C P 1989 Alternative models for the development of the pattern of spaced heterocysts in *Anabaena* (Cyanophyta). *Plant Sys Evol* 164: 27-31.

Wolk C P, Cai Y and Panoff J M 1991 Use of a transposon with luciferase as a reporter to identify environmentally responsive genes in cyanobacterium. *Proc Natl Acad Sci* USA 88: 5355-5359.

Wolk C P, Cai Y, Cardemil L, Flores E, Hohn B, Murry M, Schemetterer G, Schrautmeir B and Wilson R 1988 Isolation and complementation of mutants of *Anabaena* sp. strain PCC 7120 unable to grow aerobically on dinitrogen. *J Bacteriol* 170: 12349-1244.

Wolk C P, Ernst A and Elhai J 1994 Heterocyst metabolism and development. In: Bryant (Ed.) Molecular Biology of Cyanobacteria. Kluwer Academic Publishers, Dordrecht, Netherlands, Pp, 769-823.

Wolk C P, Thomas J, Shaffer P W, Austin S M and Galonsky A 1976 Pathway of nitrogen metabolism after fixation of ^{13}N-labelled nitrogen gas by the cyanobacterium *Anabaena cylindrica. J Biol Chem* 251: 5027-5037.

Wolk C P, Vonshak A, Kehoe P and Elhai J 1984 Construction of shuttle vectors capable of conjugative transfer from *Escherichia coli* to nitrogen fixing filamentous cyanobacteria. *Proc Natl Acad Sci* USA 81: 1561-1565.

Wolk C P and Quine M P 1975 Formation of one-dimensional patterns by stochastic processes and by filamentous blue green algae. *Develop Biol* 46: 370–382.

Wolk C P 2000 Heterocyst formation in *Anabaena*. In: Brun, Y V and Shimkets, L J (eds) *Prokaryotic Development*, Washington: American Society for Microbiology. ISBN 1-55581158-2 pp. 83–104.

Wood N B and Haselkorn R 1978 Proteinase activity during heterocyst differentiation in nitrogen fixing Cyanobacteria. In: Cohen G N and Holzer H (eds.) *Limited Proteolysis in Microorganisms.* US Departments of Health, Education and Welfare publication (NIH) 79-1591. US Government printing office. Bethesda, Maryland, pp, 159-166.

Wood N B and Haselkorn R 1980 Control of Phycobiliprotein proteolysis and heterocyst differentiation in *Anabaena. J Bacteriol* 141: 1375–1385.

Yoon H S and Golden J W 1998 Heterocyst pattern formation controlled by a diffusible peptide. *Sci* 282: 935–938.

Chapter 3

Role of Cyanobacteria in Organic-scented Rice Farming

Y.V. Singh and B.V. Singh

Contents

Glimpses of Cyanobacteria, 2006 48–66
Daya Publishing House, New Delhi

ABSTRACT

Organic farming production system aims at promoting and enhancing agro-ecosystem health, biodiversity, biological cycles and soil biological activities. Unplanned commercialization of farming, use of fertilizers and pesticides has resulted in negative effect on the environment, availability of chemical residues in food and an overall reduced quality of food which have led to a marked increase in various diseases. Organic farming has the capability to take care of these problems. Besides, the obvious, immediate and positive effects, organic farming has on the environment and quality of food, it also greatly helps a farmer to become self-sufficient in his requirements for agro-inputs and reduce his costs. Organic farming and trade of organic produce is growing rapidly worldwide and its growth is naturally heading towards what can be called a boom. As per a survey report published in 2003, almost 23 million hectares (m ha) are managed organically worldwide. Role of cyanobacteria (Blue-green algae) as biofertilizer in augmenting soil fertility and rice crop productivity is well documented in the article. Cyanobacteria can play an important role in organic rice farming where use of chemical fertilizers is prohibited. Experimental results showed increase in rice grain yield and concentration of iron/zinc in rice grains due to the application of blue green algae alone or in combination with other organic amendments. There is immense market potential of blue green algal biofertilizer in our country. Commercial production of biofertilizers which can be achieved with an effective popularization campaign is attractively profitable. Production of 'organic basmati rice' has the potential of developing huge export market in the country.

1. Introduction

1.1 What is Organic Farming?

Organic farming is a holistic production management system, which promotes and enhances agro-ecosystem health, including biodiversity, biological cycles and soil biological activity. The words such as 'biological', 'ecological', 'biodynamic', 'organic biological', 'natural', and 'sustainable' are broadly synonymous with organic farming (Lockeretz, 1990). The objective of environmental, social and economic sustainability lies at the heart of organic farming which are both a philosophy and an efficient system of agriculture.

This type of farming has economic, environmental and sustainability implications for the development of agriculture in any country.The term 'organic' is not directly related to the type of inputs used, but refers to the concept of the farm as an organism in which all the component parts *i.e.* soil minerals, organic matter, micro-organisms, plants, animals and humans-interact in a coherent manner. In many countries organic agriculture is known as biological agriculture, reflecting the reliance on ecosystem management rather than external inputs like chemical, organic, biological or otherwise.

1.2 Why to go for Organic Farming

During the last few decades, approach and outlook towards agriculture and marketing of food has seen a quantum change worldwide. Whereas earlier the seasons and the climate of an area determined what would be grown and when, today it is the "market" that determines what it wants and what should be grown. The focus is now more on quantity and outer appearance rather than intrinsic or nutritional quality. Pesticide and other chemical residues in food and an overall reduced quality of food have led to a marked increase in various diseases, mainly various forms of cancer and reduced bodily immunity.

This commercialization of farming has also had a very negative effect on the environment. The use of pesticides has led to enormous levels of chemical buildup in our environment, soil, water, air, animals and even in our own bodies. Fertilisers have a short-term effect on productivity but a longer-term negative effect on the environment where they remain for years after leaching and running off, contaminating ground water and water bodies. In the name of growing more to feed the earth, we have taken the wrong road of unsustainability. The effects already show–farmers committing suicide in growing numbers with every passing year; the horrendous effects of pesticide sprays (endosulphan) by a government-owned plantation in Kerala, India some years ago; the pesticide-contaminated bottled water and aerated beverages are only some instances. The bigger picture that rarely makes news however is that millions of people are still underfed, and where they do get enough to eat, the food they eat has the capability to eventually kill them. Yet, the picture painted for the future by agro-chemical and seed companies and governments is rosy and bright. Another negative effect of this trend has been on the fortunes of the farming communities worldwide. Despite this so-called increased productivity, farmers in practically every country around the world have seen a downturn in their fortunes.

This is where organic farming comes in. Organic farming has the capability to take care of each of these problems. Besides, the obvious, immediate and positive effects organic or natural farming has on the environment and quality of food, it also greatly helps a farmer to become self-sufficient in his requirements for agro-inputs and reduce his costs. Organic farming aspires to a combine mixtures of agronomic, environmental, social and ethic objectives (Stockdale *et al.*, (2001).

2. History of Organic Farming

Detailed description about the principle and practices of organic farming conducted in different countries are available in literature and the roots of organic farming practices can be traced in European literature back to late nineteenth/early twentieth century (Conford, 1988). An Austrian philosopher Dr Rudolf Steiner presented an alternative vision of agriculture arising from his spiritual science of Anthroposophy. Slightly later, Dr Hans Muller, working in Switzerland and later in Germany founded a movement for agricultural reforms on the principles of self-sufficiency and economic viability of farms in which soil fertility was to be maintained through crop rotation, careful management and use of animal manures (Zeichaus-Hartelt, 1991). Later Dr Hans Peter Rusch contributed important ideas relating to soil fertility and soil microbiology, which led to the further development of organic-biological agriculture in German speaking countries (Rusch, 1968).

In the U.K and other countries the world scientists like Sir George Stapledon and Sir Albert Howard were influenced by the agricultural ideas of Rudolf Steiner, although they did not adopt them directly (Howord, 1943; Conford,1988).Stapleton's work with alternate husbandry system and Howard 's work on the role of organic matter in soil and composting provided a powerful stimulus to Lady Eve Belfour (Belfour,1943). In 1939, she began the Haughley Experiment, which continued till 1969, to investigate the link between the way food is produced, food quality and human health (Belfour, 1976). Howard and Belfour's ideas emphasizing the role of a healthy, fertile soil in the production of healthy crops and livestock and the link with human health were persued in the U.S by the Rodale family, who founded the Soil and Health Foundation in 1947 (Harwood, 1990). The early organic movement focused strongly on the issues of human health and promotion of soil fertility through the use of composts and organic fertilizers. Pesticides did not become a major issue in organic agriculture until the publication of *Silent Spring* generated widespread public concern (Carson, 1963).

The formation of the International Federation of Organic Agriculture Movement (IFOAM) in 1972 gave an international framework for the discussion and codification of internationally recognized principles of organic farming. FAO and WHO have officially declared that International guidelines on organically produced food products should be considered important for consumer protection and information as they facilitate trade (Yussefi, 2003). The Codex Alimentarius Commission, a joint FAO/WHO food standards programme, the body that sets International food standards, started to develop Guidelines for the production, processing, labeling and marketing of organically produced food in 1991.The Codex guidelines are important for equivalence judgements under the rule of World Trade Organization (WTO). Modern organic farming represents a merging of a number of different streams of thought (Conford, 1988;Harwood, 1990 and Tate, 1994). There are strong evidences that organic farming systems can deliver agronomic and environmental benefits both through structural changes and the tactical daily management of the farming system.

3. Organic Farming in India and the World

As per an Food and Agriculture Organization (FAO) study of mid-2003, India has 1,426 certified organic farms producing approximately 14,000 tons of organic food/produce annually. There are a number of farms in India which have either never been chemically managed/cultivated or have converted back to organic farming because of their farmers' beliefs or purely for reason of economics. These thousands of farmers cultivating hundreds of thousands of acres of land are not classified as organic though they are. Their produce either sells in the open market along with conventionally grown produce at the same price or sells purely on goodwill and trust as organic through select outlets and regular specialist bazaars. These farmers may never opt for certification because of the costs involved as well as the extensive documentation that is required by certifiers.

Organic trade has become a growing reality and the growth is naturally heading towards what can be called a boom. According to SOEL Survey (February, 2003), almost 23 million hectares (m ha) are managed organically worldwide. Currently, the major part of this area is located in Australia (10.5 m ha), Argentina (3.2m ha) and Italy (1.2m ha) (Hussein, 2003). Recent years have seen very rapid growth in organic farming, particularly in Europe and United States but also in many other regions of the world including China, Latin America, Australia and Africa. In European Union, area under policy supported organic production increased to 2.1 per cent of total utilizable agricultural area (UAA) in 1998 from 0.1 per cent UAA in 1985 which is a 30-fold increase in 13-years (Limpkin, 1999). Several countries have now achieved 3-10 per cent of their agricultural area managed organically, and in some cases more than 30 per cent on regional basis. But many countries are still at or below 1 per cent level and India is in this category where only 0.03 area is under organic agriculture. A list of major organic agriculture adopting countries and India's position is given in Table 3.1.

Table 3.1

Sl. No.	*Country*	*Organic Hectares*	*% of Agricultural Area*	*Organic Farms (No.)*
1.	Australia	10,500,000	2.31	1,380
2.	Argentina	3,192,000	1.89	1,900
3.	Italy	1,230,000	7.49	56,440
4.	USA	950,000	0.23	9,949
5.	United Kingdom	679,631	3.96	3,981
6.	Uruguay	678,481	4.00	334
7.	Germany	632,165	3.70	14,703
8.	Spain	485,079	1.66	15,607
9.	Canada	430,600	0.58	3,236
10	France	419,750	1.40	10,364
	India	41,000	0.03	5,661
	Total (World)	**22,811,267**		**398,804**

4. The Blue-green Algae or Cyanobacteria

Algae are the heterogenous assemblage of plants that includes prokaryotes and eukaryotic organisms. They are the pioneer colonizers both in hydrosphere and xerosphere and occupy the base of the trophic pyramid. These organisms have been found to synthesize 0.8×10^{11} tonnes of organic matter, constituting about 40 per cent of the total organic matter synthesized annually on this planet (Goyal, 2002).

Blue-green algae/cyanobacteria constitute the largest, most diverse and widely distributed group of prokaryotic microscopic organisms that perform oxygenic photosynthesis. These are also known as Cyanobacteria, Myxophyceae and Cyanophyceae. These are ubiquitous in distribution, more common in tropics; and are able to withstand extremes of temperature and drought. Their ability to withstand adverse ecological conditions, capacity to thrive well in hostile environments and response to the onset of dry conditions by entering into dormant resistant state have distinguished this group of organisms as pioneers of plant succession. Nitrogen fixation is carried out in specialized cells known as heterocysts, which have thick walls and hence, physically prevent the entry of oxygen and provide necessary anaerobic conditions for the activity of the enzyme nitrogenase. Heterocysts have a high rate of respiration that scavenges the diffused oxygen, lack photoreaction II of photosynthesis due to which there is no splitting of water and evolution of oxygen during photosynthesis (Thomas, 1970). Therefore, these cells act as ideal site for nitrogen fixation under aerobic conditions. Although, nitrogenase enzyme is reported to be present in vegetative cells also, however, it remains inactive because of the presence of oxygenic photosynthesis.

The basic significance of the ecological observations on the abundance of blue-green algae in Indian rice field soils became apparent when it was recognized that heterocystous forms could fix atmospheric nitrogen that is made available to the plants during life cycle and after its death by decomposition of cells, which became available to the subsequent crops. Till the findings of Wyatt and Silvey (1969), only forms like *Anabaena. Nostoc, Cylindrospermum, Tolypothrix, Calothrix, Scytonema, Hapalosiphon, Westiellopsis* and *Stigonema* belonging to orders Nostocales and Stigonematales were considered to be capable of fixing atmospheric nitrogen. Since then, many of the non -heterocystous forms are also reported to fix nitrogen under aerobic or micro-aerophillic conditions except *Gloeocapsa* that can fix nitrogen under aerobic conditions (Stewart *et al.*, 1979). In the later case, it is accomplished because of the spatial and temporal separation of the processes of carbon and nitrogen fixation. The vegetative cells of trichomes of diazotrophic blue-green algae contain nitrogenase enzyme, which gets inactivated by oxygen. Once nitrogen is fixed, many organisms are known to incorporate ammonia into amino acids by the enzyme glutamate dehydrogenase that is either absent or present in low amounts in these organisms (Pearce *et al.*, 1969). Under the conditions of ammonia limitations, glutamine synthetase and glutamate synthetase is the major pathway for ammonia assimilation (Thomas *et al.*, 1975; Wolk *et al.*, 1976).

5. Utilization of BGA Biofertilizers for Rice Crop

A large number of reports are available in India over the impact of different biofertilizers and soil amendments on soil fertility and productivity. The positive role of organic manure and biofertilizers in augmenting soil fertility and crop productivity is well known. Presently, increasing cost of fertilizers, growing ecological and produce quality concern, harmful effect of agrochemical and pesticides and need for energy conservation have created considerable interest for the use of organics as a source of plant nutrition (Deka Medhi *et al.*, 1996). Thus, biofertilizers and organic manures can play an important role in organic farming where use of chemical fertilizers is prohibited. Utilization impact of these bio and organic sources in crops like rice or rice based cropping systems in term of nutrient supply is well documented in literature (Bhardwaj *et al.*, 1981; Singh and Singh, 1987; Mandal *et al.*, 1992; Deka Medhi *et al.*, 1996; Mahapatra, *et al.*, 1997; Singh and Mandal, 1997, 2000).

De (1939) was the first to suggest a positive role for the blue greens in the sustenance of the nitrogen status of rice fields of our country. Since then lot of information has been generated in tropics regarding improvement in the fertility status of rice soils to sustain rice yields by utilizing diazotrophic blue-green algae as biological input (Venkataraman, 1972; Singh, 1961; Singh and Bisoyi, 1989). Better crop yields under pot culture experiments were attributed to enhanced N levels due to Blue-Green Algae inoculation (De and Mandal, 1956). Singh (1939, 1942) in Uttar Pradesh attributed the enhanced nitrogen contributions to the nitrogen fixing capacity of blue-green algae and the nitrogen contribution was calculated as 48 kg N/ha/crop. They also suggested the persistence of residual effect of these organisms in the following season. Venkataraman (1966) initiated the work on the algalization of Indian rice fields and much interest was generated in this aspect. The All India Coordinated Project on the utilization of blue-green algae in rice fields throughout the country revealed that the supplementation of chemical fertilizer input with blue greens could save up to 30 per cent of the chemical fertilizer (Venkataraman, 1972, 1981). Multi-location trials conducted under varying agro-climatic conditions using different rice varieties indicated that the algal inoculation could save 30 kg N/ha that, however, this depends upon the agro-ecological conditions. (Venkataraman and Goyal, 1969). The performance of *Aulosira fertilissima* in water logged rice fields was assessed and a 14 per cent increase in soil nitrogen by Blue-Green Algae incorporation was reported (Singh, 1976; Singh and Bisoyi, 1989). A significantly positive effect of Blue-green algae and *Azolla* on rice crop has been reported in West Bengal (Singh and Singh, 1987; Singh and Mandal, 1997,2000). Singh *et al.* (1981) also reported that Blue-Green Algae, besides amending the flooded soils, released ammonium nitrogen and enhanced the level of soil carbon. Venkataraman (1980) suggested the use of local or native strains for an efficient utilization and better crop response. Blue-green algae are also considered to minimize the occurrence of weeds in the rice fields (Subramanyan, 1972) besides producing growth-promoting substances. Extracts of *Cylindrospermum* sp. (Mohan and Mukherjee, 1980) or *Hapalosiphon welwitschii* (Bongale and Bharati, 1990) accelerated the germination of seedlings. Gupta and Shukla (1967) found an increase in the shoot and root length and protein content of rice grains by pretreatment of seeds with *Phormidium* extract. Reclamation of the alkaline soils in Uttar Pradesh was demonstrated by Singh (1950, 1961) by introducing blue greens along with rice and sugarcane crops. These were reported to reduce the pH of soil and improve upon exchangeable calcium and water holding capacity. Under these situations, Na^+ ions are immobilized and the sodium clay is converted to calcium type (Kaushik, 1989; Apte and Thomas 1984). Rice fields in Kerala having acidic soils with low pH support the growth of several blue-green algae (Aiyer, 1965; Anand and Shantha Kumar, 1986). Many Cyanobacteria exhibit a wide range of pesticide tolerance (Ahmad and Venkataraman, 1973; Kaushik and

Venkataraman, 1983; Pabbi and Vaishya, 1992). These findings support the view that Blue-Green Algae can be effectively used as a nutrition source in rice.

6. Field Application of Blue-green Algae

The recommended method of application of the algal inoculum is broadcasting on standing water, about 3 to 4 days after transplantation. After the application of algal inoculum, the field should be kept water logged for about a week's time. Establishment of the algal inoculum can be observed within a week of inoculation in the form of floating algal mats, more prominently seen in the afternoon.

Blue-green algae can be used to obtain 10-15 per cent increase in rice crop yield. Recommended crop management practices for organic farming do not normally interfere with the establishment and activity of the blue-green algae in the field. Heavily fertilized rice fields generally show profuse growth of green algae that act as weeds and also reduce tillering in the rice plants. These can be differentiated from the blue-green algae by the grass green colour and fibrous nature. The green algae turn dark violet when treated with iodine but the blue-green algae remain unaffected. The green algae can be removed manually and buried in a pit and if their growth is intense, copper sulphate @ 4.0 kg/ha can be used which will selectively kill the green algae. Blue-Green Algae biofertilizer should be used in every rice crop as a kind of insurance to the crop yield as well as to prevent the deterioration of soil physico-chemical properties.

7. Algal Biofertilizer versus Chemical Fertilizer

Biofertilizers are renewable sources of plant nutrients to supplement chemical fertilizers in sustainable agricultural system. Chemical fertilizer is the major supplier of nutrients besides organic and green manures. Despite sizeable increase in the use of chemical fertilizers, the gap between the nutrient removal and replenishment is quite high. The role of biofertilizers in organic agriculture assumes special significance particularly in present context of high cost of chemical fertilizers. The effect of biofertilizer is not as instantaneously obvious as in the case of chemical fertilizers. Their effect can hardly be seen with the naked eye but one must have the patience to realize the potential of this inexpensive input, which is self-regenerating and has sustained cumulative effect. As the crop plants utilize the products of biological activities of a biofertilizer, there is always a possibility of their population build up in the soil that may exhibit the benefits after a few superimposed applications. These are also known to play an important role in soil conditioning and may also be source of growth regulators and organic matter.

Algal biofertilizers are much cheaper than the chemical fertilizers available in the market. The crop plants are able to utilize more nutrients from the soil in presence of the algal inoculation because of slow release of the fixed and metabolizable nitrogen. The chemical fertilizers provide the nutrients in a cyclic manner but the blue-green algal biofertilizers provide the nutrients in a gradual and linear manner. This results in the utilization of the applied nutrients more effectively. The diazotrophic Cyanobacteria are more compatible with nitrate nitrogen than ammonium nitrogen. This leads to rapid multiplication of blue-green algae in the photic zone of rice field where oxidizing conditions tend to keep the level of nitrate nitrogen high. In presence of combined nitrogen the blue-green algae not only continue to grow but show concentration dependent increase in the uptake of added chemical nitrogen. This metabolized nitrogen which is prevented from being lost through denitrification, percolation and run off, is slowly released through excretion and autolysis (Suseela and Goyal, 1995).

Further, the algal nitrogenase has 'switch on' mechanism, which is activated when the level of combined nitrogen falls below a threshold value, which seems to be around 40 ppm. This enables the increased algal biomass to contribute more of the biologically fixed nitrogen when the level of fertilizer nitrogen in the ecosystem is reduced due to progressive utilization and loss. (Goyal, 2002).

8. Organic Farming of Scented Rice

A number of organic farming practices like use of biofertilizers and other organic amendments can improve the carbon sequestration prevent the soil loss due to erosion and nutrient loss into the water bodies and the environment. One of the basic principles of soil fertility management in organic farming system is the plant nutrition based upon biologically derived nutrients. These nutrients are not readily available to the crop plant and thus, require the activity of soil microbial flora.

Scented (Basmati) rice has a good market locally as well as export potential. This market may be further exploited if scented rice is grown organically. With this background, field experiments were conducted at IARI, New Delhi during 2003-05 on organic farming in scented rice to improve the quality of the produce and soil and long-term environmental sustainability. Pusa Basmati-1 variety of scented rice was taken. In the experiment pulse crop was grown in summer season and its biomass was incorporated in the soil after picking the pods; which enriched the soil with organic matter. In the experiments, organic inputs like *Azolla* @ 1.0t/ha, Blue-Green Algae @15kg/ha, FYM@ 5.0t/ha and Vermicompost @ 5.0 t/ha were applied. For comparison absolute control ($N_0P_0K_0$) and recommended fertilizer dose ($N_{80}P_{30}K_{30}$) were also taken. A significant increase (14.3 to 35.7 per cent) in grain yield of rice was recorded over absolute control due to the application of different biofertilizers and organic amendments when applied alone (Table 3.2). The increase in yield was further enhanced when these organic amendments were applied in combination of two or three bio-inputs and was in the range of 65.8 to 96.4 per cent and 100 to 109.2 per cent, respectively over absolute control. Grain yield (4.25 t/ha) obtained under combined application of four organic amendments was at *par* with the yield recorded under recommended dose of chemical fertilizer application. An interesting observation recorded was that there was no serious attack of any insect pest or disease in organically grown rice or wheat crop.

Microbial population (Actinomycetes, Bacteria, Fungi and Blue-Green Algae) was enhanced due to the application of organic amendments in comparison to control and recommended fertilizer application which accordingly resulted in a notable enhancement in dehydrogenase enzyme activity (Table 3.3). Rice grain analysis for Fe and Zn contents showed a significant increase in Fe contents in the treatments having 2 or more organic amendment added altogether over control. Zn content in grain also increased due to organic treatments but the increment was not significant over control.

9. Future of Cyanobacteria vis-à-vis Organic Farming in India

About 65 per cent of India's cropped area is not irrigated and it can be safely assumed that high-input demanding crops are not grown on these lands. Fertilizer use on dry lands is always low, as chemical fertilizers require sufficient water to respond. Pesticide use also in these lands is low, as the economics of these hardy or "not so profitable" crops will not permit expensive inputs. These areas are at least "relatively organic" or perhaps even "organic by default". While neither of these terms necessarily denotes a healthy farm or a recommended agriculture system, it would at least imply a non-chemical farm that can be converted very easily to an organic one providing excellent yields and without the necessity and effort of a conversion period.

Table 3.2: Effect of Different Organic Treatments on Grain Yield and Zinc and Iron Concentration in Grains of Scented Rice During Kharif, 2003 and 2004

Tr. No.	*Treatments*	*Rice Grain Yield (t/ha)*		*Concentration in Grain*	
		2003[a]	*2004*[b]	*Zinc (ppm)*	*Iron(ppm)*
1	*Azolla*(A)*	2.87	1.81	1.28	2.54
2	BGA (B)	2.70	1.85	1.34	2.46
3	FYM (F)	2.69	1.75	1.32	2.24
4	Vermicompost(V)	2.90	1.94	1.36	2.66
5	A+B	3.35	2.04	1.38	3.25
6	A+F	3.70	1.93	1.40	3.42
7	A+V	4.08	2.10	1.45	3.85
8	B+F	3.33	2.15	1.44	3.26
9	B+V	3.91	2.11	1.45	3.50
10	F+V	3.75	2.17	1.43	3.58
11	A+B+F	4.15	2.20	1.47	3.96
12	A+F+V	4.28	2.15	1.48	4.10
13	B+F+V	4.10	2.22	1.46	3.92
14	A+B+F+V	4.33	2.26	1.50	4.25
15	$N_{80}P_{30}K_{30}$	4.43	1.56	1.21	4.38
16	$N_0P_0K_0$	2.02	1.87	1.24	1.96
	C.D (at 5 per cent)	0.35	0.42	0.18	0.48

Rate of application/ha*: Azolla1.0 tonne;* Blue-Green Algae 30 kg; FYM 5.0 tonne; Vermicompost 5.0 tonne.

a: Pusa Sugandh 2; b: Pusa Basmati 1.

Table 3.3: Effect of Different Organic Treatments on Microbial Population (CFU/gm of soil) and Enzymatic Activity at Mid Season Stage of Rice Crop During kharif, 2003

Tr.No.	*Treatments*	*Actinomy-cetes × 10³*	*Bacteria × 10³*	*Fungi × 10³*	*BGA × 10³*	*Activity* (mgm)*
1	Azolla [A]	307	390	36	56	122.71
2	BGA [B]	331	363	60	74	119.19
3	FYM [F]	250	301	44	63	114.08
4	Vermicompost [V]	282	393	36	27	113.96
5	A+B	300	379	25	17	134.20
6	A+F	289	390	26	33	141.78
7	A+V	192	300	23	51	118.34
8	B+F	255	393	29	47	113.85
9	B+V	235	393	32	60	127.42
10	F+V	280	387	23	55	118.65
11	A+B+F	267	383	49	71	142.34
12	A+F+V	367	440	80	93	126.74
13	B+F+V	350	401	89	91	130.58
14	A+B+F+V	311	360	34	27	127.98
15	$N_{80}P_{30}K_{30}$	197	371	44	13	107.75
16	$N_0P_0K_0$	201	356	27	2	114.53
	C.D at 5 per cent	75	25	8	19	7.58

*Dehydrogenase enzyme activity

There is immense market potential of blue green algal biofertilizer considering the vast area of about 43 mha under rice cultivation in our country. Use of these cheaper and easily manageable inputs can make the agriculture increasingly eco-friendly and induce long-term sustainability. Commercial production of biofertilizers and organic farm produce, which can be achieved with an effective popularization campaign, is attractively profitable and appears to be the appropriate option for popularization and large-scale adoption of the technology. Production of 'organic basmati rice' has the potential of developing huge export market in the country. Those who would like to be involved in the production, promotion and extensive use of biofertilizers and organic farming, must be trained for the purpose. The gap between the researchers, extension workers and farmers need to be bridged through revamped extension system that ensures minimum technology dissemination loss. There is a need to create a viable, remunerative and enabling environment for larger participation of the private sector and farming community in order to realize our goal of maximum Blue-Green Algae biofertilizer production and its application in rice in general and organic scented rice in particular.

10. References

Ahmed M H and Venkataraman G S 1973 Tolerance of *Aulosira fertilissima* to pesticides. *Curr Sci* 42: 108.

Aiyer R S 1965 Comparative algological studies in rice fields in Kerala state. *Agric Res J* Kerala 3: 100.

Anand N and Shanthakumar H 1986 Blue-green algae from several rice fields in Kerala State, India. *Hydrobiologia* 144: 223-227.

Apte S K and Thomas J 1984 Sodium requirement and metabolism in nitrogen fixing Cyanobacteria. *J Bio Sci* 6: 771-794.

Belfour E B 1943 *The Living Soil.* Faber and Faber, London.

Belfour E B 1976 In: *The Living Soil and the Haughley Experiment.* Universe Books, New York.

Bognale U D and Bharathi S G 1990 Algal flora of cultivated soils of Karnataka State, India. *Phykos* 19: 95-113.

Carson R 1963 *Silent Spring.* Humish Hamilton, London.

Chakraborty P K, Mandal L N and Majumdar, A 1988 Organic and chemical sources of nitrogen: its effect on nitrogen transformation and rice productivity under submerged conditions. *J agric Sci* UK 11: 91-94.

Conford P Ed 1988 *An Anthology of Writings on Organic Farming 1900-1950* In: *The Organic Tradition.* Green Classics, Green books, Bideford, Devon, UK.

De P K 1939 The role of blue-green algae in nitrogen fixation in rice fields. *Proc Roy Soc* London, Ser B 127: 121-139.

De P K and Mandal L N 1956 Fixation of nitrogen by algae in rice fields. *Soil Sci* 81: 453.

Deka M B, Borthakur H P and Borthakur S N 1996 Effect of organic and inorganic source of nitrogen on the nutrients in soil and soil solution and growth of rice. *J Indian Soc Soil Sci* 44: 263-266.

Foster C and Lampkin N 1999 In: *Organic Farming in Europe: Economics and Policy*. University of Hohenheim, Hohenheim.

Goyal S K 2002 A profile on algal biofertilizer. In: Kannaiyan, S (Ed) *Biotechnology of Biofertilizers* TNAU, Coimbatore, Tamilnadu, India. pp 250-258.

Goyal S K, Singh B V, Nagpal Veena and Marwaha T S 1997 Improved method for production of algal biofertilizer. *Indian J Agric Sciences* 67: 314-315.

Gupta A B and Shukla A C 1967 Studies on nature of algal growth promoting substance and their influence on growth, yield and protein content of rice plants. *J Sci Tech* Kanpur 5: 162-163.

Harwood R R 1990 A history of sustainable agriculture. In: Edwards C A, Lal R, Maddee P, Miller R H and House G (eds) *"Sustainable Agriculture Systems"*. pp 3-9.

Howord A 1943 *An Agricultural Testament*. Oxford University Press, London.

Kaushik B D and Venkataraman G S 1983 Response of Cyanobacterial nitrogen fixation to insecticides. *Curr Sci* 52: 321-323.

Kaushik B D 1989 Reclamation potential of Cyanobacteria in slat affected soils. *Phykos* 28: 101-109.

Lampkin N H 1999 Organic farming in the European Union-overview, policies and perspective. In: Organic farming in the European Union-Perspective for the 21st century. Proceedings EU commission/Austrian Government conference. Baden/ Vienna. May 1999. Euro Tech Management, Vienna.

Lockeretz W 1990 Major issues confronting sustainable agriculture. In: Francis C A, Flora C B and King L D (eds) *Sustainable Agriculture in Temperate Zone,* John Wiley and Sons, New York.

Mahapatra P, Panda M M, Chalam A B, Chakraborty S P and Mohanty S K 1997 Effect of green manuring and N fertilization on the yield and nitrogen use efficiency of wetland rice. *J Indian Soc Soil Sci* 45: 95-99.

Mandal B K, Subudhi U K and Singh Y V 1992 Response of rice (*Oryza sativa*) to green leaf manure and prilled urea. *Indian J. of Agric. Sci.* 62: 432-435.

Mohan M and Mukerji K G 1980 Effect of extra-cellular products of blue-green algae (N_2 fixer and Non N_2 fixer) on the growth of *Oryza sativa* C.V. Pusa 2-21. National Workshop on algal systems. Indian Society of Biotechnology. IIT. New Delhi 32: 31-35.

Pabbi S and Vaishya A K 1992 Effect of Insecticides on Cyanobacterial growth and nitrogen fixation. In: Proc Natl Symp Cyanobactl Nitrogen Fixation. Kaushik B D (Ed.).

Pearce J, Leach C K and Carr N G 1969 The Incomplete tricarboxylic acid cycle in the blue-green algae, *Anabaena variabilis*. *J Gen Microbiol* 55: 371-378.

Rusch H P 1968 In: Bodenfruchtbarkeit: Eine Studie Biologischen Denkens Verlag. K F Haug, Heidelberg.

Singh B V and Dhar D W 2004 Blue-green algae-a potential biofertilizer for rice crop. In: Nehra, Sampat (Ed.), *Plant Microbes and Biotechnology.* Deptt of Botany, University of Rajasthan, Jaipur, India. pp 259-269.

Singh R N 1950 Reclamation of Usar lands in India through blue-green algae. *Nature* 165: 325-326.

Singh P K and Bisoyi R N 1989 Blue-green algae in rice fields. *Phykos* 28: 181-195.

Singh A L and Singh P K 1987 Nitrogen fixation and balance studies of rice soils. *Biology and Fertility of Soil* 4: 15-19.

Singh Y V and Mandal B K 1997 Nutrition of rice (*Oryza sativa*) through *Azolla*, organic materials and urea. *Indian J Agronomy* 42: 626-633.

Singh Y V and Mandal B K 2000 Rate of mineralization of *Azolla*, other organic materials and urea in water logged soils. *Tropical Agriculture* (Trinidad) 77: 119-122.

Singh P K 1976 Algal inoculation and its growth in water-logged rice fields. *Phykos* 15: 5-10.

Singh R N 1939 An investigation in to the algal flora of paddy field soils of the United Provinces. *Indian J Agric Sci* 9: 55–77.

Singh R N 1942 The fixation of elementary nitrogen by some of the commonest blue-green algae from the paddy soils of the United Province and Bihar. *Indian J Agric Sci* 12: 743.

Singh R N 1961 *Role of Blue-green Algae in Nitrogen Economy of Indian Agriculture.* Indian Council of Agricultural Research, New Delhi.

Stewart D W P, Rowell P, Ladha J K and Sampio M J A 1979 Blue-green algae (Cyanobacteria)–some aspects related to their role as sources of fixed nitrogen in paddy soils. In: *Nitrogen and Rice,* IRRI, Los Banos, pp 263.

Stockdale E A, Lampkin N H, Hovi M, Keatinge R, Lennartsson E K M, Macdonald D W, Padel S, Tattersall F H, Wolfe M S and Subrahmanyan R 1972 Some observation on utilization of blue green algal mixtures in rice cultivation in India. In: Desikachary T V (Ed) *Taxonomy and Biology of Blue-green Algae,* University of Madras, India.

Suseela M R and Goyal S K 1995 Growth and nitrogen fixing potential of Cyanobacteria. *Phykos* 34: 131-134.

Tate W B 1994 The development of the organic industry and market: an international perspective. In: Lampkin N H and Padel S (eds)*The Economics of Organic Farming: An International Perspective.* CAB International, Wallingford. Oxon. UK pp 11-25.

Thomas J 1970 Absence of pigments of Photosystem II of Photosynthesis in heterocysts of blue-green algae. *Nature* 228: 181-183.

Thomas J, Wolk C P, Shaffer P W, Austin S M and Galonsky A 1975 The initial organic products of fixation of 15N–labelled nitrogen gas by the blue-green alga *Anabaena cylindrica. Biochem Biophys Res Commun* 67: 501-507.

Tripathi U, Pabbi S and Singh P K 2001 Status of blue-green algal biofertilizers in India. In: Singh P K, Dhar D W, Pabbi S, Prasanna R and Arora A (eds). *Recent Advances in the Exploitation of Blue-green Algae and Azolla.* IARI, New Delhi pp 45-57.

Venkataraman G S 1966 Algalization. *Phykos* 5:164.

Venkataraman G S 1972 *Algal Biofertilizers and Rice Cultivation.* Today and Tomorrow Printers and Publishers, India.

Venkataraman G S 1980 Algal biofertilizer: Potential and problems. In: Seshadri C V, Thomas S and Jeeji B N (eds) *Proc Natl Workshop on Algal Systems,* Indian Society of Biotechnology, IIT, New Delhi pp1-10.

Venkataraman G S 1981 Blue-green algae. A possible remedy to nitrogen scarcity. *Curr Sci* 56: 549-550.

Venkataraman G S and Goyal S K 1969 Influence of blue-green algae on high yielding paddy variety IR8. *Sci cult* 25: 58.

Walk C P, Thomas J, Shaffer, P W, Austin S M and Galonsky A 1976 Pathway of nitrogen metabolism after fixation of 15N–labelled nitrogen gas by the Cyanobacterium. *J Biol Chem* 251: 5027-5234.

Watson C A 2001 Agronomic and environmental implications of organic farming systems. *Advances in Agronomy* 70: 261-311.

Wyatt J T and Silvey J K G 1969 Nitrogen fixation by *Gloeocapsa. Science* 165: 908-909.

Yussefi M 2003 The world of organic future agriculture 2003-Statistics and prospects.

Zeichaus-Hartelt C 1991 Bioland-ein Verbandentwickelt sich. *Bioland* 2: 13-14.

Chapter 4

Studies on Application of Some Potential Cyanobacterial Strains in Scale-Up Production of Ethnomedicinal Plants of Southern Rajasthan

Usha Pandey

Contents

ABSTRACT

The present investigation was designed to study the scale-up production of two Ethno-medicinal plants *Andrographis*

Glimpses of Cyanobacteria, 2006 67–77
Daya Publishing House, New Delhi

paniculata and *Bacteria cristata* using two Cyanobacteria- *Spirulina subsalsa* and *Nostoc piscinale* as biofertilizers The post soils containing seedlings of medicinal plants were inoculated with 10g of fresh cultured strains of these Cyanobacteria. Plants height, number of layers, total leaf area, total leaf biomass and total plant biomass increased significantly in plants supplemented with inoculums. Relative growth, net assimilation rate, leaf production rate, specific leaf area etc. also increased in treated plants. These ethno medicinal plants in response to Cyanobacterial biofertilizer application, allocated greater proportion of their photoassimilates for growth and development of photosynthetic organs (Leaves and stem). This has relevance because in these trees, leaves and stems are the main sources of extracting chemical principles of medicinal significance

1. Introduction

Cyanobacteria are oxygenic photosynthetic prokaryotes Although occur predominantly under water logged conditions, they are ubiquitous in distribution They have been implicated in maintaining the fertility of agriculture lands. The extensive research during last few decades has been strengthened our knowledge towards economic use of this group of organisms. They are widely used as biofertilizers. Their growth not only adds nitrogen to the soil but also helps reducing soil erosion, decreasing soil compaction, adding organic matter liberating growth regulator (Venkataraman, 1986, Goyal, 1993). Certain Cyanobacteria perform the process of O_2 sensitive N_2 fixation and O_2 evolving photosynthesis simultaneously (Stewart, 1969). An Oxygenic free atmosphere is essential for N_2 fixation by nitrogenase (Stewart 1970) as its component, Fe-protein and Mo-Fe protein are inactivated equally by molecules oxygen (Stewart *et al.*, 1985). The ability to fix molecular nitrogen has led this group of organism to be used as biofertilizers. During past two decades the practice of utilizing Cyanobacteria in scale-up production of rice has been widely advocated and adopted in our country (Venkatraman, 1981; Kannaiyan, 1990)

The global concern about the scale-up production of agriculture and horticulture crops to meet the demand of fast growing human population has forced scientist to utilize Cyanobacterial potentials in a cost effective and ecofriendly manner. This has particular concern because synthetic fertilizers not only pollute the environment but also reduce the soil fertility in long run. Use of Cyanobacterial biofertilizers in sustained production of rice has been practicised from several years (Venkatraman, 1975). However, the importance of these organisms in scale-up production of medicinal plants are not known (Pandey, 2004). The southern part of Rajasthan in rich in plant biodiversity including those of medicinal plants living in close association with wild plants, the tribals of this belt developed sound knowledge of utilization of medicinal plants in treatment of various diseases. Realizing the side effects of allopathic medicines and growing need of plant based medicines and medicinal products, the scale-up production of ethno-medicinal plants using cost effective and eco-friendly practices has becomes desirable. Therefore, the present investigation was undertaken to study the scale-up production of two medicinal plants namely *Andrographis paniculata* and *Barleria cristata* using two Cyanobacteria, *Spirulina subsalsa* and *Nostoc piscinale* as a sourse of biofertilizer.

2. Materials and Methods

2.1 Propagation of Medicinal Plants

The medicinal plants species chosen for the present study include *Andrographis paniculata cristata*. One month old seedings of these species were collected from Govt. Banki Nursery, Udaipur and planted (one per pot) in 30 cm diameter × 35cm long earthern pots filled with garden soil [sandy loam, pale brown colour, pH 7.35; organic carbon 0.90 per cent; total nitrogen 0.07 per cent; available P 0.005 per cent; exchangeable K 0.09 per cent and cation exchange capacity 16.1m eq(%)]. Seedling was uniform watered at desired intervals.

2.2 Cyanobacteria Bioassays

Both the strain chosen for the present study were isolated from agriculture field of Maharana Pratap Agriculture University, Udaipur and grown in BG-11 median (Rippka *et al*,1979) medium. Test cultures were maintained at 28±2°C in an orbital shaker, illuminated with fluorescent lamps providing a light intensity of 1800 lux around the culture vessels for 16:8 light/dark cycle. Concentration of protein, carbohydrates and lipids were determines following Lowry's *et al.*, (1951), Spiro (1986) and Hellebust and Craige (1978) respectively.

Biochemical analysis of 10 day harvested biomass was done in triplicate for chlorophyll, carotenoids and phycobiliproteins (Jensen, 1978). α-amylase was determined as per Oser (1965) and acid phosphate was estimated by the method of Patni and Aaronoson (1978). Ammonia excretion was estimated at intervals of 10 days according to Solarzano (1969).

2.3 Soil Inoculation

The pot soils containing seedling were inoculated with 10 gm of fresh cultured strains of following a factorial randomized block designs. The plants with Cyanobacterial innoculums are hereafter referred to as treated plants.

2.4 Plants Sampling And Analysis

Samples were collected from the control and each set of treatment at 15 days intervals. However, only the data for days 45 and 60 are presented For each treatment three replicate pots were maintained. Three plants from each set were harvested at each sampling date. Plant height and number of leaves and flowers were recorded. Leaf area was determined with leaf area meter (LI-CoR Lincoln, USA). Chlorophyll pigments were extracted in 80 per cent acetone The optical densities of the extracts were measured at 645 and 663nm using an UV-visible spectrophotometer Chlorophyll concentrations were calculated using the formula given by Maclachlan and Zalike (1963).

For biomass determination, plants were separated into leaves, stem and root and oven dried separately at 80°C to constant weight. Relative growth rate (RGR), net assimilation rate (NAR), leaf area ratio (LAR), leaf production rate (LPR) etc were computed from formulae given by Hunt (1982). Three-way analysis of variance and the t-test were performed to test the level of significant difference

3. Results

The biochemical characterization of both Cyanobacterial strains are present in Table 4.1. Concentrations of chlorophyll and phycobiliproteins were high in *Nostoc piscinale* However, concentrations of carotenoids, proteins and carbohydrates remained higher in *Spirulina subsalsa*. Further, *Nostoc piscinale* showed superiority with respect to the concentrations of lipids and α-amylase and acid phosphatase activity (Table 4.1).

Plant height increased throughout the experiment period in both control and treated plants (Table 4.2) The increase in plant height was maximum in *Barleria cristata*Plants inoculated with *Nostoc piscinale* compared with control, the number of leaves increased in treated plants, the increase being maximum in plant, inoculated with *Nostoc piscinale*. In both the treatments, the number of flowers increased with respect to control (Table 4.2). Root and shoot biomass increased in treated plants throughout the study. However, increase in shoot biomass showed precedence over root biomass. Total leaf area and total leaf biomass increased in treated plants throughout the study. Total biomass gain in *Spirulina subsalsa* and *Nostoc piscinale* inoculated *A paniculata* and *B cristata* plants were 22.4 per cent, 20 per cent and 30 per cent and 22.5 per cent respectively for 65 days old plants. Chlorophyll concentration increased significantly in plants having Cyanobacterial inoculum (Table 4.2). However, the increase in chlorophyll concentration was maximum at initial stage of plant growth. Variations in total biomass and total chlorophyll were significant with respect to different treatments plant age and their interactions (Table 4.3)

Table 4.1: Pigments and Biochemical Properties of Cyanobacteria after 10 days

Sl.No.	*Pigment and Biochemical properties*	*Spirulina subsalsa*	*Nostoc piscinale*
1.	Chlorophylls(mg g^{-1} dry wt)	8.45±0.45	9.40±1.00
2.	Carotenoids (mg g^{-1} dry wt)	1.98±2.00	1.60±0.20
3.	Phycobiliproteins (mg^{-1}dry wt)	130.15±14.8	145.0±20.50
4.	Proteins (mg g^{-1} dry wt)	450.0±4.165	390.0±32.0
5.	Carbohydrate (mg g^{-1} dry wt)	85.0±9.05	80.0±9.15
6.	Lipid (mg g^{-1} dry wt)	12.15±1.40	130.5±1.10
7.	α-amylase (μg ml^{-1})	190.68±20.0	210.0±22.10
8.	Acid phosphatase (μg ml^{-1})	2.90±0.25	3.15±0.45
9.	Ammonia excretion (μg ml^{-1})	0.082±0.15	13.50±1.20

Relative growth rate (RGR) remaining higher in treated plants throughout the study (Table 4.4). Similar trends were observed in NAR and LPR, the maximum leaf production was observed in plants inoculated with *N. piscinale* (Table 4.4). The root production rate (RPR) also increased in treated plants, but increase was significantly

Table 4.2: Effect of Cyanobacterial innoculum on Growth of *Andrographis paniculata* (A) and *Barleria cristata* (B) Plants

Measurement	*45 Days*						*60 Days*					
	Control		*S. subsalsa*		*N. piscinale*		*Control*		*S. subsalsa*		*N. piscinale*	
	A	*B*	*A*	*B*	*A*	*B*	*A.*	*B*	*A*	*B*	*A*	*B*
Height(cm)	13	19	23*	30	26*	34*	18	27	30*	41*	34*	46***
Leaves per plant (No)	25	13	33	17	38**	18*	32	16	40*	20	45**	22***
Leaves per plant (gdw)	1.25	0.69	1.63*	086*	196**	105*	170	090	210**	112*	2.30**	1.16*
Flower per plant (gdw)	–	–	–	–	–	–	48	36	59*	45*	64**	50
Shoot per plant (gdw)	3.28	2.05	4.86*	260*	5.61**	301**	490	270	590*	310**	6.21**	3.25**
Root per plant (gdw)	1.20	0.65	148*	080	1.76**	093*	154	082	1.96*	1.02	2.15**	1.04**
Total leaf area (cm²)	80	76	109*	105*	126**	112**	98	101	1.32**	126**	1.46***	1.40***
Biomass per plant (gdw)	448	2.70	634**	3.40**	7.37**	394*	640	350	7.86**	4.12**	8.36**	4.29**
Total chlorophyll (mg^{-1} gdw)	460	4/52	6.59*	6.95**	7.18**	7.62**	4.48	4.45	6.35*	6.80*	7.0**	7.39**

Level of significance $^{*}p<0.05$, $^{**}p<0.01$, $^{***}p<0.001$.

lower than those of LPR. Specific leaf area (SLA) and leaf area ratio (LAR) and leaf area duration (LAD) increased significantly in plants inoculated with Cyanobacteria. Leaf weight ratio (LWR) increased throughout the study. However, root: shoot ratio (RSR) decreased in treated plants and magnitude of reduction was maximum in plants inoculated with *N. piscinale.*

Table 4.3: Mean Square Estimates of Chlorophyll and Biomass of *Andrographis paniculata (A)* and *Barleria cristata (B)* as Affected by Cyanobacterial Innoculum

Sl.No.	*Sources*	*df*	*Means of Square*	
			Chloro	*Biomass*
1.	SS	1	1.2170***	12.0362***
2.	NP	1	1.2205***	12.1895***
3.	Age	2	0.0942***	52.0765***
4.	Replicate	2	0.0010^{NS}	0.1630^{NS}
5.	SS × Age	2	0.0490*	6.9210***
6.	Np × Age	2	0.0510*	6.9875***

ss: *Spriulina subsalsa;* NP: *Nostoc piscinale,****,**,*, NS: significant at p<0.005, 0.025, 0.100 and not significant, respectively df degree of freedom.

4. Discussion

The beneficial effects of Cyanobacterial inoculum on plant growth observed in the present study are consistent with reports for rice (Meelu, 1992; Ganter *et al.*, 1995 Goyal 1993) and ethnomedicinal plants *Vitex negundo* (Pandey 2004). Inoculation of soil with *S. subsalsa* and *N. piscinale* increased plant height, number of leaves and biomass Analysis of variance showed significant differences in biomass accumulation and chlorophyll concentrations in plants inoculated with Cyanobacteria.

Flowering was slightly delayed in the innoculum-supplemented plants, indicating that biofertilizers application stimulates vegetative growth but leads to slower transition from vegetative to reproductive phase. However, treated plants produced more flowers, which suggest that greater amount of accumulated photosynthate ultimately diverted to these reproductive sinks. Increase in leaf area, leaf dry weight and leaf production rate in plants supplemented cyanobacteria innoculum further suggest that the plant respond to

Table 4.4: Effects of Cyanobacteria Innoculum on Growth Characteristics (dry weight basis) of *Andrographis paniculata* (A) and *Barleria cristata*(B) Plants

Measurement	45 Days						60 Days					
	Control		*S. subsalsa*		*N. piscinale*		*Control*		*S. subsalsa*		*N. piscinale*	
	A	*B*	*A*	*B*	*A*	*B*	*A*	*B*	*A*	*B*	*A*	*B*
RGR ($gg^{-1}d^{-1}$)	0.034	0030	0062***	0060***	0067***	0063***	0015	0013	0038***	0034***	0040***	0.036***
NAR ($mg\ cm^{-2}d^{-1}$)	0.425	0400	0530*	0500*	00542**	0512**	0400	0381	0500*	0468**	0510**	0.480**
LPR ($gg^{-1}d^{-1}$)	4.65	410	710**	700**	728***	716**	362	341	490*	476*	520**	4.90**
RPR ($gg^{-1}d^{-1}$)	2.48	220	310*	300*	324**	310**	215	202	248	232	298**	2.68**
LAD (m^2d^{-1})	4.80	436	595**	524**	625***	570**	610	578	765*	690*	798**	7.45**
SLA ($mg\ cm^{-2}$)	2.80	270	310	295	329*	308	290	276	336*	320	352**	3.28
LAR ($cm^2\ g^{-1}$)	96.25	8430	10565	9730	11200*	10450	10135	9460	10675	10120	11330	107.00
LWR ($g\ g^{-1}$)	0.27	0261	0340*	0320	0355**	0334	0290	0284	00 360	0339*	0368**	0.348
RSR	0.25	0.23	0.20	0.21	0.19*	0.20	0.24	0.23	0.19	0.20	0.18*	0.20

RGR:relative growth rate; NAR: net assimilation rate; LPR: leaf production rate; RPR: root production rate; LAD:leaf area duration; SLA:specific leaf area; LAR: leaf area ratio; LWR: leaf weight ratio; RSR: root shoot ratio.

Level of significance: *$p<005$, **$p<001$, ***$p<0001$.

these biofertilizers by increasing and allocating greater proportion of its photo assimilates for growth and developed of photosynthetic organs. This effect was more pronounced in plant inoculated with *N. piscinale.* This diazotrophic Cyanobacteria is well known for its nitrogen fixation potential (Venkataraman 1975). Increase in specific leaf area (SLA) indicated high photosynthetic efficiency and rich biochemical status of plants. The stimulatory effect of these biofertilizers on RGR may be ascribed to an increased in LAR and NAR. RGR is the product of LAR and NAR. Increase in LAR throughout the study period indicateD consistent increase in plants photosynthetic efficiency due to biofertilizers application. This bear particular significance since consistent increase in photosynthetic efficiency could also support growth of those plant parts which have great medicinal significance. Supplementing soils with nitrogenous fertilizers are well known to stimulate leaf growth and increase plant photosynthetic efficiency (Meelu 1992, Ganter *et al.,* 1995). *Spirulina subsalsa* and *Nostoc piscinale* are well known N-rich biofertilizers

The study indicates that supplementing soil with CyanobacteriaL, inoculum have fertilizing effects on *A. paniculata* and *B. cristata* plants, promoting vegetative growth as well as the flower yields. Treated plants, although showed stimulating effects at all parts allocated greater proportion of its photoassimilate for growth and developed of shoot and photosynthetic organs This has particular significance because for these plants and stems are used as the source for extracting main chemical principles of medicinal significance. The study suggests that the Cyanobacterial inoculum provide eco-friendly and cost effective source of biofertilizer for scale-up production of ethnomedicinal plants.

5. Acknowledgement

The author is thankful to Principal, GN Girl's PG College for providing Laboratory facilities.

6. References

Gantar M, Kerby, N W, Rowell P, Obrent and Scrimgeour C 1995 Colonization of wheat (*Triticum vulgare L*) by nitrogen fixing Cyanobacteria: IV dark nitrogenase activity and effects of Cyanobacteria on natural super ^{15}N Abundance in the plants *New Phytol* 129: 337-343.

Goyal S K 1993. Algal biofertilizers for vital soil and free nitrogen. *Proc Indian Natn Sci Acad*. 3:295-302.

Hellebust J A and Craige J W 1978 *Handbook of Physiological and Biochemical Methods* Cambridge University Press Cambridge. 85-90.

Hunt R 1982 *Plant Growth Curves*. Edward Arnold, London.

Jensen A 1978 Chlorophylls and carotenoids (pigments) In: Hellebust J A and Craige J S (eds) *Handbook of Physiological and Biochemical Methods*. Cambridge University Press. 59-70.

Kannaiyan S, Aruna S J, Merina P K S, Hall D O 1987 Immobilized Cyanobacteria as a biofertilizer for rice crop *J Appl Phycol* 7: 1-9.

Lowry O H, Rosenbrough N J, Larr A C and Randall R J 1951 Protein measurement with folin phenol reagent *J Biol Chem* 93: 265-275.

Maclachlan S and Zalik S 1968 Plastid structure, chlorophyll concentration and free amino acid composition of a chlorophyll mutant of barley. *Can J Bot* 41:1053-1062.

Meelu O P 1992 Biofertilizer and their potential in crop production, 281-286 In: Dhaliwal G S, Hansra B S, Jerath N (eds) *Changing Scenario of our Environment*. Punjab Agricultural University, Ludhiana, India.

Oser B L 1965 Determination of serum amylase activity In: *Hawk's Physiological Chemistry* Tata McGraw Hill Publishing Company Bombay.

Pandey U 2004 Effects of Cyanobacteria biofertilizer application on growth of ethnomedicinal plant *Vitex negundo*. *Indian Journal of Applied and Pure Biol* 19: 2-6.

Patni N J and Aaronsom S 1978 Extracellular acid phosphate of *Ochromonas danica* In: Hellebust, J A and Craige J S (eds) *Handbook of Physiological and Biochemical Methods*. Cambridge Univ Press Cambridge 263-266.

Rippka R, Deruelles J, Waterbury J B, Herdman M and Stanier R Y 1979 Generic assignment, strain histories and properties of pure cultures of Cyanobacteria. *J Gen Microbiol* 111: 1-61.

Solarzano L 1969 Determination of ammonia of nature water by the phenol hypochlorite method. *Limnol Oceanogr* 799-801.

Spiro R G 1966 Analysis of sugars found in glycoproteins *Methods in Enzymology* 8: 3-26.

Stewart W D P 1969 Biological and ecological aspects of nitrogen fixation by free-living microorganisms. *Proc R Soc London B*. 112: 376-388.

Stewart W D P 1970 Algal fixation of atmospheric nitrogen *Plant and Soil* 32: 555-558.

Stewart W D P, Rowell P, Cossar J D and Kerby N W 1985 Physiological studies on nitrogen fixing Cyanobacteria In: Ludden P W, Burris G J, (eds) *Nitrogen Fixation and CO_2 Metabolism* A Stanbock symposium in honor of Robert H Burris 269-282.

Venkataraman G S 1975 The role of Cyanobacteria in tropical rice cultivation In: Stewart W D P (eds) *Nitrogen Fixation by Free-living Microorganisms*. Cambridge Univ Press, 207-260.

Venkataraman G S 1981 Blue-green algae A possible remedy to nitrogen scarcity: *Curr Sci* 50: 253-256.

Venkataraman L V 1986 *Cyanobacteria as Biofertilizers* In: Handbook of microalgal Mass Culture. CRC Press455-471.

Chapter 5

Cyanobacteria as a Source of Antimicrobial Compounds

V.D. Pandey, R.K. Gupta, R.C. Gupta, A.K. Mishra and S.K. Singh

Contents

ABSTRACT

Cyanobacteria produce a wide variety of novel antibacterial and antifungal compounds, which could be advantageous for their survival in natural environment. They

Glimpses of Cyanobacteria, 2006 78–86
Daya Publishing House, New Delhi

are suitable candidates for the discovery of novel antibiotics, and could be used as biocontrol agents of bacterial and fungal phytopathogens.

1. Introduction

Cyanobacteria (blue-green algae or cyanophytes) are a large, diverse and widely distributed group of prokaryotic organisms, resembling gram-negative bacteria in cellular organization and green plants in oxygenic photosynthesis (Stanier and Cohen-Bazire, 1977). Endowed with the remarkable capacities to adapt to varying environmental conditions (Tandeau de Marsac and Houmard, 1993), they colonize almost all kinds of terrestrial and aquatic ecosystems. Cyanobacteria have been the subject of both basic and applied research for years. More recently, they have received attention as potential sources of bioactive compounds (biologically active compounds), which may refer to the compounds that at low concentrations affect the growth, development or physiological processes of a cell or an organism in beneficial or detrimental manner. In addition to the best known neurotoxins and hepatotoxins, they are known to produce a wide variety of novel compounds with diverse biological activities. These include antibacterial compounds (Pergament and Carmeli, 1994; Ishida *et al.*, 1997), antifungal compounds (Moon *et al.*, 1992, Frankmölle *et al.*, 1992a, b), antiviral agents (Knübel *et al.*, 1990; Gustafson *et al.*, 1989), algicides (Mason *et al.*, 1982; Bagchi, 1995), anticancerous/antineoplastic agents (Trimurtulu *et al.*, 1994; Barchi *et al.*, 1984), muscle-relaxant (Harada *et al.*, 1995), antimitotic agents (White *et al.*, 1995; Yoo and Gerwick, 1995) and enzyme inhibitors (Murakami *et al.*, 1994; Okino *et al.*,1997). Many of these cyanobacterial bioactive compounds hold immense potential for the development of pharmaceuticals.

The present article briefly focuses on antimicrobial (antibacterial and antifungal) compounds produced by cyanobacteria.

2. Antimicrobial Compounds

Antimicrobial screening aimed at discovering antibacterial and antifungal antibiotics has largely been carried out with non-cyanobacterial microorganisms, particularly streptomycetes and fungi, which are the source for majority of antibiotics currently known. Photoautotrophic organisms like cyanobacteria have not yet received much attention. However, systematic screening and chemical investigation of cyanophytes over the past two decades have revealed the presence of antimicrobial activities in many species. A number of potent antibacterial and antifungal compounds have been isolated with chemical characterization.

Scytophycins, the potent antifungal and cytotoxic agents, are congeneric macrolides, which are produced by certain species of the genera *Scytonema* and *Tolypothrix* of the family Scytonemataceae (Ishibashi *et al.*, 1986; Carmeli *et al.*, 1990). Tolytoxin (6-hydroxy-7-0-methyl-scytophycin b), a well-characterized scytophycin, was originally isolated from *Tolypothrix conglutinata* var. *colorata* in 1977 (Moore, 1981). The cyanophytes, which are now known to produce scytophycins and tolytoxin include *Scytonema pseudohofmanni, S.mirabile, S.burmanicum* and *S.ocellatum* (Ishibashi *et al.*, 1986; Carmeli *et al.*, 1990). Scytophycins and tolytoxin, in addition to antifungal activity, exhibit strong cytotoxicity towards cancerous cell lines (Moore *et al.*, 1986; Patterson and Carmeli, 1992).

Many cyclic peptides and depsipeptides from blue-green algae have been reported as antimicrobial agents. These include majusculamide C (antifungal agent) from the marine cyanobacterium *Lyngbya majuscula* (Carter *et al.*, 1984), hormothamnins (antibacterial and antifungal compounds) from the marine cyanobacterium *Hormothamnion enteromorphoides* (Gerwick *et al.*, 1989), calophycin (fungicide) from the terrestrial blue-green alga *Calothrix fusca* (Moon *et al.*, 1992), laxaphycins (antifungal agents) from the terrestrial cyanophyte *Anabaena laxa* (Frankmölle *et al.*, 1992a,b), schizotrin A (antifungal and antibacterial compound) from *Schizothrix* sp. (Pergament and Carmeli, 1994), tenuecyclamides (antibacterial and cytotoxic agent) from lithophytic cyanobacterium *Nostoc spongiaeforme* var. *tenue* (Banker and Carmeli, 1998), kawaguchipeptin B (antibacterial agent) from the toxic cyanobacterium *Microcystis aeruginosa* (Ishida *et al.*, 1997) and tolybyssidins (antifungal compounds) from the epilithic cyanobacterium *Tolypothrix byssoidea* (Jaki *et al.*, 2001).

Tubercidin and toyocamycin are fungicidal and cytotoxic nucleosides produced by *Plectonema radiosum* and *Tolypothrix tenuis,* respectively (Stewart *et al.*, 1988). Hapalindoles (indole alkaloids) are antibacterial and antifungal agents isolated from an edaphic cyanobacterium *Hapalosiphon fontinalis* (Moore *et al.*, 1987). The terrestrial cyanophyte *Fischerella muscicola* has been reported to produce an antifungal compound, fischerindole L, which is chemically related to hapalindoles (Park *et al.*, 1992). Ambiguine isonitriles are fungicidal hapalindole-type alkaloids known to be produced by cyanobacteria *Fischerella ambigua*, *Hapalosiphon hibernicus* and *Westiellopsis prolifica* of the family Stigonemataceae (Smitka *et al.*, 1992). The antifungal compounds tjipanazoles, which are chemically indolocarbazoles, have been isolated from the terrestrial cyanobacterium *Tolypothrix tjipanasensis* (Bonjouklian *et al.*, 1991). Ambigols, from terrestrial cyanobacterium *Fischerella ambigua*, are polychlorinated aromatic compounds, which exhibit antibacterial and antifungal activities (Falch *et al.*, 1993). The antibacterial oxazole peptide alkaloid, muscoride A, is produced by the fresh water cyanobacterium *Nostoc muscorum* (Nagatsu *et al.*, 1995). Malyngamides, amides of the fatty acid (-)-7(S)-methoxytetradec-4 (E)-enoate, are antibacterial principles of the marine cyanobacterium *Lyngbya majuscula* (Gerwick *et al.*, 1987). Tanikolide, an antifungal lactone, is another antimicrobial compound produced by *L. majuscula* (Singh *et al.*, 1999).

3. Release and Regulation

Antimicrobial and other bioactive compounds produced by cyanobacteria may be released extracellularly to surroundings, either actively or passively or by autolysis (Metting and Pyne, 1986). The biosynthetic pathways leading to the production of antimicrobial compounds in cyanobacteria and the factors regulating their production are not completely known. There are only few reports to indicate that the production of antimicrobial compounds in these organisms is regulated by culture conditions and nutrients (Patterson and Bolis, 1993, 1995; Fish and Codd, 1994).

4. Ecological Implication

Ecologically, antimicrobial compounds of cyanobacteria are important because of their possible role in chemical interactions that occur in communities of microorganisms. Such compounds may confer selective advantage to cyanobacteria to grow and survive in natural habitats. Chrost (1975) reported considerable reduction in bacterial population in lakes during the occurrence of cyanobacterial water blooms. This phenomenon may partly or completely be attributed to inhibitors/antibacterial compounds produced and released by cyanobacteria.

5. Conclusion

The development of resistance in pathogenic microorganisms against commonly used antibiotics has necessitated a search for new antimicrobial substances from sources other than the traditional microorganisms (Trias and Gordon, 1997). The presence of antimicrobial compounds in cyanobacteria open up exciting new possibilities for the exploitation of these organisms as potential source of antibiotics in pharmaceutical industries. Because of rapid growth rate, simple growth requirements, amenability to controlled laboratory culture and ubiquitous distribution, they appear to be suitable candidates for the purpose. In agriculture, they would be suitable biocontrol agents of phytopathogenic bacteria and fungi.

The pharmaceutical potential of cyanobacteria has yet to be fully realized. Increased research effort must therefore be devoted to extensive screening, purification and characterization of antimicrobial compounds, strain improvement, optimization of growth media and culture conditions, and high-density culture.

6. References

Bagchi S N 1995 Structure and site of action of an algicide from a cyanobacterium, *Oscillatoria late-virens*. *J Plant Physiol* 146: 372-374.

Banker R and Carmeli S 1998 Tenuecyclamides A-D, cyclic hexapeptides from the cyanobacterium *Nostoc spongiaeforme* var *tenue*. *J Nat Prod* 61: 1248-1251.

Barchi J J Jr, Moore R E and Patterson G M L1984 Acutiphycin and 20, 21-didehydroacutiphycin, new antineoplastic agents from the cyanophyte *Oscillatoria acutissima*. *J Am Chem Soc* 106: 8193-8197.

Bonjouklian R, Smitka T A, Doolin L E, Molloy R M, Debono M, Shaffer S A, Moore R E, Stewart J B and Patterson G M L1991 Tjipanazoles, new antifungal agents from the blue-green alga *Tolypothrix tjipanasensis*. *Tetrahedron* 47: 7739-7750.

Carmeli S, Moore R E and Patterson G M L 1990 Tolytoxin and new scytophycins from three species of *Scytonema*. *J Nat Prod* 53:1533-1542.

Carter D C, Moore R E, Mynderse J S, Niemczura W P and Todd J S 1984 Structure of majusculamide C, a cyclic depsipeptide from *Lyngbya majuscule*. *J Org Chem* 49: 236-241.

Chrost R J 1975 Inhibitors produced by algae as ecological factor-affecting bacteria in water ecosystems. *Acta Microbial. Pol Ser* B7: 125-133.

Falch B S, König G M, Wright A D, Sticher O, Rüegger H and Bernardinelli G 1993 Ambigol A and B: new biologically active polychlorinated aromatic compounds from the terrestrial blue-green alga, *Fischeralla ambigua*. *J Org Chem* 58:6570-6575.

Fish S A and Codd G A. 1994 Analysis of culture conditions controlling the yield of bioactive material produced by the thermo tolerant cyanobacterium (blue-green alga) *Phormidium*. *Eur. J Phycol* 29:261-266.

Frankmölle W P, Larsen L K, Caplan F R, Patterson G M L, Knübel G, Levine I A, and Moore R E 1992a Antifungal cyclic peptides from the terrestrial blue-green agla *Anabaena laxa*. I. Isolation and biological properties. *J Antibiotics* 45: 1451-1457.

Frankmölle W P, Knübel G, Moore R E and Patterson G M L 1992b Antifungal cyclic peptides from the terrestrial blue-green agla *Anabaena laxa*. II. Structures of laxaphycins A, B, D and E. *J Antibiotics* 45: 1458-1466.

Gerwick W H, Reyes S and Alvarado B 1987 Two malyngamides from the Carribean Cyanobacterium *Lyngbya majuscula*. *Phytochem* 26: 1701-1704.

Gerwick W H, Mrozek Ch, Moghaddam M F and Agarwal S K 1989 Novel cytotoxic peptides from the tropical marine Cyanobacterium *Hormothamnion enteromorphoides*. I. Discovery, isolation and initial chemical and biological characterization of the hormothamnins from wild and cultured material. *Experientia*, 45:115-121.

Gustafson K R, Cardellina II J H, Fuller R W, Weislow O S, Kiser R F, Snader K M, Patterson G M L and Boyd M R 1989 AIDS–antiviral sulfolipids from Cyanobacteria (blue-green algae). *J Nat Cancer Inst*, 81:1254-1258.

Harada K I, Fujii K, Shimada T, Suzuki M, Sano H, Adachi K and Carmichael W W 1995 Two cyclic peptides, anabaenopeptins, a third group of bioactive compounds from the Cyanobacterium *Anabaena flos-aquae* NRC 525-17. *Tetrahedron Lett* 36:1511-1514.

Ishibashi M, Moore R E, Patterson G M L, Xu C and Clardy J 1986 Scytophycins, cytotoxic and antimycotic agents from the cyanophyte *Scytonema pseudohofmanni*. *J Org Chem* 51: 5300-5306.

Ishida K, Matsuda H, Murakami M and Yamaguchi K 1997 Kawaguchipeptin B, an antibacterial cyclic undecapeptide from the Cyanobacterium *Microcystis aeruginosa*. *J Nat Prod* 60:724-726.

Jaki B, Zerbe O, Heilmann J and Sticher O 2001 Two novel cyclic peptides with antifungal activity from the cyanobacterium, *Tolypothrix byssoidea* (EAWAG 195) *J Nat Prod* 64: 154-158.

Knübel G, Larsen L K, Moore R E, Levine I A and Patterson G M L 1990 Cytotoxic, antiviral indolocarbazoles from a blue-green alga belonging to the Nostocaceae. *J Antibiotics* 43: 1236-1239.

Mason C P, Edwards K R, Carlson R E, Pignatello J, Gleason F K and Wood J M 1982 Isolation of chlorine-containing antibiotic from the freshwater Cyanobacterium *Scytonema hofmanni*. *Science* 215: 400-402.

Metting B and Pyne J W 986 Biologically active compounds from microalgae. *Enzyme Microb Technol* 8: 386-394.

Moon S S, Chen J L, Moore R E and Patterson G M L 1992 Calophycin, a fungicidal cyclic decapeptide from the terrestrial blue-green alga *Calothrix fusca*. *J Org Chem* 57:1097-1103.

Moore R E 1981 Constituents of blue-green algae. In: Scheuer P J (Ed) *Marine Natural Products*, Vol 4, Academic press, New York, pp, 1-52.

Moore R E, Patterson G M L, Mynderse J S, Barchi J J Jr, Norton T R, Furusawa E and Furusawa S 1986 Toxins from cyanophytes belonging to the Scytonemataceae. *Pure Appl Chem* 5: 263-271.

Moore R E, Cheuk C, Yang X-Q G, Patterson G M L, Bonjouklian R Smitka T A, Mynderse J S, Foster R S, Jones N D, Swartzendruber J K and Deeter J B 1987 Hapalindoles, antibacterial and antimycotic alkaloids from the Cyanophyte *Hapalosiphon fontinalis*. *J Org Chem* 52: 1036-1043.

Murakami M, Okita Y, Matsuda H, Okino T and Yamaguchi K 1994 Aeruginosin 298-A, a thrombin and trypsin inhibitor from the blue-green alga *Microcystis aeruginosa* (NIES-298). *Tetrahedron Lett* 35: 3129-3152.

Nagatsu A, Kajitani H and Sakakibara J 1995 Muscoride A: a new oxazole peptide alkaloid from freshwater Cyanobacterium *Nostoc muscorum*. *Tetrahedron Lett*, 36: 4097-4100.

Okino T, Qi S, Matsuda H, Murakami M and Yamaguchi K 1997 Nostopeptins A and B, elastase inhibitors from the Cyanobacterium *Nostoc minutum J Nat Prod* 60: 158-161.

Park A, Moore R E and Patterson G M L 1992 Fischerindole L. a new isonitrile from the terrestrial blue-green alga *Fischerella muscicola*. *Tetrahedron Lett* 33: 3257-3260.

Patterson G M L and Carmeli S 1992 Biological effects of tolytoxin (6-hydroxy-7-0-methyl-scytophycin b), a potent bioactive metabolite from Cyanobacteria. *Arch Microbiol* 157: 406-410.

Patterson G M L and Bolis C M 1993 Regulation of scytophycin accumulation in cultures of *Scytonema ocellatum*. I. Physical factors. *Appl Microbiol Biotechnol* 40: 375-381.

Patterson G M L and Bolis C M 1995 Regulation of scytophycin accumulation in cultures of *Scytonema ocellatum*. II. Nutrient requirements. *Appl Microbiol Biotechnol* 43: 692-700.

Pergament I and Carmeli S 1994 Schizotrin A: a novel antimicrobial cyclic peptide from a cyanobacterium. *Tetrahedron Lett* 35: 8473-8476.

Singh I P, Milligan K E and Gerwick W H 1999 Tanikolide, a toxic and antifungal lactone from the marine cyanobacterium, *Lyngbya majuscula*. *J Nat Prod* 62: 1333-1335.

Smitka T A, Bonjouklian R, Doolin L, Jones N D, Deeter J B, Yoshida W Y, Prinsep M R, Moore R E and Patterson G M L 1992 Ambiguine isonitriles, fungicidal hapalindole-type alkaloids from three genera of blue-green algae belonging to the Stigonemataceae. *J Org Chem* 57: 857-861.

Stanier R Y and Cohen-Bazire G 1977 Phototrophic prokaryotes: the Cyanobacteria. *Ann Rev Microbiol* 31: 225-274.

Stewart J B, Bornemann V, Chen J L, Moore R E, Caplan F R, Karuso H, Larsen L K and Patterson G M L 1988 Cytotoxic, fungicidal nucleosides from blue-green algae belonging to the Scytonemataceae. *J Antibiotics* 41:1048-1056.

Tandeau de Marsac, N and Houmard, J. 1993 Adaptation of Cyanobacteria to environmental stimuli: new steps towards molecular mechanisms. FEMS Microbiol Rev 104:119-190.

Trias J and Gordon E M 1997 Innovative approaches to novel antibacterial drug discovery. *Curr Opin Biotech* 8: 757-762.

Trimurtulu G, Ohtani I, Patterson G M L, Moore R E, Corbett T H, Valeriote F A and Demchik L 1994 Total structures of cryptophycins, potent antitumor depsipeptides from the blue-green alga *Nostoc* sp strain GSV224. *J Am Chem Soc* 116:4729-4737.

White J D, Kim T-S and Nambu M 1995 Synthesis of Curacin A: a powerful antimitotic from the cyanobacterium, *Lyngbya majuscula*. *J Am Chem Soc* 117: 5612-5613.

Yoo H-D and Gerwick W H 1995 Curacins B and C, new antimitotic natural products from the marine cyanobacterium, *Lyngbya majuscula*. *J Nat Prod* 58:1961-1965.

Chapter 6

Distributional Patterns of Cyanobacteria

M. Kumar and G.S. Paliwal

Contents

ABSTRACT

Cyanobacteria is a group of unique, polymorphic and ubiquitous prokaryotes. They are also termed as Cyanophytes, blue-green algae and Cyanoprokaryotes. They possess simple cellular organization and have a capacity to grow in all possible

Glimpses of Cyanobacteria, 2006 87–106
Daya Publishing House, New Delhi

habitats including terrestrial, aquatic, lithophytic, endophytic, epizoic, epiphytic etc. They prefer to grow in freshwater ecosystems. They are cosmopolitan in distribution and dominated in the Proterozoic era (between 2.5 and 0.5 billion years ago), which is called the ' golden era of Cyanobacteria'. They were pioneers to bare rocks and are therefore very important for ecological point of view. There have been identified many beneficial and harmful effects of Cyanobacteria on human life.

Many classifications have been proposed for the identification of Cyanobacteria. Most of them are based on morphological characters. The classification given by Fristch (1945) is universally accepted by almost all the workers Cyanobacteria live successfully as hydrophytes (aquatic), edaphophytes (terrestrial), aerophytes, gyophytes (cryophytes), endophytes, endozoophytes, parasites and eluviatite. Cyanobacteria are found in lentic as well as lotic freshwater and marine reservoirs in planktonic or reserves forms. They are most abundant in freshwater ecosystems They can survive in very harsh conditions of environment and are able to tolerate the temperatures below the freezing point (*e.g. Nostoc commune*) to 85°C (*Conferva thermalis Birdwoodii*). Their growth, development and distribution are greatly influenced by the extent of some ecological factors like light, temperature, pH, UV radiations, soil etc.

1. Introduction

Cyanobacteria, also known as Blue-green algae or Cyanoprokaryotes constitute the most diverse group of plant kingdom. They occur in a variety of habitats including lithophytic, aquatic, epiphytic etc. These have the ability to survive in diverse ecological conditions. They are polymorphic, prokaryotic microorganisms with single cell, colonial, and trichome organization. They resemble with gram-negative bacteria in cellular organization and green plants in oxygenic photosynthesis (Stanier and Cohen Bazire, 1977) and are named variously *i.e.* Cyanophytes, Cyanophyceae and most recently Cyanoprokaryotes (Fabbro and Mc Gregor, 2003).

The members of this group are capable of splitting water during photosynthesis and to fix atmospheric CO_2 by using water as the reductant and leaving O_2 as a bioproduct. In fact, they belong to a large, highly diverse, widely distributed, and ecologically important oldest group of microorganisms. They are the most archaic organisms on the earth that grow during the first three billion years of the earth's history *i.e.* during the Precambrian period, in this period, these algae absorbed CO_2 from the earth's atmosphere and released oxygen, thus helped this green earth planet hospitable to other organisms. The tiny members of this group dominated the biota in the Proterozoic Era, an Era between 2.5 and 0.5 billion years ago. This Era is known as the 'Age of Cyanophyceae' (van den Hoek *et al.*, 1993). The Era is also termed as "Golden Era of Cyanophyceae".

They lack the conventional mode of sexual reproduction but show great capacity to differentiate various cells, facilitating the asexual reproduction. This group belongs to structurally simple organisms than other land plants (Graham and Wilcox, 2000). Cyanobacteria are devoid of true nucleus (the structure that encloses the DNA) and chloroplast (the structure that encloses the photosynthetic membranes and are evident in photosynthetic true algae). In fact, blue-green algae are more akin to bacteria which have similar biochemical and structural characteristics. Cyanobacteria or blue-green algae do not show much relation in the way of life, but they are remarkable organisms. The microscopic threads of algae weave together and form carpet on the earth. Each layer of this carpet or mat, depending on its exposure to light and oxygen, performs a unique function. Endowed with the remarkable capacities to adapt to varying environmental conditions (Tandeau de Marsac and Houmard, 1993), they colonize almost all kinds of terrestrial and aquatic ecosystems, often in some specific environments where no other vegetation can exist.

The blue-green algae contain most simple cellular organization of the plant kingdom. They can grow in almost all possible habitats including full sunlight and complete darkness. However, majority of them are aerobic photoautotrophs, their life processes require only oxygen, light and inorganic substances. A number of species of blue-green algae can tolerate temperatures as high as 72°C in the thermal pools of Yellow Stone National Park (USA), some species are halophilic or salt tolerant (as high as 27 per cent, for comparison, the salt in sea water is 3 per cent). They were characterized as the first plants to colonize on bare areas of rocks and soil, as an example subsequent to cataclysmic volcanic explosion (at Krakatoa Indonesia in 1883). The blue-greens that form symbiotic relationship with a wide range of other life forms, can convert nitrogen gas into ammonium (Chapman and Chapman, 1973; Lee, 1999). At the onset of adverse environmental conditions some members of Cyanobacteria develop modified cells, called an akinetes containing large amount of reserve food rich in carbohydrates. They can tolerate adverse ecological conditions such as complete drying of a pond and extreme cold; as a consequence, akinetes serve as "seeds" for the growth of juvenile filaments when favourable conditions return.

Biochemically, their cell wall is made up of peptidoglycan and contains diaminopimelic acid, which is common in the bacterial cell walls. It is suggested that Cyanophyceae and bacteria, both have originated from the same genetic stock hence the BGA are also termed as 'Cyanobacteria'. The presence of gas vacuoles in cells is a common feature in planktonic forms of the blue-green algae (Walsby, 1987) Some Cyanobacteria occur as endozoic *i.e.* inside the body of various animals. According to Langeron (1923), about 14 species of Oscillatoriaceae have been recorded from the digestive and respiratory tract of vertebrates. Cyanobacteria also cause precipitation of calcium and magnesium salts at the bottom of the springs to constitute, 'travertine'. A review of the literature shows that higher pH of water scarcity of CO_2 and higher amount of phosphorus is conducive to the production of the biomass of Cyanobacteria (Carmichael, 1994; Gajdek, 2000).

Cyanobacteria is highly successful group of plants, which enjoys a wide distribution along with wide variety of habitats *ie* freshwater, marine, epilithic, epipelic, epiphytic, endophytic, thermophilic etc, but are often abundant in freshwater, sea-water and in terrestrial environments (Fogg *et al.*, 1973) At present, Cyanophyceae records 160 genera and 1500 species from all over the world. Out of which, only 98 genera, and 833 species of blue-green algae have been reported so far from India. Sarma and Khan, (1980). Montagne (1849) for the first time described a blue-green alga, *Calothrix indica* from Assam, India. The members of Cyanobacteria from dominant plant life in frigid rivers of the foot-hills of Garhwal Himalayas, where they create an extensive benthic mat or scum under/over water ponds, streams and over the surface of moist soils. Sometimes the entire surface of water body becomes blue-green due to enriched growth of Cyanobacteria. Most of

Cyanobacteria occur as hydrophytes but other forms like epilithic, epipelic, saprophytic etc are not uncommon.

Classification of Fritsch (1945) is well accepted by most of the workers, as he divided this group into five orders namely, Chroococcales, Chaemosiphonales, Pleurocapsales, Nostocales and Stigonematales. Morphologically, this group includes simple unicellular/multicellular and unbranched/branched filamentous or spherical organisms. They may be true or false branched. Most of the members are motile and trichome forming hormogonia. They show gliding movement based on the torsional waves of the fibrillae (*e.g. Oscillatoria*).

Cyanobacteria have considerable and increasing economic importance including both, beneficial and harmful effects on human life. The beneficial uses of this group are in pharmaceutical as well as therapeutic industries. Due to the production of secondary metabolites antimicrobial compounds, antibacterial *e.g.* lithophytic Cyanobacterium *Nostoc spongiaeforme* var *tenue* (Banker and Carmeli, 1998), *Microcystis aeruginosa* (Ishida *et al.*, 1997), antifungal *e.g.* terrestrial blue-green alga *Calothrix fusca* (Moon *et al.*, 1992) etc, as food due to good source of proteins, vitamins and minerals *e.g. Spirulina,* as biofertilizers by fixing 78 per cent molecular atmospheric nitrogen through some filamentous and heterocystous Cyanobacterial species of *Anabaena, Nostoc, Aulosira* etc, in reclamation of usar lands and increasing the water holding capacity and in hydrogen production *e.g.* strains of *Synechococcus, Microcystis* and *Oscillatoria* (Prescott, 1969; Kumar, 1990; Kumar and Singh, 1995).

The harmful effects of Cyanobacteria are due to the production of toxins by some members of this group *e.g. Microcystis aeruginosa, Anabaena flos-aquae, Lyngbya majuscula, Oscillatoria nigrovirdis* etc. By the production of toxins, Cyanobacterial blooms alter the physical and chemical characteristics of water upto the deterioration of water quality. Thus causes unpotability of water, skin disorders, diseases by eating fishes from affected waters, in irrigation with effected water, in washing of clothes and dishes with toxic water, in water supplies of tanks, in stock watering, in swimming and recreation, in nuisance and undesirable condition, poisonous conditions etc. These are some major impacts caused by Cyanobacteria on human society.

2. Classification of Cyanobacteria

The classification of the blue-green algae have been reviewed and evaluated by Geitler (1925), Elenkin (1934; 1936), Fritsch (1942, 1945) and Papenfuss (1955). The classification given by Fritsch based only on the morphological characteristics is fully accepted by almost all the workers. The outline classification given by Fritsch (1945) is as follows:

2.1 Outline Classification as Proposed by Fritsch (1945)

Fritsch divided Blue-green algae into five orders namely Chrooccocales, Chaemosiphonales, Pleurocapsales, Nostocales, Stigonematatales (Table 6.1).

Classification As Proposed By Fritsch (1945) Modified After Desikachary (1959)

(1) Chroococcales

It is divided into two families:

(a) Chroococcaceae

1. Cells spherical, ellipsoidal, cylindrical, seldom spindle shaped, single or forming colonies, membrane thick, mucilaginous, often lamellated.
2. Colonies shapeless, spherical, ellipsoidal, tubular or hemispherical.
3. Common genera of this family are *Synechocystis, Microcystis, Gloeocapsa, Aphanocapsa, Chroococcus, Aphanothece, Gloeothece, Merismopedia, Dactylococcopsis, Eucapsis, Synechococcus, Coelosphaerium, Rhabdoderma and Gomphosphaeria.*

(b) Entophysalidaceae

1. Thallus generally attached, while some are free floating.
2. Cells arranged in erect radial rows, sometimes with a prominent one-sided development of sheath, which forms mucilaginous filamentous stalk.
3. Cells spherical or ellipsoidal, seldom cylindrical.
4. Common genera of this family are *Johannesbaptistia, Chorogloea, Entophysalis, Placoma.*

Table 6.1: The Taxonomic Classification of Blue-green Algae Adopted by Fristch (1945)

Order	*General Characteristics*	*Families*	*Representative Genera*
Chroococcales	Plant unicellular or colonial; colonies not showing polarity; multiplication by binary fission or by endospores; no heterocysts.	Chroococceae Cyanochloridaceae* Entophysalidaceae	*Choococcus, Microcystis* *Placoma, Entophysalis*
Chaemosiphonales	Plant unicellular or colonial; colonies showing distinct polarity; multiplication by endospores or exospores; no heterocysts.	Dermocarpaceae Chamaesiphonaceae Endonemataceae Siphononemataceae	*Dermocarpa,Stichosiphon* *Chamaesiphon* *Endonema* *Siphononema*
Pleurocapsales	Plant filamentous; multiplication by endospores; no heterocysts.	Pleurocapsaceae Hyellaceae	*Pleurocapsa, Oncobyrsa* *Hyella, Solentia*
Nostocales	Plant filamentous; but no division into prostrate and upright filaments; multiplication by short motile filaments called hormogonia, with or without heterocysts.	Oscillatoriaceae Nostocaceae Mircochaetaceae Rivulariaceae Scytonemataceae Brachytricchiaceae	*Microcoleus, Oscillatoria, Phormidium, Spirulina* *Ananaena, Anabaenpsis,* *Nostoc, Wollea Microchaete* *Calothrix, Dichothrix, Gloeotrichia* *Scytonema, Tolypothrix,* *Brachytrichia*
Stigonematales	Plants filamentous; showing distinct prostrate and upright systems; multiplication by hormogonia, or rarely by akinetes; heterocysts present.	Pulvinulariaceae* Capsosiraceae Nostochopsidaceae Loefgreniaceae* Stigonemataceae	*Pulvinularia* *Capsosira* *Mastigololeus, Nostochopsis* *Loefgrenia* *Fischerella, Hapalosiphon, Westiella*

* It seems to us that the members placed by Fritsch in this family are not true blue-green algae.

(2) Chaemosiphonales

It is divided into three families:

(a) Cyanidiceae

1. Cells spherical.
2. Plants free living, not differentiated into base and apex.
3. Cells single or in groups.
4. Reproduction by endospores and pseudoendospores.
5. Only have single genus *Chroococcidiopsis.*

(b) Chameosiphonaceae

1. Plants attached, differentiated into base and apex.
2. Form colony by the development of exospores.
3. Young cells are club shaped which later elongate to form sporangium.
4. Single genus *Chaemosiphon* is known.

(c) Dermocarpaceae

1. Plants attached, differentiated into base and apex.
2. The entire contents divide to form endospores.
3. Sporangia spherical, ellipsoidal, club shaped, pyriform or cylindrical with a basal mucilaginous stalks or disc.
4. Endospores 2 to many, mostly 8-32.
5. Common genera of this family are *Dermocarpa, Stichosiphon.*

(3) Pleurocapsales

It is divided into two families:

(a) Pleurocapsaceae

1. Thallus do not form filament.
2. Cells with a firm or gelatinous membrane.
3. Form colonies, which are attached.
4. Single genus *Myxosarcina* is known.

(b) Hyellaceae

1. Thallus form filaments. It is made of prostrate region with

radially growing filaments a discoid thallus and erect filaments with often endolithic conditions.

2. Filaments contain one to many rows of cells.
3. Filaments have apical growth, branched or unbranched.
4. Common genera of this family are Hyella, Hydrococcus, *Scopulonema, Xenococcus.*

(4) Nostocales

It is divided into three families:

(a) Oscillatoriaceae

1. Trichome false branched or unbranched without heterocysts and spores.
2. Trichome straight or regularly or irregularly coiled.
3. Common genera of this family are *Crinalium, Hydrocoleum, Trichodesmium, Microcoleus, Schizothrix, Oscillatoria, Sirocoleus, Porphyrasiphon, Spirulina, Polychlamydum, Katagnymene, Dasygloea, Lyngbya, Symploca, Phormidium.*

(b) Nostocaceae

1. Trichomes free or in common mucilage.
2. Cells in a single row, sheath present.
3. Heterocyst and spores present or absent if present it is intercalary.
4. Common genera of this family are *Richelia, Wollea, Anabaenopsis, Nostoc, Cylindrospermum, Raphideopsis, Aphanizomenon, Pseudoanabaena, Anabaena, Hormothamnion, Aulosira, Nodularia.*

(c) Scytonemataceae

1. Filaments with thick firm sheath.
2. Trichomes false branched, with single row of cells, heterocysts, and spores present.
3. Common genera of this family are *Plectonema, Pseudoscytonema, Hydrocoryne, Camptylonemopsis, Scytonematopsis, Scytonema, Tolypothrix.*

(*d*) Microchaetaceae

1. Trichomes with single row of cells, unbranced or false branched.
2. Heterocysts and spores present.
3. Common genera are *Microchaete, Fortiea.*

(*e*) Rivulariaceae

1. Trichomes with single row of cells.
2. Apices tapering in a hair.
3. Heterocysts present or absent, if present it is basal commonly.
4. Spores present or absent.
5. Common genera are Calothrix, Gloeotrichia, Dichothrix, Leptochaete, Rivularia and Homeothrix.

(5) Stigonematales

It is divided into five families:

(*a*) Capsosiraceae

1. Thallus attached.
2. Filaments free, erect irregularly subdichotomously or laterally branched.
3. Spores and heterocysts present or absent.
4. Single genus *Stauromatonema* is known.

(*b*) Nostochopsidaceae

1. Branching true, Lateral or reverse 'V' shaped.
2. Pedicellate heterocyst present.
3. Common genera are *Mastigocoleus and Nostochopsis.*

(*c*) Mastigocladopsidaceae

1. Filaments sheathed and branched.
2. Branching both reverse 'V' shaped and lateral.
3. Heterocyst present.
4. Single genus *Mastigocladopsis* is known.

(*d*) Mastigocladaceae

1. Trichomes with reverse 'V' shaped branching.
2. Heterocyst and spores present.
3. Common genera are *Brachytrichia* and *Mastigocladus*.

(*e*) Stigonemataceae

1. Branching true and lateral.
2. Main filament prostrate, often forming spores or in a chroococcaceous stage growth of filament by the division of an apical cell.
3. Heterocysts present.
4. Common genera are *Iyengariella, Fischerellopsis, Camptylonema, Westiella, Fischerella, Hapalosiphon, Stigonema, Westiellopsis.*

2.2 Stanier *et al.* (1971) have also Classified the Blue-green Algal Groups on the Basis of their Morphological Characters

A. Reproduction by repeated binary fission in a single plane, frequently resulting in the formation of short chains of cells.	1 Group
1. Cells cylindrical, ellipsoidal or spherical immediately after division; cells do not contain retractile polar granules.	Group IA
2. Cells cylindrical or ellipsoidal after division; cell always contains retractile polar granules.	Group IB
B. Reproduction by binary fission in two or three successive planes at right angles to one another; cells spherical.	
1. Cell divisions occur at regular intervals; never form parenchymatous masses of cells.	Group II
a) Cells not unsheathed; do not contain gas-vacuoles.	Group IIA
b) Cells unsheathed; do not contain gas-vacuoles.	Group IIB
c) Cells not unsheathed; contain gas-vacuoles.	Group IIC

2 Cell division irregular; growth leads to the formation of parenchymatous, tightly packed masses of cells. Group III

2.3 Rippka *et al.* (1971) after Stanier, has also given the Classification of Cyanobacteria which has been Described as Follows

Rippka divides the Cyanobacteria into five sections:

Section I (Unicellular)

Unicelluar Cyanobacteria that reproduce by binary fission, *e.g. Synechococcus, Synechocystis, Gloeothece, Gloeocapsa,* Gloeobacter, Chaemosiphon.

Section II (Unicellular)

Unicelluar Cyanobacteria that reproduce by multiple fission, *e.g., Dermocarpa, Chroococcidiopsis, Xenococcus, Pleurocapsa, Dermocarpella, Myxosarcina.*

Section III (Filamentous)

Filamentous non-heterocystous. Cyanobacteria that divide in only one plane, *e.g. Spriulina, Pseudoanabaena, Oscillatoria.*

Section IV (Filamentous)

Filamentous heterocystous. Cyanobacteria that divide in only one plane, *e.g. Anabaena, Nostoc, Nodularia, Scytonema, Cylindrospermum, Calothrix.*

Section V (Filamentous)

Filamentous heterocystous cyanobacteria that divide in more than one plane, *e.g. Chloroglocopsis, Fischerella.*

2.4 Practically almost all the workers accept the classification given by Fritsch.

3. Distribution of Cyanobacteria

Cyanobacteria enjoy a worldwide distribution and the majority of species are cosmopolitan. Such worldwide distribution is presumably related both to the antiquity of the group and to the high degree of resistance which is shown by most of its members towards adverse conditions. In the range of habitats on the one hand they predominate in the flora of arid deserts and are abundant on rocks and buildings exposed to the full intensity of tropical sunlight while, on the other different species shows massive growth in temperate freshwater lakes. Blue-green algae are the dominant plant life in the frigid lakes of the Antarctic. They are abundant on rocky seashores, common in moist soil, on barks of trees and various other habitats.

3.1 Distribution According to Habitats

Some Cyanobacteria are also found in drier conditions. They are found on the trunks of trees, on wires, on rocks, on walls, in hot springs and several other habitats. According to their habitats they are designed into following groups:

3.1.1 Hydrophytes

They are more or less completely submerged or free floating on the surface of the water. They may be subdivided into following groups:

Benthophytes

Those, which are found, attached to the bottom of reservoirs like pools, lakes, rivers etc.

Epactiphytes

These genera grow along the shores of lakes and ponds *e.g. Scytonema, Rivularia.*

Thermophytes

These Cyanobacteria can tolerate the temperature upto 70° C or more. The members of family Chroococcaceae and Oscillatoriaceae may survive upto 85°C. This supports that blue-green algae are primitive *e.g. Synechococus*.

Planktophytes

Those Cyanobacteria that float on water *e.g. Nostoc, Rivularia, Triponema, Microspora, Cylindrospermum, Gloeotrichia* etc.

Halophytes

Such Cyanobacteria occurs in saline water *e.g. Oscillatoria.*

Epiphytes

They are found upon other living plants *e.g. Microspora.*

3.1.2 Edaphophytes

They are called terrestrial algae found upon or inside the surface of the earth. They can be:

Cryptophytes

They are subterranean in habit and occur inside the soil.

Saphophytes

They are surface algae. They are found upon the surface of the soil *e.g. Nostoc, Aulosira.*

3.1.3 Aerophytes

These algae are aerial in habitat. They are found upon the trunks of trees, walls, fencing wires, rocks and animals and many other aerial substrata. They can be subdivided as:

Epiphyllophytes

Such Cyanobacteria are epiphytic upon leaves of trees *e.g. Phyllosiphon.*

Epiphloephytes

These Cyanobacteria grow on the bark of trees mined with many mosses and liverworts *e.g. Phormidinum, Scytonema, Haplosiphon and Schizothrix* .

Epizoophytes

These Cyanobacteria are found even on the bodies of land animals. Certain Chaetophorales are found even on the hairs of sloth.

Lithophytes

These Cyanobacteria grow on the rocks and walls *e.g. Scytonema.*

3.1.4 Gyophytes

These Cyanobacteria are found on ice and snow and cause 'red snow', 'yellow snow' and 'green snow'. The 'Red snow' is caused by species of *Gloeocapsa* and diatoms. Certain species of *Pleurococcus* and *Nostoc* cause 'yellow or yellow green snow'.

3.1.5 Symbionts or endophytes

Cyanobacteria grow in symbiotic association of other plants. The most striking example of symbiosis is Lichen. Various

Cyanophyceae as *Chroococcus, Nostoc, Microcystis, Gloeocapsa, Scytonema, and Rivularia* etc have been separated from lichens *Anabaena azollae* is found inside the leaves of *Azolla* (A Pteridophyte), *Anabaena cycadae* is found in the corolloid roots of Cycas (a Gymnosperm), *Nostoc* is found in the leaves of *Sphagnum* (Bryophytes), *Nostoc* has also been reported from the tissue of *Anthoceros and Notothylas* (Bryophytes).

3.1.6 Endozoophytes

Certain Cyanobacteria occur inside the body of animals. According to Langeron (1923), about 14 species of Oscillatoriaceae are found in the digestive and respiratory tracts of various vertebrates.

3.1.7 Parasites

Some Cyanobacteria live as parasite upon other plants.

3.1.8 Eluviatite algae

Such Cyanobacteria are found in rapidly flowing water.

3.2 Distribution According to Presence in Water

Most of the part of the land is covered with water. The water can be classified as freshwater and seawater. Cyanobacteria are found both in freshwater and marine environments on the basis of their presence in water they are classified as Freshwater Cyanobacteria and Marine Cyanobacteria.

3.2.1 Freshwater Cyanobacteria

Cyanobacteria are widespread and often abundant in freshwater. Cyanobacteria can be divided into two forms *i.e.* Planktonic and Attached freshwater forms.

Planktonic Freshwater Forms

These belong mostly to the Chroococcales and Nostocales. In chroococcales, genus *Coelosphaerium, Gomphosphaeria, Merismopedia* and *Microcystis* are the most commonly represented. The common filamentous forms of order Nostocales belong to the genera *Arthrospira, Oscillatoria, Lyngbya, Spirulina, Anabaena, Anabaenopsis, Aphanizomenon* and *Gloeotrichia*. The species of *Anabaena* are the main phytoplankton organisms in Crater Lake, Oregon (Utterback *et al.*, 1942).

Attached Freshwater Forms

The number of attached freshwater forms is more than the planktonic freshwater form. The nature of the attached freshwater flora is largely determined by the general nature of the substratum and is classified into following forms:

1. Epilithic: growing on rock and stones.
2. Epiphytic: growing on larger plants.
3. Epipelic: growing on, or in, bottom sediments.

The striking development of BGA is seen in Antarctic lakes.

3.2.2 Marine Cyanobacteria

They are widely distributed in marine ecosystems although they are seldom. The dominant epilithic algae on temperate rocky shores are some species of *Calothrix, Entophysalis, Gloeocapsa, Phormidium, Plectonema, and Oscillatoria*. Common endolithic marine forms are *Hyella caespitosa, Mastigocoleus testarum, Gomontia polyrhiza, Phormidium molle and Gloeocapsa* sp.

Cyanobacteria are rare in the supralittoral and littoral regions of tropical shores because of the severe desiccation, which results during low tide.

3.2.3 Terrestrial Cyanobacteria

Cyanobacteria are found abundantly in terrestrial habitats. They occur predominantly on the surface of soils, stones, rocks, and trees, particularly in moist, neutral, or alkaline areas.

4. Factors Affecting their Distribution

The most important factor favouring the worldwide distribution is their ability to survive in the extreme conditions of environment. Along with this light, pH, temperature, and other physical factors also play an important role in the distribution of Cyanobacteria.

4.1 Light

Cyanobacteria grow in abundance in the presence of high light. Rarely or no Cyanobacteria are observed to grow in dark, because most of the Cyanobacteria are obligate photoautotrophs *i.e.* they grow in the presence of light.

4.2 Temperature

The optimum temperature required for the growth of blue-green algae is nearly 35°C. It has the ability to grow in wide range of temperatures than other algal groups. Certain thermophilic blue green algae have the ability to tolerate upper extremes of temperature. They require a temperature of about 30-45°C. Thermophilic species *Synechococcus* species may not survive at temperatures below 50°C. The maximum temperature at which a Cyanobacterium has been recorded is about 73°C and *Synechococcus* is the single genus reported to grow in such a high temperature (Ward and Castenholtz, 2000).

4.3 pH

Cyanobacteria prefer neutral to alkaline medium for their abundant growth. Various observations show that Cyanobacteria grow abundantly at higher pH *i.e.* upto 11.

4.4 UVB and C Radiations

Some species of Cyanobacteria have the ability to tolerate UV B and C radiations that make this group most widely distributed in all the area.

4.5 Soil

Cyanobacteria grow well in all types of soil especially ranging from loam to clay.

All the above mentioned factors are conducive for the growth of cyanobacteria and hence are also favourable for their distribution.

5. Conclusion

A variety of habitats are represented on our planet, which favour the growth and accumulation of Cyanobacterial taxa. That is why this algal group enjoys a worldwide distribution and is cosmopolitan. Changing environment, due to human activities is resulting in the appearance of several new species and disappearance of the others. It is paradoxical that while the underdeveloped countries need to encourage the growth of blue-green algae, the countries of west are adopting measures to control their growth because they take it as creating nuisance in the environment. They are equally important in several aspects such as biofertilizers and as soil conditioners etc. Thus the blue-green algae, which were probably predominant among the aboriginal organisms of the earth, are still quantitatively and qualitatively important in the biosphere.

6. Acknowledgement

Authors are thankful to the UGC, New Delhi for providing financial support through a Major Research Project.

7. References

Banker R and Carmeli S 1998 Tenuiecyclamides A-D cyclic hexapeptides from the cyanobacterium *Nostoc spongiaeforme var tenue. J Nat Prod* 61: 1248-1251.

Carmichael W W 1994 *The toxins of cyanobacteria. Scientific American* 270: 78-86.

Chapman V J and Chapman D J 1973 *The Algae.* Macmillan, London.

Desikachray T V 1959 *Cyanophyta.* Indian Council of Agricultural Research, New Delhi, India.

Elenkin A A 1934 *Sur le principles fonda mentaux due system les Cyanophyceae Soc Bot.* 1934: 51–83.

Elenkin A A 1936 Monographic algarum cyanophycearum aquidulcuem et lessestruim infinbus URSS. enventrum Vol 1, Acad Nauk URSS, Moscow.

Fabbro L and Mc Gregor G 2003 *Blue–green Algae General Information NRM Facts Water Series–W3*, Queenslad Govt Natural Resources and Mines, Australia.

Fogg G E, Stewart W D P, Fay P and Walsby A E 1973 *The Blue-Green Algae.* Academic Press, London.

Fritsch F E 1942 The interrelations and classification of the Myxophyceae. *New Phytol*, 41: 134-148.

Fritsch F E 1945 *The Structure and Reproduction of the Algae.* Vol-II Cambridge Univ Press, London.

Gajdek P 2000 Mikrocystyny sinic W Zbiornikach Wodnych Wiadomo ci chemiczne. 54: 637-650.

Geitler L 1925 Synoptische Darstellung der cyanophycean in morphologische and systematischer Hinsicht *Berh Bot Cbl II.* 41:163-294.

Graham Le and Wilcox L W 2000 *Algae*. Prentice Hall, Upper saddle River, NJ 07458 USA.

Ishida K, Matsuda H, Kurakami M and Yamaguchi K 1997 Kawagudhipeptin B, an antibacterial cyclic undecapeptide from the Cyanobacterium *Microcystis aeruginowa*. *J Nat Prod* 60: 724-726.

Kumar H C 1990 *Introductory Phycology*. Affiliated East West Press New Delhi.

Kumar H D and Singh H N 1995 *A Textbook of Algae*. Affiliated East West Press, New Delhi.

Langeron M 1923 Les Oscillariees parasites du tube digestif de phomme st des animaux. *Ann paurasitol lee maine et comp* 1: 75-89, 113-123.

Lee R E 1999 *Phycology*. Cambridge Univ Press, London.

Montagne C 1849 Sixiene centuric de plantes cellulaires nouvelles tant di indigences qu exotiques. *Ann Sci Nat Bot* 12: 285-320.

Moon S S, Chen J L, Moore R E and Patterson G M L 1992 Calophycin, a fungicidal cyclic decapetide from the terrestrial blue-green alga *Calothrix fusca*. *J Org Chem* 57:1097-1103.

Papenfuss G F 1955 Classification of Algae in A century of Progress in Natural Sciences, California. *Acad Sci*115-224.

Prescott G W 1969 *The Algae: a Review*, London; Nelson.

Sarma Y S R K and Khan M 1980 *Algal Taxonomy in India. Bot Reco and Monograpgh–2*, Todays and Tomorrow's Print and Publ, New Delhi.

Stanier R Y and Cohen-Bajire G 1977 *Phototrophic prokaryotes: The Cyanobacteria:* pp 225-227 In: Starr M P, Ingraham J L and Balows A (eds) Palo Alto C A: Annual Review of Microbiology: Resource competition during seasonal succession towards dominance by Cyanobacteria *Ecology* 70: 229-245.

Tandeau de Marsac N and Houmard J 1993 *Adaptation of Cyanobacteria to environmental stimuli: New steps towards molecular mechanisms*. 104: 119-190.

van den Hoek C, Mann D G and Johns H M 1993 *Algae: An Introduction to Phycology*. 2nd ed Cambridge Univ Press, Cambridge.

Walsby A E 1987 Mechanism of buoyancy regulation in planktonic Cyanobacteria with gas vesicles. pp 377-392 In: Fay P and Van Ballen C (Eds), *The Cyanobacteria*. Elsevier Science Publ, Ansterdam.

Utterback D L, Phifer LD and Robinson R J 1942 Some chemical, physical and optical properties of crater lake. *Ecology* 23: 97-103.

Chapter 7

Biotechnological Potential of a Cyanobacterium: *Spirulina*

Yashveer Singh

Contents

© R.K. Gupta, M. Kumar & G.S. Paliwal (*Editors*)
Glimpses of Cyanobacteria, 2006 107–121
Daya Publishing House, New Delhi

ABSTRACT

The filamentous blue-green microalga, *Spirulina* is a rich source of good quality protein (60-70 per cent of dry weight) for malnourished people. In addition, *Spirulina* species have high contents of iron, pro-vitamin A, Vitamin B_{12} and iodine. *Spirulina* based formulation for skin and body care are extremely effective in cosmetics as they activate skin metabolism and stimulate the enzymes required for the same. *Spirulina* contains well balanced amino-acid composition and regarding lipids, galactosyldiglycerides, phosphatidylglycerol and free fatty acids are dominant. The lipopolysaccharide comprised about 2 per cent of the dry weight and total carbohydrate content of this cyanobacterium is 16 per cent of the dry weight and hydrolysis yielded glucose, sucrose, levulose, glycerol and several polyols. Two heaped tablespoons of *Spirulina* powder (20 g) provide all the daily body requirements of vitamin B_{12} as well as significant quantity of other B-complex with vitamins including 70 per cent of recommended daily allowances (RDA) for vitamin B_1 (thiamine), 50 per cent for B_2 (riboflavin and 12 per cent for B_3 (niacin). The presence of unsaturated essential fatty acid, linolenic acid in *Spirulina* species can help in arthritis, obesity and heart diseases in human.

1. Introduction

Malnutrition has become a major problem in the developing countries, which is further aggravated by the rapid increase in human population. In view of the increasing cost of animal protein, unavailability of suitable cheap plant protein and growing tendency of man of prefer plant products, it has become necessary for phyco-biotechnologists to look for alternative sources of protein. Regarding the cheap sources of protein and lipids, variety of microalgae is known today. The most promising microalga, *Spirulina*, which has great potential as protein supplement for vegetarian people may afford easily and manageable sources of cheap protein, minerals, essential amino acids and fatty acids, and vitamins. The term single cell protein (SCP) were coined to designate the protein obtained from the cyanobacterium, *Spirulina*.

The four most prevalent deficiency diseases of public health importance in developing countries, including India, are protein-calorie malnutrition, nutritional anaemia, xerophthalmia and endemic goiter (Ramalingaswami, 1977). The filamentous, blue-green microalga, *Spirulina* is a rich source of good quality protein (60-70 per cent dry weight) for malnourished people (Venkataraman and Becker, 1982). It is useful in the context of the above problems in view of its high contents of protein, iron, vitamin B_{12}, pro-vitamin A and iodine. Thus, it is a good source of non-conventional vegetarian protein for food and feed. It is also a relatively extensively cultured microalga for biomass production (Vonshak and Richmond, 1988). For the use as food, decoloration of dry *Spirulina* may be obtained by extraction with ethanol and acetone, yielding a pale-yellow meal with 80 per cent efficiency. It is practically odourless and contains 84.2 per cent protein. Now *Spirulina* is also being used for skin care, controlling blood sugar and for nursing mothers. Today there is an increasing faith in natural herbs and their formulation for skin and body care all over the world. Totally natural in composition, *Spirulina* is fast replacing synthetic contents in the preparation of cosmetics. *Spirulina* based formulation are extremely effective in cosmetics as they activate skin metabolism and stimulate the enzymes required for the same.

In may cases nursing mothers may not produce sufficient milk owing to lack of nutritive food, infection or even due to stress factors. In Vietnam, this problem is overcome by incorporation of *Spirulina* tablets to nursing mothers. Now a days, *Spirulina* tablets with various trade names, containing high protein content and vitamins are available in the Indian markets and being sold by chemists and drugists. 'Zyrulina', a *Spirulina* tablet (500 mg) is being marketed by Alidac Genetics and Pharmaceutical; and manufactured by Cadila Health Care Ltd., Ahmedabad. In India, many small entrepreneur and big manufacturers like Thapars and Dabur have taken up the production of *Spirulina*. Commercial preparations like 'Sunnova' and 'Spiruvit' as capsules containing dry powder of *Spirulina* are available in the market. There is ample scope for increasing the demand of this wonderful alga in the domestic market atleast for enriching the fish and poultry feed. In U.S.A., *Spirulina* is declared as 'Food for Tomorrow' and given to the athletes and soldiers as protein source.

2. Physiology of Cyanobacteria

Cyanobacteria, the oldest group of oxygen-evolving photosynthetic prokaryotes, are playing a key role in the biological nitrogen fixation especially in paddy crops. Many Cyanobacteria (blue-green algae) are still untapped resource for the production of lipid, protein and other fine chemicals. On other hand, algae as seaweeds are the largest primary producer in aquatic ecosystem on the earth and as a phytoplankton; it makes the basis of biotic existence in water. The ability of blue-green algae to survive through earlier harsh environment made them in widespread adaptability to the extremes of radiations (Sinha and Hader, 1996), illumination, desiccation (Tripathi, 1998) temperature, pH, toxicant (Rai *et al.*, 1981) and nutrient availability (Goyal and Goyal, 1998). These characteristics retained through the long evolutionary sequences conferred cosmopolitan nature on Cyanobacteria. Blue-green algae have striking morphological and physiological similarities with the bacteria. All Cyanobacteria are mostly Grams-ve and possess chlorophyll *a*.

3. Chemical Composition of *Spirulina*

The chemical composition of *Spirulina* reflects its potential as human food, animal feed and a source of natural products. The protein content varies from 60 per cent to 70 per cent of dry weight depending upon growth conditions and the percentage of ash. In acid-washed *Spirulina,* the protein content is 70 per cent . An amino acid analysis of spray-dried *Spirulina* was produced in Mexico. Although the amino acid composition was well balanced, it was low in sulphur-containing amino acid and in tryptophan.

Tornabene *et al.* (1985) have examined the lipid and lipopolysaccharide constituents of *Spirulina*. The dominant lipids in *Spirulina* were mono-,di- and higher galactosyldiglycerides and phosphatidylglycerol. The dominant component in the nonpolar lipid fraction was free fatty acids. The lipopolysaccharide comprised 1.8 per cent of the dry weight. The carbohydrate content of the lipopolysaccharide consisted of hexose, heptose and glucosamine. The carbohydrate content of *Spirulina* is about 16 per cent of the dry weight and hydrolysis yielded glucose, sucrose, levulose, glycerol and several polyols. Glucosan and particularly rhamnosan both containing glucosamine, were found to be the main carbohydrate storage products. The presence of sterols is related to the anti-microbial activity of this alga. From a nutritional standpoint, the most fatty acid components are linolenic acid and linoleic acid, which in *Spirulina* produced in Mexico, were 1.04 per cent and 1.24 per cent of dry weight, respectively.

4. Nutritional Potential

As a rule, algal proteins are poorly utilized when intact cells are fed to monogastric animals or humans due to their digestibility. Blue-green algae in general, and *Spirulina* in particular, are unique in that they are highly digestible and thus do not require special processing. According to Venkataram and Becker (1982) only a very small differences were observed between the digestibility of fresh *Spirulina* (82 per cent) and sun-dried and freeze-dried *Spirulina*, which yielded 65 per cent to 70 per cent digestibility, respectively.

Table 7.1: Vitamin Content of *Spirulina platensis*

Vitamin		*Content (mg/kg dry weight)*
ß-carotene (pro-vitamin A)	:	1700
Cyanocobalamine (B_{12})	:	1.6
d-ca-pantothenate	:	11
Folic acid	:	0.5
Inositol	:	350
Niacin(B_3)	:	118
Pyridoxine (B_6)	:	03
Thiamine (B_1)	:	55
Tocopherol (E)	:	190

Spirulina is very suitable as a dietary supplement in dairy and poultry farms. It has been demonstrated to be a satisfactory source of protein for many animals like cow, buffalo, goat and pig. *Spirulina* increases the protein content in the milk of cow, buffalo and eggs of hen. It has also a cholesterol-lowering effect. In human, its supplementation in diet is used for the treatment of pancreatitis, hepatitis and constipation.

The nutritive value of *Spirulina* is amplified in that it is extremely high in vitamin A and B_{12} and its lipids are made up of unsaturated fatty acids that do not form cholesterol. *Spirulina* have non-toxic, easily digestible, mucoprotein cell walls. These make *Spirulina* a potential food item for persons suffering from coronary illness and obesity. In addition to being exceptionally high in protein, *Spirulina* appears to have the highest vitamin B_{12} content of any unprocessed plant or anima! food, representing a boon to vegetarian diets. Two heaped tablespoons of *Spirulina* (20g), provide all the daily body requirements of vitamin B_{12}, as well as significant quantities of other B-complex vitamins, including 70 per cent of the recommended daily allowance (RDA) for vitamin B_1 (thiamine), 50 per cent for B_2 (riboflavin) and 12 per cent for B_3 (niacin). Other nutritional attributes of *Spirulina* include vitamin A and unsaturated fatty acids. Extraction methods could provide high protein power of decoloured *Spirulina* suitable for widespread uses.

The effect of *Spirulina* tablets as a supplement to a reducing diet on clinical-chemical parameters and the body weight of patients was recently conducted in Germany. During the four weeks that *Spirulina* tablets were taken (14 tablets before each meal), body weight fell significantly by 1.4 kg. Various therapeutic properties of *Spirulina* have been reported and a study showed that *Spirulina* and its enzymic hydrolysates promote skin metabolism and prevent keratinization. The water-soluble blue pigment, phycocyanin may stimulate the immune system, providing protection from a variety of diseases. Epidemiological studies also suggested that high dietary pro-vitamin A (β-carotene) intake decreases cancer risks (Underwood, 1984). By virtue of its high vitamin A and B_{12} content *Spirulina* can serve as a food supplement in preventing night blindness in children and pre and post-natal anaemias in women. Linolenic acid (an essential amino acid) stimulates prostaglandin synthesis. Clinical studies demonstrate that dietary linolenic acid can help arthritis, heart disease and obesity. *Spirulina* extracts restored a significant amount of the colinestarase activity in human erythrocytes (RBC). The consumers of *Spirulina* have reported several consistent effects. These include the ability to postpone or skip a meal with quantities as little as three *Spirulina* pills, reduction of pre-menstrual stress and freedom from hangovers when 3 gm. of *Spirulina* are ingested the previous evening. Phycocyanin has also been found application in biopigments, ointments, creams and lotions. Intake of food rich in β-carotene a precursor of vitamin A reduces vitamin A deficiency and acts as antioxidants that further reduce the toxic effects of free radicals (Verma *et al.*, 1998).

5. Morphological Features

Spirulina is a multicellular Cyanobacterium having blue-green filaments of 2 to 12 µm diameter cylindrical cells in unbranched, helicoidal trichomes. The filaments are motile, gliding along their axis and have no heterocysts. *Spirulina* can grow rapidly, reaching high filament densities in warm, shallow brackish lakes. It is one of the most common and abundant algae in many alkaline-saline lakes in Africa and America. In India, it is found in Sambhar Salt Lake, Rajasthan and in ponds in Varanasi (Singh *et al.*, 1994). The helical shape of the trichome is characteristic of *Spirulina*, but helix pitch, length and dimensions vary with the species. The helical shape is maintained only in liquid media, the filaments becoming true spirals on solid media. Jeeji Bai and Seshadri (1986) recognized three distinct morphological variants of *Spirulina fusiformis* from Madurai (T.N.): (i) an S-type variant with trichomes having more or less regular and distant coils, (ii) a C-type variant with trichomes having a distinct spindle shape with close coils; and (iii) an H-type variant with trichomes forming a dumb-bell shape with very close to tight coils (Figure 7.1). Under high light intensity and nutrient concentration these distinct types readily transformed from one type to another. Some common species of *Spirulina* are-*S.platensis, S.major, S. geitleri, S.jenneri* and *S.maxima*.

6. Ecology

Spirulina, a ubiquitous organism, was first isolated by Turpin (1827) from a fresh water stream and can be found in widely differing environments such as soils, marshes, fresh, brackish and sea waters and thermal springs. *Spirulina* may colonize extreme environments that are unsuitable for many other algae. This protein-rich alga has been eaten by people in the Lake Chad region of Africa for many generations. Also the Aztecs in Mexico were consuming *S. maxima* as a major source of protein in the 16th century (Ciferri, 1983).

Typical examples are *S.platensis* and *S. maxima*, which grow profusely in certain alkaline lakes in Central America and Africa. The example of ideal growth and ecology of *Spirulina* is found in Lake Chad of Africa. Here *Spirulina* may become the dominant species

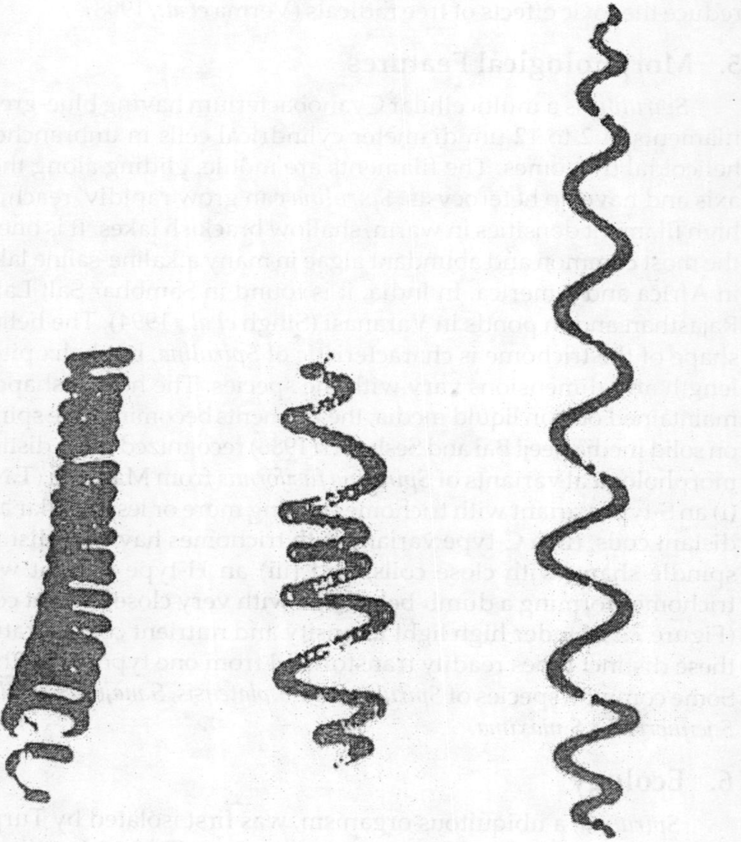

Figure 7.1 : Morphological Variants of *S. fusiformis*
Right : S-type trichome; Middle : C-type trichome, and Left : H-type trichome.

and it has been used as food since time immemorial. The Lake Chad is shallow with an average depth of about 3.5 m and the lake area is about 20,000 Km^2. The climate is subdesertic, with a long dry season alternating with a rainy season of about four months. The average yearly water temperature is 27°C with two maxima in June and September. The water pH rages from 7.2 to 9.0 and the water chemistry is dominated by HCO_3^-, with SO_4^{--} and Cl^- occurring only in small quantities. Lake with salt concentrations above 30 gl^{-1} supported monospecific population of *Spirulina*. It is known that the higher the salt content and pH of water, the predominance of *Spirulina*. Adaptation of strain of *S. platensis* to grow in cobalt and iodine-enriched media are reported by Singh and Kumar (1994) and showed that this adapted strain have higher uptake of these two trace elements than the controls in lab conditions.

7. Biomass Production

Since nearly two decades *Spirulina* has enjoyed a leading position as a high-value protein source and technologically suitable for mass production (Venkataraman, 1983). In most countries only *S. maxima* and *S. platensis* have been considered as good strains for biomass production. To make the technology available at village level to the poor and to achieve economy in production, albeit to a limited extent, low-cost technologies and substitute for costly chemicals have been developed and tried with success (Seshadri and Thomas, 1979). The pot culture technology for the rural poor has been developed by Jeeji Bai (1986). Although the daily yield and cost are more favourable in the ponds system but pot culture technology is more suitable as it is easier to maintain by a single rural woman without any outside dependence. The methodology of pot culture shall be more simple and accessible to the poor with further improvements as the villagers still depend on a supply of stock nutrient medium (Zarouk or Jeeji Bai medium) and culture. The difficulty of providing medium may be overcome by developing a fertilizer base nutrient medium as fertilizers (N, P, K) are available in all villages. In urban area wastewater can be used for *Spirulina* production during the treatment or purification of water with algae (Saxena *et al.*, 1983). Microalgae biomass production methods and their uses have been reported by Shelef and Soeder (1980).

8. Mass Cultivation

Mass production of *Spirulina* requires inexpensive yet reliable enclosures for growing the cultures. In practice, the design of these enclosures represents a compromise between the cost of the investment in relation to the expected returns and the desire to establish conditions for the highest possible output rate.

The basic design presently employed for mass cultivation of *Spirulina* consists of an oblong, shallow raceway stirred with paddles and lined with some plastic sheeting. Commercial pond areas range between 500 and 5000 m^2. These ponds may be open and exposed to the atmosphere or covered with a transparent material. Another form of closed reactor consists of transparent, polyethylene tubes. The tubes are placed on the ground and arranged in the form of raceway, the algal-laden medium being circulated by a pump. The high evaporation increases the costs of water and pumping, but more importantly, it increases the salinity of the medium. The advantage of closed system is a reduction in the amount of dirt and insects contaminating the algal product. The disadvantage of closed system is that light penetration is reduced because most of the commercially available materials used for tubing or for covering are not completely transparent, In addition, dust also accumulates on the outer surface of the covers.

The main environmental factors governing the production of *Spirulina* biomass are nutrients, temperature and light.

8.1 Nutrition

High alkalinity is mandatory for the growth of *Spirulina*, as reflected in the pH optimum for its growth, which ranges from 8.2 to 11.0. *Spirulina* can tolerate progressive changes in pH; however, the culture may rapidly deteriorate when pH is changed abruptly, as it may happen in a growth medium that is not well buffered. 0.2M $NaHCO_3$ provides good buffering capacity for the growth medium. Nitrates are the main nitrogen source assimilated by *Spirulina*, but ammonium salts may be used as long as the NH_4^+ concentration is less than 100 mg l^{-1}. Urea can be used with no ill effect at pH 8.4 as long as its concentration is below 1.5 gl^{-1}. The addition of 0.1 per cent (w/v) glucose to the growth medium enhances growth rate and cell yield. Both Na^+ and K^+ are indispensable in the *Spirulina* growth medium (Jeeji Bai and Seshadri 1986). Inhibition of growth may occurs when the ratio K^+: Na^+ is >5. Growth is uninhibited even at very high concentration of Na^+ (18gl^{-1}).

8.2 Temperature

Spirulina is a mesophilic alga. The optimum temperature for growth is in between 35 and 39°C. In outdoor cultures, however, a rise in temperature to 40°C for few hours does not cause any harmful effects. The minimum temperature that still permits growth in *Spirulina* species is about 18°C. *Spirulina* can tolerate relatively low night temperatures in winter.

8.3 Light

When nutrients and temperature are not limiting the growth of *Spirulina*, light availability to the average cell become the dominant limiting factor. The availability of light to each cell in the cultures is a function of the intensity and duration of light irradiance as well as the population density that affect the extent of mutual shading. The latter is due to the absorbance of the incident light by the algal cells closed to the illuminated surface, thereby decreasing the amount of light available to the cells below. Population density has been found to be a major factor in the production of *Spirulina*. The higher the density the lower the potential photosynthetic rate. In ponds solar irradiation is almost completely absorbed in the upper 4 to 5 cm of the cultures, leaving some 60 per cent of the cells in the pond in complete darkness. In dense mass cultures, the cells are exposed to intermittent radiation, which imposes a light-dark cycle to which the average *Spirulina* filament is continuously exposed. The relative effect of decreasing the population density, and thus of increasing the rate of light utilization per cell, is most pronounced in summer.

An important point to bear in mind is that, under artificial pond conditions, growth limitation by either temperature or light alone takes place only at its extremes. In practice, the growth rate of microalgae is governed by both temperature and light, and may also be influenced by nutritional factors. It is also reported that the effects of light and temperature on algal growth rates are interrelated.

8.4 Maintenance of Mass Cultures

The relationship between population density and the output rate of *Spirulina* has been investigated and found that a decline in output rate was always associated with a population density over 500 to 600 mg l^{-1}. A basic factor that modifies the effects of population density in mass cultures of *Spirulina* is mixing. Mixing has several functions, the most important of which is to ensure that a favourable regime of light intermittance is maintained. The most common device used to create a flow in shallow, open raceways is the paddle wheel. In a closed system, a suitable pump could be used for air-lift to stir *Spirulina* cultures. A crucial challenge for the commercial production of *Spirulina* is to maintain, throughout the year, a monoalgal culture. The basic need is to provide growth conditions that will not vary too much from the optimum. Since nutrition deficiencies can be controlled easily in mass cultures, the most important factor that affects species competition is temperature. The more the temperature declines from the optimum for *Spirulina*, the slower the growth rate and the easier it is for contaminants to take over the culture. *Chlorella*, a green microalga, seems to be the main organism contaminating *Spirulina* cultures. *Spirulina* is an exceptionally good cyanobacterium for biomass production and is suitable for use in SCP production due to its high growth rate, pH and temperature tolerance, yield and airtime requirement. The result obtained by Singh *et al.* (1994) showed that *S. major* is as good as *S. platensis* for biomass production. *Spirulina major* was isolated from a pond in Varanasi city in which people take bath.

8.5 Harvesting of *Spirulina*

The filaments of *Spirulina* are harvested from the growth medium by filtration. In some production plants today, a vibrating screen is used for harvesting *Spirulina* biomass. This machine, however, causes some cell damage due to bruising of the delicate cells. Micro straining is another method that may be used for harvesting *Spirulina* from pond.

8.6 Drying of *Spirulina*

In commercial *Spirulina* production, drying poses a problem of major economic importance, in that it may constitute about 20 per cent of the production cost. The various systems for drying differ both in the extent of capital investment and in energy requirements, and a marked effect on the food value and the taste of the products. Experience with *Spirulina* production is that the harvested slurry should be well rinsed in acid water at pH 4.0, to remove the adsorbed carbonates. It can be stored at 0 to 2°C for several days. The usual method for drying of *Spirulina* is spray drying, which yields a very good product from which pills may readily be formed. Drum drying also yields a useful product, in the form of flakes of special flavour. Direct drying of the *Spirulina* slurry in the sun is also feasible. Sun drying is not recommended for preparing an algal product instead for human consumption, for two reasons:

1. A rather unpleasant odour is associated with sun drying, due to the slowness of the dehydration process, which enables degradative processes to set in before drying is complete.
2. Sun drying does not include exposure to high heat (upto 120°C for a few seconds as in drum-drying) and may thus exhibit a higher bacterial count.

For the production of animal feed, sun drying may be an acceptable solution. In the laboratory, fish feed made of *Spirulina* that had been vacuum dried to 20 per cent dry matter is mixed with different proportions of corn meal and dried in the sun. Commercial production of *Spirulina* takes place in Mexico, U.S.A., Japan, Taiwan, Thailand, India and Israel is estimated to produce over 200 tons/year biomass of *Spirulina* (Yanagimoto and Saitoh, 1982). In India scientist of the institutes like National Botanical Research Institute (NBRI) Lucknow, Central Food Technology Research Institute (CFTRI) Mysore and Shri A.M.M. Murigappa Chettiar Research Centre, Chennai are engaged in the development of low cost technology for the mass cultivation of *Spirulina*.

9. Acknowledgement

The author thanks Dr H D Kumar, Ex-Professor of Botany, BHU Varanasi for suggestions and to the Principal, Govt Raza P G College, Rampur for library facility and moral support. UGC is highly acknowledged for financial assistance to a minor research project.

10. References

Cifferi O 1983 *Spirulina*, the edible microorganism. *Microbiol Rev* 47: 51-578.

Goyal D and Goyal S K 1998 Biotechnological Potential of Microalgae. *In:* Verma B N, Kargupta A N and Goyal S K (eds) *Advances in Phycology* APC Pub New Delhi.

Jeeji Bai N 1986 Mud pot culture of the alga *Spirulina (Arthrospira) fusiformis* for rural households. Monograph Series on Engineering of Photosynthetic Systems Vol:19 MCRC Madras.

Jeeji Bai N and Seshadri C V 1986 Nutritional Augmentation using home grown *Spirulina*. *In*: Proc International Symp "Proteins in Food, Health and Industry" Loyola College Madras.

Rai L C, Gaur J P and Kumar H D 1981 Phycology and heavy metal pollution. *Biol Rev* 56:99-151.

Ramalingaswami V 1977 Knowledge and action in the control of vitamin A deficiency. Annals New York *Acad Sc* 300:210-220.

Saxena P N, Ahmad M R, Shyam R and Amla D V 1983 Cultivation of *Spirulina* in sewage for poultry feed. *Experientia* 39:1077-1083.

Seshadri C V and Thomas S 1979 Mass culture of *Spirulina* using low-cost nutrients. *Biotechol Lett* 1: 287-291.

Shelef G and Soeder C J 1980 *Algae Biomass: Production and Use.* Elsevier/North Holl and Biomed Press Amsterdam.

Singh Y and Kumar H D 1994 Adaptation of a strain of *Spirulina platensis* to grow in cobalt and iodine enriched media. *J Appl Bacteriol* 76: 149-154.

Singh D V, Jha M and Kumar H D 1994 Biomass production of *Spirulina major*. *In*: Kashyap A K and Kumar H D (eds) *Recent Advances in Phycology*. Rastogi Pub Meerut.

Sinha R P and Hader D P 1996 Response of a rice field cyanobacterium *Anabaena* sp. to physiological stress. *Env Exptl Bot* 36:147-155.

Tripathi S N 1998 Desiccation tolerance in terrestrial cyanobacteria. *In*: Verma B N, Kargupta A N and Goyal S K (eds) *Advances in Phycology*. APC Pub New Delhi.

Turpin P J F 1827 *Spirulina oscillarioides.* In: FG De Levrault (Ed) *Dictionnaire des Sciences Naturelles* Paris 50:309-310.

Underwood B A 1984 (Ed) *Vitamin A and Cancer Prevention Conference* National Institute of Health Maryland USA.

Venkataraman L V and Becker E W 1982 *Biotechnology and Utilization of Algae: The Indian Experience.* Department of Science and Technology, New Delhi.

Venkataraman L V 1983 A Monograph on *Spirulina platensis*. Biotechnology and Application. All India Co-ordinated Project on Algae DST CFTRI Mysore.

Vonshak A and Richmond A 1988 Mass production of the blue-green alga, *Spirulina*: An overview. *Biomass* 15:233-247.

Chapter 8

Toxicity of Different Insecticides to Rice Field Cyanobacteria

Mihir K. Das and S.P. Adhikary

Contents

ABSTRACT

Diazotrophic cyanobacteria have been found to play a vital role in sustaining the yield of rice without the application of nitrogen fertilizer because of their ability to fix nitrogen.

Glimpses of Cyanobacteria, 2006 122–134
Daya Publishing House, New Delhi

However, the cyanobacteria in rice fields are subjected to variety of biotic and abiotic stresses. Use of pesticides has become imperative for achieving higher production. Indiscriminate use of pesticides are known to affect the cyanobacterial population in the rice fields. Insecticides, which constitute about 83 per cent of total pesticides affect cyanobacteria under laboratory conditions differently, either by changing the growth rate and metabolic rates or alter the population equilibria. Their effect on growth of these organisms may be stimulatory, inhibitory or neutral. In this review, the range of insecticide–induced biological responses, induction of biochemical changes in cyanobacteria, the tolerance of cyanobacteria to insecticides and interaction with other stress conditions are highlighted.

1. Introduction

One of the paradoxes in nature is the bountiful of nitrogen in the atmosphere and its relative non-availability to plants and animals. Organized crop production has, therefore, to depend largely on synthetic nitrogen fertilizers, the production of which is based on non-renewable fossil fuel resources. The spiraling cost of fossil fuels, the widening gap between demand and supply for this nutrient and the prevailing low purchasing power in the densely populated developing countries like India have made it imperative to think of an alternate, low cost technology, ecologically friendly and economically viable renewable organic resource to meet at least a part of the nitrogen demand of crops. Any saving in the consumption of chemical nitrogen without affecting the crop productivity will, therefore, be not only an economic advantage, but also a strategic necessity. The present day global interest in biological nitrogen fixation is a direct consequence of this necessity to provide relief to small and marginal farmers.

In most of the developing countries, the potential nitrogen requirement is often one-third or one-half of that of the developed nations and actual requirement is even less. In terms of nitrogen consumption per hectare of agricultural land, developing countries consume only an average of 8 kg N/ha (Venkataraman, 1981). This indicates the necessity of improving the soil nitrogen status. Although nitrogen occurs in the atmosphere in the form of molecular nitrogen in a concentration of 77 volume per cent, agriculture is in a state of poverty amidst plenty with regard to this nutrient. Nitrogen fixation is a high budgeted process requiring an input of about 147 K cal to reduce one molecule of nitrogen and this is the reason why higher plants have not evolved nitrogen-fixing mechanisms; they simply may not be able to afford the energy. Cyanobacteria are the unique group of organisms which carryout both oxygen evolving photosynthesis with the oxygen sensitive nitrogen fixation, the energy bill for the latter process being paid by the sun. Rice is perhaps the only major cereal crop to whose nitrogen economy nitrogen-fixing cyanobacteria make a significant contribution (Roger and Kulasooriya, 1980).

2. Cyanobacteria

Cyanobacteria (Cyanophytes/blue-green algae) constitute the largest, most diverse and widely distributed group of photosynthetic prokaryotes. They have the ability to invade other-wise uninhabitable sections of the environment, namely, hot springs, ice lands volcanic soils and bare rocks where the combined nitrogen content is either absent or the temperature is either of the extremes or a variety of natural or artificial chemicals are present (Whitton, 1992). In addition, these organisms multiply rapidly in the floodwater in neutral to alkaline soils and release nitrogen slowly for meeting the nutrient requirement by rice (Tirtol *et al.*, 1982; Whitton *et al.*, 1988). Many genera are aerobic nitrogen-fixers and some others are known to grow on molecular nitrogen in the anaerobic and symbiotic state (Fogg *et al.*, 1973). Agronomic potential of cyanobacteria in rice cultivation was first recognized by De (1939), who attributed the natural fertility of tropical rice fields to biological nitrogen fixation by these organisms. Since then, long term fertility experiments and nitrogen fixation measurements have confirmed the importance of cyanobacteria in maintaining a moderate, but constant rice production in fields receiving no fertilizer (Roger, 1988) and many trials have been conducted to increase rice yields by inoculation of Cyanobacteria (Venkataraman, 1972; 1975; 1981).

3. Pesticides

3.1 Pesticide and Agriculture

Modern high-yielding rice varieties that are highly responsive to fertilizers have become an integral part of the cultivation practices. With higher fertilizer input, disease and insect/pest problems have become prominent which endangered the yield necessitating the large-scale use of pesticides. Until 1850s, the use of pesticides against pests was not in practice. Before 1940, they were used in relatively small amount and considered as "Safe Pesticides" which were derivatives, of various plants. But from 1940 onward, the development of numerous synthetic organic compounds or so-called "second generation pesticides" completely changed the basis and strategies of pest control. Thus, during 1950s and 1960s, huge amount of organic pesticides were widely used in agriculture and horticulture to facilitate the production and protection of food, feed and fibre in greater quantity. During 1960s, it was realized that any one single pesticide did not completely eliminate all the species of a target pest and also the concept of resistance was recognized. This has produced a number of trends in use and development of new pesticides for effective control. As a result, the number of chemical pesticides has increased during the past 40 years. In India, the use of pesticides steadily increased from a meagre 4000 tonnes of technical grade agrochemical in 1960 to more than 82,000 tonnes in 1991, out of which 68,000 tonnes are insecticides (SPPL, Bombay, 1992).

3.2 Pesticide Pollution

Though pesticides were the miracle drugs for the plants during 1950s, doubts were raised concerning the hazardous effects of these chemicals on the environment from the early 1960s. Richael Carson was one of the first to become aware of the danger of pesticide pollution and in her famous book entitled 'Silent Spring", she produced evidences of toxic effects of pesticides upon wild life and potential hazards to man (Misra and Mani 1994). Despite increasing research efforts towards crop improvement by methods obviating the use of pesticides (Day, 1987), agriculture remains heavily dependent on these agrochemicals (Gadkari, 1988). The soil, being the ultimate sink in the rice fields, harbours many beneficial cyanobacteria, which are potentially susceptible to pesticides. Cyanobacteria are known to contribute to the maintenance of soil fertility (Singh, 1961; Venkataraman, 1972). It is therefore, imperative that regular monitoring be undertaken to evaluate the influence of pesticides on these organisms.

In general, pesticide toxicity is regarded only as the lethal effects of the chemicals upon a particular organism. The direct effects concern changes in growth rate and changes on specific metabolic rates *i.e.* photosynthesis and respiration. Secondly, pesticides have toxicity, which is selective for certain groups of microorganisms (Padhy, 1985). The effects of this kind may alter population equilibrium in an indirect manner by changing the competitive efficiency of one or other group. Thirdly, pesticides may promote either directly or indirectly, the growth of one or more types of organisms. The type favoured, could be either beneficial or undesirable in paddy field ecosystem. Finally, toxic action of pesticides causing death of certain groups of organisms would result in the release of nutrients by the decay of those organisms, thus inducing a corresponding increase in the number of pesticide tolerant forms.

Many works have been carried out in various laboratories of the world on the responses of pesticides/herbicides to cyanobacteria through *in vitro* determinations and/or under *in vivo* conditions. Most of the works pertain to the determinations of lethal doses of pesticides on different species/strains of cyanobacteria. The more recent studies have been biochemical and physiological in nature than the earlier ones, departing from examination of pesticide-induced qualitative and quantitative fluctuations in soil microbial populations and concentrating more on factors such as the impairment of nitrogen-fixation (Kapoor and Sharma, 1980; Kaushik and Venkataraman, 1983). Although studies on the interaction of cyanobacteria with agrochemicals continue to be widely conducted, there has been an increase in the amount of attention paid to pesticides, particularly insecticides. Since 1979, emphasis has also been laid on the examination of the response of cyanobacteria to mixtures of pesticides (Stratton, 1983; Tandon *et al.*, 1988). This is encouraging, since in rare instances, would any pesticide be presenting isolation in the soil.

4. Insecticides

Insecticides, which constitute 83 per cent of total pesticides produced, have been studied quite extensively with respect to their effect on cyanobacteria.

4.1 Organochlorine Insecticides

As a group, organochlorine insecticides are considered hazardous to these organisms due to their toxicity and persistence (Lal and Saxena, 1980). It appears that exposure of microorganisms, altered call membrane permeability and integrity and interfered with the synthesis of DNA, RNA and proteins (Lal and Saxena, 1980). One of the organochlorines examined by a number of workers for its effect on soil cyanobacteria is BHC and its isomer, Lindane or the commercial preparation of Gammexane (5 per cent active ingredient). Kar and Singh (1979a) in a study on the detoxification of HCH (as gammexane) by *Nostoc muscorum* and *Wollea bharadwajae* found that 4 ppm HCH was algistatic to these two organisms. The toxicity of HCH to *N. muscorum* was reduced by increasing the concentration of individual nutrient (dibasic potassium phosphate from 10 to 40 ppm and calcium chloride from 55 to 330 ppm) (Kar and Singh, 1979b). When higher concentrations of these two nutrients were used in combination in the growth medium, HCH became more toxic to cyanobacteria. The authors suggested that individual nutrients at suitable concentrations may in some way have obstructed the movement of HCH in to the cells or interact with it to form a less toxic product and these effects were lessened by antagonism when the nutrients were used in combinations (Kar and Singh, 1979b). Sharma and Gour (1981), using *Anabaena doliolum, Aulosira fertilissima* and *Nostoc* species isolated from paddy fields found that maximum levels of tolerance of these organisms to Lindane was 9, 15 and 10 ppm respectively. Sardeshpande and Goyal (1982) noted significant Endosulfan-induced reduction in the growth of *Anabaena iyengarii, Hapalosiphon intricatus* and *Calothrix bharadwajae*. Tandon *et al.* (1988) also observed an inhibitory effect of 1 ppm Endosulfan on an *Anabaena* sp. and also on *Aulosira fertilissima*. Photosynthesis and nitrogenase activity were impaired by Endosulfan in *A. fertilissima*, but not in *Anabaena* sp. they also pointed out the seriousness of such observations considering that organochlorine pesticide, Endosulfan is widely used in countries such as India following the ban of Endrin and restricted use of DDT and BHC. Sahu *et al.* (1992) observed that Endotaf at 10 ppm was toxic to almost all test cyanobacteria, except *Lyngbya major*, *Scytonema pascheri and Calothrix parietina,* which could tolerate upto 250 ppm. The mechanism of resistance of the ensheathed to pesticide is not known; the presence of a sheath outside the cell wall of these organisms might be protecting them from adverse effects of the toxic chemicals. Das and Adhikary (1996) also observed that EC_{50} level of Hildan decreased the growth and chlorophyll-a content of ten test cyanobacteria more intensely than Roger and Sevin. Reddy *et al.* (1999) conducted short-term experiments to study the effect of HCH on *Synechococcus aeruginosus* and found that HCH inhibited photosynthesis and respiration, which enhanced with concentration and exposure duration. Besides, HCH also affected the quantities of chlorophyll, proteins, carbohydrates, nucleic acids and enzymes like α-amylase and acid phosphatase. Significant reduction in the cellular metabolites of this alga reflects the impact of HCH on their metabolic pathways.

4.2 Carbamate Insecticides

Carbamate group of insecticides has also been used for testing their influence on cyanobacteriain paddy field ecosystem. Kar and Singh (1979a), working with Carbofuran (Furadan, 3 per cent active ingredient) observed that its 30-ppm was algistatic to *Nostoc muscorum* and *Wollea bharadwajae*. So far the influence of various environmental factors are concerned, Kar and Singh (1978) found that the insecticide at 100-500 ppm was the most toxic at p^H 7.5 to 10; the toxicity increased with decreasing light intensity and at critical population levels. The toxicity of Carbofuran was reduced by increasing the concentration of dibasic potassium phosphate (10 to 40ppm) and calcium chloride (55 to 330 ppm). These findings further illustrate the complexity of pesticide interactions with other soil chemicals and in general, the involvement of environmental factors in determining their efficacy. Adhikary *et al.* (1984) found that Sevin (50 per cent w/v) at 10 mg/ml increased the survival, growth and nitrogen fixation while higher concentrations (50 mg/ml) showed an inhibitory effect on *Anabaena* sp. and *Westiellopsis prolifica*. Since the recommended doses of Sevin in the rice field was less than the lethal doses, Adhikary *et al.* (1984) suggested that excessive use of the insecticides may affect the overall nitrogen economy of rice field soils. Rath and Adhikary (1994) screened the relative tolerance of twenty two species of cyanobacteria belonging to seven different genera, *viz. Anabaena, Nostoc, Cylindrospermum, Calothrix, Scytonema, Aulosira* and *Westiellopsis* to commercial grade Furadan containing 3 per cent active ingredient of carbofuran and observed varied responses among the different species, tolerance values ranged from 15 to 70 mg/ml of the pesticide. All the test species of *Calothrix* showed maximum tolerance up to the higher doses of Carbofuran (70 mg/ml), while *Anabaena fertilissima* and *Aulosira* sp. UU 1437 were proved to be quite susceptible to the pesticides, indicating the ensheathed forms to be the ideal strains for using as biofertilizer. *Nostoc muscorum* was susceptible to carbaryl at 120 ppm (Vaishampayan, 1985). Carbaryl acted on *N. muscorum* by causing cell wall rupture and cell lysis with the induction of various mutations. The major concern was that of paddy field pesticides such as carbaryl can elicit toxicity and/or mutagenicity in paddy field cyanobacteria, this may also be the case with the main crop, leading to a possible carcinogenic hazard to mankind (Vaishampayan, 1985). Begum and Islam (1993) studied the effect of Basudin and Furadan on the growth and nitrogen fixing capacity of *Nostoc commune* and *Calothrix marchica*, which are very common in rice fields of Bangladesh. Recommended dose (for field) of Furadan showed inhibitory effect in the growth and nitrogen fixation of both the algae. On the contrary, Basudin did not generally affect these activities in both the organisms.

Sahu *et al.*, 1992, while studying the effect of pesticide on 17 cyanobacterial species, observed that in the presence of Furadan and Sevin upto 100-ppm in the test soil, the occurrence of micro algae remained almost unaffected. With increase in pesticide level, a proportionate decrease in the occurrence of cyanobacterial forms was noted. At the 500-ppm level of the carbamate insecticides tested, most of the unicellular and filamentous heterocystous forms could not grow in the soils. Only the algal species *Gloeocapsa atrata, Lyngbya major, Calothrix parietina* and *Scytonema pascheri*, which possess well-detined sheaths around the cells/trichomes, could tolerate 1000 ppm of Sevin and Furadon. To the contrary, Das and Adhikary (1996), in another set of experiment, found that *Nostoc muscorum* which does not possess a thick slime layer, tolerated higher concentration of Sevin which implies that the presence of only additional surface features around the trichome of the trichome can not be the only factor for tolerance to higher levels of pesticide.

4.3 Organophosphate Insecticides

Extensive research on the interaction of rice field cyanobacteria with organophosphate insecticides have been undertaken in various laboratories of India because of their wide application in the tropical rice fields. These chemicals, being biodegradable and non-persistent in the soil, are preferred by farmers for pest management practices. From among the organophosphate insectides, effect of Malathion and Phorate have been studied by Gangawane (1979), Sardeshpande and Goyal (1982), Kaushik and Venkataraman (1983), Tandon *et al.*, 1988. Malathion was found to have a differential toxicity to *Anabaena* sp. and *Aulosira fertilissima*; former survived up to 500 ppm, where as letter recovered to control levels at 10 ppm of the pesticide (Tandon *et al.*, 1988). Sardeshpande and Goyal (1982) reported the inhibition of growth and nitrogen fixation in *Hapalosiphon* sp. by 1.0 and 0.5-ppm phorate respectively. However, 1-ppm Phorate stimulated the growth of *Anabaena iyengarii* and *Calothrix membranacea*. This level of Phorate was not tolerated by *Nostoc* sp. in laboratory culture (Gangawane, 1979). Subramanian *et al.* (1994) while studying the effect of 50 ppm of Monocrotophos and Malathion on ten strains of filamentous heterocystous cyanobacteria found that *Aulosira fertilissima* ARM 68 and *Nostoc muscorum* ARM 221 grew maximally with the pesticide even in the absence of inorganic phosphate in the medium. This suggested that these organisms used the pesticides as a sole source of phosphorous in the absence of inorganic phosphate in the medium and as an additional source of phosphorous when inorganic phosphorous was available in the medium. Another interesting observation was that phosphorous starvation increased alkaline phosphatase activity while the pesticides induced acid phosphatase activity in the test cyanobacteria (Subramanian *et al.*, 1994)

5. Fungicides and Herbicides

Fungicides and herbicides also have serious consequences on the overall productivity of soil by interfering with the activity of cyanobacteria (Vyas, 1988). Many workers have published their findings (Moore, 1967; Venkataraman and Rajyalakshmi, 1971, 1972; Gangawane and Saler, 1979; Gangawane and Kulkarni, 1979, Gangawane, 1980; Kapoor and Sharma, 1980, Khalil *et al.*, 1980, Kashyap and Pandey, 1982, Vaishampayan, 1984 a, Panigrahi, 1984; Singh and Tiwari, 1988a, 1988b; Selvakumar *et al.*, 2001) on the interaction of fungicides and herbicides with cyanobacteria.

6. Conclusion

The above reports on the interaction of pesticides (Insecticides) with cyanobacteria have shown that:

1. Agrochemicals affect differently the growth and nitrogen fixation of different cyanobacterial species,
2. There exists an inter-and intra specific strain variation for pesticide toxicity.

3. Recommended field dose of many pesticides did not generally affect the biological activity of these organisms, however, excessive application had an adverse effect on nitrogen fixing cyanobacteria, thereby decreasing crop productivity.

7. References

Adhikary S P, Dash P and Pattnaik H 1984 Effect of the carbamate insecticide Sevin on *Anabaena* sp and *Westiellopsis prolifica*. *Acta Microb Hung* 31:335-338.

Begum N Tahmida and Islam M M, Sazzadul 1993 Study on the effect of basudin and furadan on *Nostoc commune* Vaucher and *Calothrix marchica* Lemm. pp103-109 *In*: Tilzer M M and Khondker M (eds) *Hypertrophic and Polluted Freshwater Ecosystems: Ecological Bases for Water Resource Management*. Proc Int Symp Limnol.

Das M K and Adhikary S P 1996 Toxicity of three pesticides to several rice field cyanobacteria. *Trop Agric* 73: 155-157.

Day P R 1987 Crop improvement constraints and challenges. pp 3-12 *In*: *British Crop Protection Conference on Weeds* BCPC Publ, Surrey, UK.

De P K 1939 *The Role of Blue-green Algae in Nitrogen Fixation in Rice Fields.* Proc Roy Soc, London 127: 121-134.

Fogg G E, Steward W D P, Fay P and Walsby A E 1973 *The Blue-Green Algae*. Academic Press, London.

Gadkari D 1988 Assessment of the effects of photosynthesis–inhibiting herbicides Diuron, DCMU, Metamitron and Metribuzin on growth and nitrogenase activity of *Nostoc muscorum* and a new cyanobacterial isolate strain G4. *Biol Fertil Soils* 6: 50-54.

Gangawane L V 1979. Tolerance of Thimet by nitrogen-fixing blue-green algae. *Pesticides* 13: 33-34.

Gangawane L V 1980 Tolerance of nitrogen-fixing blue-green algae to Brassicol, Bavistin and Fytolan. *J Indian Bot Sci* 59:157-160.

Gangawane L V and Kulkarni L 1979 Tolerance of certain fungicides by nitrogen-fixing blue-green algae and their side effects on rice cultivars. *Pesticides* 13: 37.

Gangawane L V and Saler R S 1979 Tolerance of certain pesticides by nitrogen-fixing blue-green algae. *Curr Sci* 48: 306-308.

Kapoor K and Sharma V K 1980 Effect of certain pesticides on survival, growth and nitrogen fixation of blue-green algae *Anabaena doliolum* Bharadwaja. *Z Allg Mikrobiol* 20: 465-469.

Kar S and Singh P K 1978 Toxicity of carbofuran to blue-green alga *Nostoc muscorum*. *Bull Environ Contam Toxicol* 22: 707-714.

Kar S and Singh P K 1979a Detoxification of pesticides carbofuran and hexachlorocyclohexane by blue-green algae, *Nostoc muscorum* and *Wollea bharadwajee*. *Microb Lett* 10: 111-114.

Kar S and Singh P K 1979b Effect of nutrients on the toxicity of pesticides carbofuran and hexachlorocyclohexane to blue-green alga, *Nostoc muscorum*. *Z Allg Mikrobiol* 19: 467-472.

Kashyap A K and Pandey K D 1982 Inhibitory effects of rice field herbicide Machete on *Anabaena doliolum* Bharadwaja and protection by nitrogen sources. *Z Pflanzenphysiol* 107: 339-345.

Kaushik B D and Venkataraman G S 1983 Response of cyanobacterial nitrogen fixation to insecticides. *Curr Sci* 52: 321-323.

Khalil K, Chaporkar C B and Gangawane L V 1980 Tolerance of blue-green algae to herbicides. pp 36-39 *In:* Proc Natl Work on "Algal systems" *Indian Soc Biotchnology* IIT, New Delhi.

Lal R and Saxena D M 1980 Cytological and biochemical effects of pesticides on micro-organisms. *Residue Review* 73: 49-86.

Misra S G and Mani D 1994 *Agricultural Pollution* Vol.II Ashish Publ House, New Delhi.

Moore R B 1967 Algae as biological indicators of pesticides. *J Phycol* 3: 4.

Padhy R N 1985 Cyanobacteria and Pesticides. *Residue Reviews* 95: 1-44.

Panigrahi K C 1984 *Physiological and Genetical Effect of Pesticides on Blue-Green Algae: Effect of Carbamate Pesticides*. PhD Thesis, Berhampur University, Berhampur (Orissa) India.

Rath B and Adhikary S P 1994 Relative tolerance of several nitrogen-fixing cyanobacteria to commercial grade Furadan (carbofuran, 3 per cent). *Ind J Exptl Biol* 32: 213-215.

Reddy T R K, Jyothi B L B and Reddy G R K 1999 Influence of hexachlorocyclohexane on metabolic activity and cell contents of microalgae. *Phykos* 38: 43-54.

Roger P A 1988 Reconsidering the utilization of blue-green algae in wetland rice cultivation. pp119-137 *In*: Dutt S K and Sloger C (eds) *Biological N_2-Fixation Associated with Rice Production*. Oxford and IBH Publ Co Pvt Ltd, New Delhi.

Roger P A and Kulasooriya S A 1980 *Blue-green Algae and Rice*; International Rice Research Institute, Los Banos, Manila.

Sahu J, Das M K and Adhikary S P 1992 Reaction of blue-green algae of rice field soils to pesticide application. *Trop Agric* 69: 362-364.

Sardeshpande J S and Goyal S K 1982 Effect of insecticides on the growth and nitrogen fixation by blue-green algae. pp588-605 *In*: Kaushik B D (ed) *Proc Natl Sym Biological N_2 fixation*, Today and Tomorrow's Printers and Publishers, New Delhi.

Selvakumar G, Gopalaswami G and Kannaiyan S 2001 Isolation and characterization of cyanobacterial isolates from herbicides applied rice soil. *Phykos* 40: 129-133.

Sharma V K and Gaur Y S 1981 Nitrogen fixation by pesticide-adapted strains of paddy field cyanophytes. Int *J Ecol Environ Sci* 7: 117-122.

Singh L J and Tiwari D N 1988a Some important parameters in the evaluation of herbicide toxicity in diazotrophic cyanobacteria. *J Appl Bacteriol* 64: 365-376.

Singh L J and Tiwari D N 1988b Effects of selected rice field herbicides on photosynthesis, respiration and nitrogen assimilating enzyme systems at paddy soil diazotrophic cyanobacteria. *Pestic Biochem Physiol* 31: 120-128.

Singh R N 1961 *Role of Blue-Green Algae in Nitrogen Economy of Indian Agriculture*, ICAR, New Delhi.

SPPL Bombay 1992 *Indian Pesticide Directory* (Ed. III) 225-367 Scientia Publ Pvt Ltd, Bombay.

Stratton G W 1983 Interaction effects of permethrin and atrazine combinations towards several non-target micro-organisms. *Bull Environ Contam Toxicol* 31: 297-303.

Subramanian G, Sekar S and Sampoornam S 1994 Biodegradation and utilization of organophosphorous pesticides by cyanobacteria. *Int Biodet Biodeg* 33: 129-143.

Tandon R S, Lal R and Rao V V S N 1988 Interaction of Endosulfan and Malathion with blue-green algae *Anabaena* and *Aulosira fertilissima*. *Environ Poll* 52: 1-9.

Tirol A C, Roger P A and Watanabe I 1982 Fate of nitrogen from a blue–green alga in a flooded rice soil. *Soil Sci Plant Nutr* 28: 559-569.

Vaishampayan A 1984a Biological effects of a herbicide on a nitrogen-fixing Cyanobacterium (blue-green alga); an attempt for introducing herbicide resistance. *New Phytol* 96: 7-11.

Vaishampayan A 1985 Mutagenic activity of Alachlor, Butachlor and carbaryl to a nitrogen-fixing Cyanobacterium, *Nostoc muscorum*. *J Agric Sci* 104:571-576.

Venkataraman G S 1972 *Algal Biofertilizer and Rice Cultivation*. Today and Tomorrows Printers and Publishers, New Delhi.

Venkataraman G S 1975 The role of blue-green algae in tropical rice cultivation. pp207-218 *In*: Stewart, W D P (Ed) *Nitrogen Fixation by Free Living Micro-organisms*, Cambridge Univ Press.

Venkataraman G S 1981 Blue-green algae: a possible remedy to nitrogen scarcity. *Curr Sci* 50: 253-256.

Venkataraman G S and Rajyalakshmi B 1971 Tolerance of blue-green algae to pesticides. *Curr Sci* 40: 143-144.

Venkataraman G S and Rajyalakshmi B 1972 Relative tolerance of nitrogen-fixing blue-green algae to pesticides, *Indian J Agric Sci* 42: 119-121.

Vyas S C 1988 *Non-target Effects of Agricultural Fungicides*. CRC Press, Florida, USA.

Whitton B A 1992 Diversity, Ecology and Taxonomy of the cyanobacteria. pp1-51 *In*: Mann N H and Carr N G (eds) *Photosynthetic Prokaryotes* Plenum Press, New York.

Whitton B A, Aziz A, Francis P, Rother J A, Simon J W and Tahmida Z N 1988 Ecology of deep water rice fields in Bangladesh. I. Physical and chemical environment. *Hydrobiol* 169: 3-67.

Chapter 9

Stress Response of Mutant Strain of *Nostoc calcicola* to High Concentration of $NaHCO_3$

P. Jaiswal, S. Roy and A.K. Kashyap

Contents

ABSTRACT

Cyanobacteria synthesize new proteins under stress conditions irrespective of the nature of stress. In present communication we report a mutant strain (HCO_3^{-R}) of *Nostoc calcicola* (an 'usar land isolate), which can tolerate $NaHCO_3$

Glimpses of Cyanobacteria, 2006 135–151
Daya Publishing House, New Delhi

concentration as high as 1000mM (4 folds higher than the tolerance limit of wild type). Though growth of the mutant strain was very low at this concentration. To check the possibility of any stress response by the mutant strain, soluble protein profile of mutant strain was investigated through polyacrylamide gel electrophoresis. Results indicated that it was able to synthesize a 35KD protein during the initial hour of transfer to high $NaHCO_3$ and synthesis of other proteins was absent or low. After 24 h almost all the proteins present in control reappeared indicating that high $NaHCO_3$ concentration acts as a severe ionic stress for HCO_3^R mutant in the initial period, but it maintains the synthesis of some proteins, which help it to overcome the stress until it adapts itself to the condition. Also growth and tolerance level of *N. calcicola* and its mutant differed in different carbon source *viz.* $NaHCO_3$, $KHCO_3$ and CO_2.

1. Introduction

Cyanobacteria can be considered to be a very successful group of prokaryotes, which have existed on earth for at least 2 billion years (Schopf and Walter, 1982; Buick, 1992, Summons *et al.*, 1999). They inhabit vast range of aquatic and damp environments including arctic, freshwater, hyper saline, deep-sea (euphotic) coastal soda lakes, soil, lichen symbiosis etc. They are by far most diverse group of photosynthetic prokaryotes in structural and developmental respect. Like all living cell Cyanobacteria are open system interacting with their environment by the constant interchange of matter, energy and information. For each environmental factor there exists a range, within which cells are capable of growth, however cells are capable of survival over a larger range of variation. The nutrient status and growth condition in the environment is equally critical on modulating the physiological and biochemical responses of Cyanobacteria. The chemical nutrient, which increase the growth at low level limits the metabolic activity at fairly higher concentration and behaves like stress.

Stress responses are not uncommon in Cyanobacteria. These organisms originated millions of years ago and have successfully lived through several environmental stresses on our planet. So they offer an appropriate model system for stress studies. The stress responses in Cyanobacteria starts with destabilization of cellular metabolism due to loss of membrane permeability, denaturation of proteins, changes in metabolic balances etc. leading to growth inhibition (Rai, 1997). This is followed by an acclimation phase initiated by change in gene expression or genetic mechanism, which trigger the synthesis of low molecular stress metabolites, modify the membrane composition and induce certain transport process.

2. Materials and Methods

2.1 Organism and Growth Condition

Nostoc calcicola Breb, an 'usar' land isolate was maintained in our laboratory for last twenty years (Kashyap *et al.*, 1982). The Cyanobacterium was cultivated and maintained in combined nitrogen-free medium with A_5 trace elements and Fe-EDTA because as source of iron (Allen and Arnon, 1955). For all the purpose of cultivation the medium was diluted two folds except the trace elements and Fe-EDTA because the organisms could grow better in diluted medium than the original concentrations indicated by Allen and Arnon (1955). The cultures were incubated at 25 ± 1°C and illuminated with day light fluorescent tubes with photon flux density of 25 $\mu Em^{-2}s^{-1}$ on the surface of culture vessels. The pH of the medium was maintained initially at 7.8 by using 10mM HEPES buffer. The bicarbonate resistant mutant of the Cyanobacterium was isolated by chemical mutagenesis using MNNG (Jaiswal and Kashyap,2002).

2.2 Protein Estimation

Protein content of the Cyanobacterial cells was determined by using Folin-ciocalteau reagent (Lowry *et al.*,1951) with a modification suggested by Herbert *et al.*, 1971).

2.3 Estimation of Photosynthetic Pigments

Acetone soluble pigments were estimated calorimetrically by using method of Myers and Kratz (1955), and quantitative estimation of water soluble pigment was done by following method of Brody and Brody, (1961).

2.4 SDS-PAGE Gel Electrophoresis

SDS polyacrylamide gel electrophoresis was carried out in a vertical system according to the method of Lammli (1970). [^{35}S] Methionine labelling and autoradiography: Radioactive metionine

(^{35}S) was used in experiments on response of bicarbonate resistant mutant at higher level of bicarbonate. Mutant cells grown at ambient CO_2 was harvested by centrifugation (5000 x g, 10 min) and suspended in fresh medium (buffered) in two sets, (equal in inoculum size. One containing 1000mM $NaHCO_3$ and other devoid of bicarbonate. Cultures were incubated in light for 4 h and pulsed with ^{35}S (30μ Ci ml^{-1}) for last 3 h. After the period of incubation the radioactive methionine was removed by centrifugation and washing with cold methionine. Subsequently the pellet was subject to protein extraction. After SDS-PAGE, the gels were dried and exposed to x-ray film at -80°C for appropriate period.

3. Results

3.1 Effect of $NaHCO_3$

In order to ascertain the differences between wild type and HCO_3^{-R} mutant of *N. calcicola* growth was followed in graded concentrations of $NaHCO_3$ and results have been summarized in Figure 9.1. It is evident from the results that *N. calcicola* showed a concentration dependent growth upto 75mM $NaHCO_3$. Further increasing the concentration of $NaHCO_3$ suppressed the growth of the Cyanobacterium. It tolerated upto 250mM $NaHCO_3$ in the medium. The enhancement of growth at optimal growth promoting concentration *i.e.* 75mM was more than 2 folds in terms of protein content and 3 folds in terms of specific growth rate. Two ways ANOVA showed that the increasing concentration of $NaHCO_3$ showed statistically significant effect on growth of *N. calcicola*. $F_{(6,30)} = 175.35$, $p<0.001$

Similarly effect of different concentration of $NaHCO_3$ on growth of bicarbonate resistant mutant (HCO_3^{-R}) of *N. calcicola* has been presented in Figure 9.2. Growth of the mutant was very poor in medium devoid of bicarbonate (control). The increasing concentration of $NaHCO_3$ supported the growth upto 250mM, and growth at this concentration was almost 2.5 folds higher than control condition. The mutant tolerated upto 1000mM concentration of $NaHCO_3$ in the medium however very little growth was observed at this concentration. Thus 250mM is the optimal growth promoting concentration for the mutant and concentration of $NaHCO_3$ above this decreased the growth of the mutant. Thus it can be said that the mutant is bicarbonate resistant or in other words it is a bicarbonate

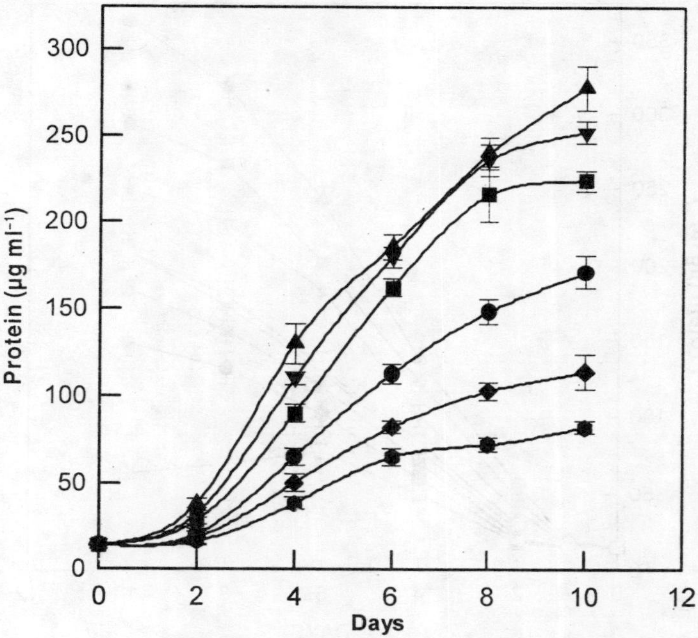

Figure 9.1: Diazotrophic Growth of *Nostoc calcicola* in Elevated Levels of $NaHCO_3$ Supplemented Medium. [5mM(■), 50mM(▲), 75mM(▼), 100mM (◆) and 250mM (●)]. Medium devoid of $NaHCO_3$ served as control (●). pH of the medium maintained by HEPES buffer (10mM, pH 7.8). Vertical bars represent ± SD.

stress tolerant mutant of cyanobacteria reported so ever. Two way ANOVA showed that there is significant effect of $NaHCO_3$ concentrations on growth of the mutant. $F_{(7,28)} = 38.38$, $p<0.001$

3.2 Effect of $KHCO_3$

An experiment similar to $NaHCO_3$ was performed using $KHCO_3$ as bicarbonate source (Figure 9.3). $KHCO_3$ supported the growth of *N. calcicola* upto 20mM and enhancement in growth was about 2 folds in terms protein content (10th day). The Cyanobacterium was able to tolerate upto 100mM concentration of $KHCO_3$ and growth at

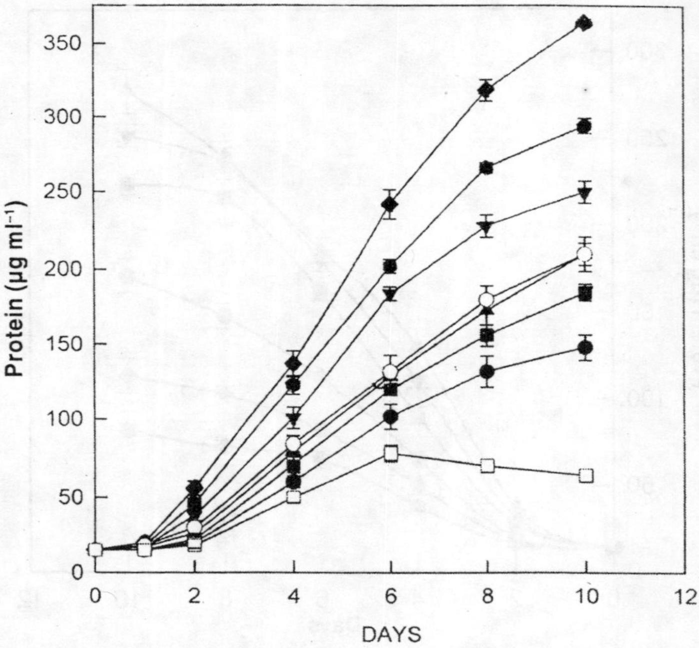

Figure 9.2: Diazotrophic Growth of Bicarbonate Resistant Mutant of *Nostoc calcicola* in Elevated Levels of $NaHCO_3$ Supplemented Medium. [5mM(■), 50mM(▲), 75mM(▼), 100mM (◆) and 250mM (⬢)]. Medium devoid of $NaHCO_3$ served as control (●). pH of the medium maintained by HEPES buffer (10mM, pH 7.8). Vertical bars represent ± SD.

this concentration was almost one third of the control. Statistical analysis of data revealed that growth of *N. calcicola* differed significantly due to different concentrations of $KHCO_3$. $F_{(5,20)} = 105.9$, $p<0.001$.

In order to observe the resistance of HCO_3^{-R} mutant to $KHCO_3$ muatnt strain was grown in varying concentrations of $KHCO_3$ (0-75mM), and results have been presented in Figure 9.4. The mutant showed 1.6 folds increase in in growth (on day 9) compared to control when the medium was supplemented with 10mM $KHCO_3$. Further increase in $KHCO_3$ concentrations suppressed the growth of the mutant in concentration dependent manner. The maximum $KHCO_3$

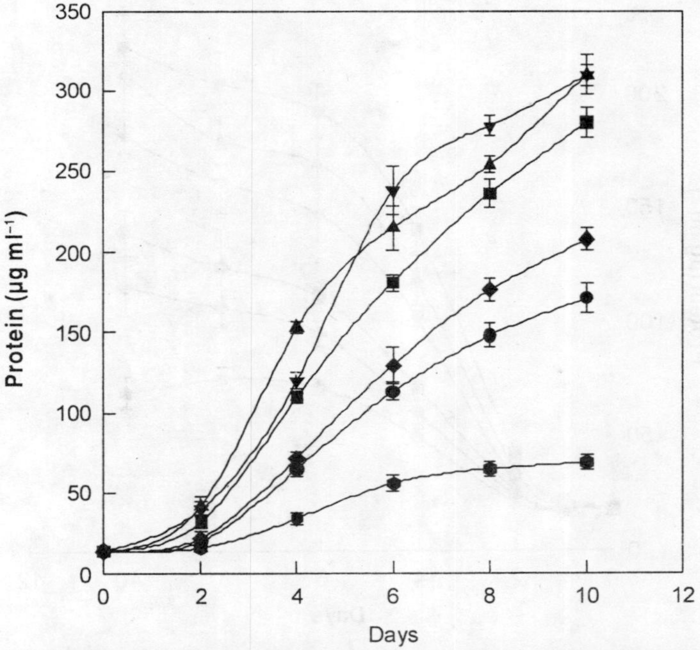

Figure 9.3: Growth Pattern of *Nostoc calcicola* in Elevated Levels of $KHCO_3$ Supplemented Medium. [10mM(■), 20mM(▲), 30mM(▼), 50mM (◆) and 100mM (●)]. Medium devoid of $KHCO_3$ served as control (●). pH of the medium maintained by HEPES buffer (10mM, pH 7.8). Vertical bars represent ± SD.

concentration tolerated by mutant strain was 75mM. Further increase in $KHCO_3$ concentration resulted in death of the mutant strain indicating that the mutant was relatively sensitive to $KHCO_3$ in comparison to wild type *N. calcicola*. Two way analysis of variance showed growth of HCO_3^{-R} mutant at different time intervals was significantly different at different levels of $KHCO_3$ supplementation. $F_{(5,25)} = 24.76, p < 0.001$.

3.3 Effect of CO_2

In order to observe the effect of gaseous form of inorganic carbon on growth and pigment content of Cyanobacterial strains, experiment was performed in airtight culture vessel, where different

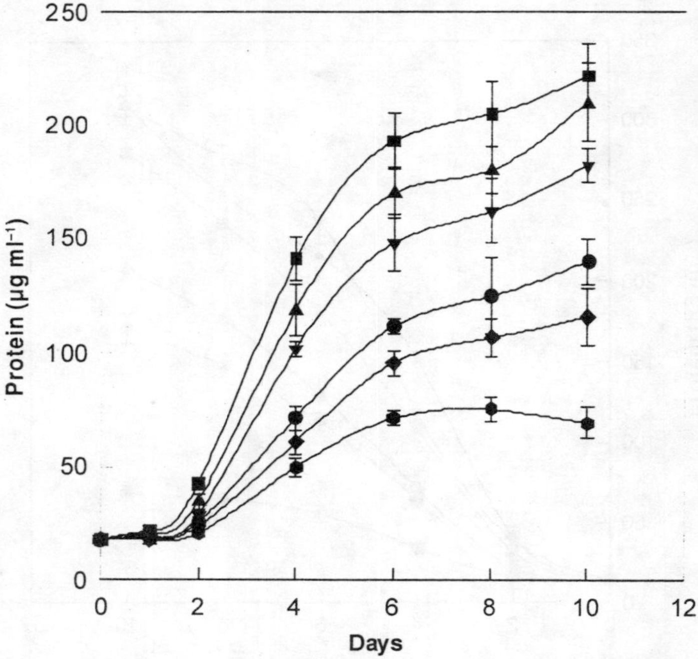

Figure 9.4: Growth Pattern of Bicarbonate Resistant Mutant of *Nostoc calcicola* at Elevated Levelss of $KHCO_3$ Supplemented Medium. [10mM(■), 20mM(▲), 30mM(▼), 50mM (◆) and 75mM (⬢)]. Medium devoid of $KHCO_3$ served as control (●). pH of the medium maintained by HEPES buffer (10mM, pH 7.8). Vertical bars represent ± SD.

concentrations (per cent, v/v) of CO_2 was maintained after replacing equal amount of air. *N. calcicola* and its HCO_3^{-R} mutant were grown with desired concentrations of CO_2 for 10 days and results has been presented in Tables 9.1 and 9.2, respectively. Increasing concentrations of CO_2 supported the growth of *N. calcicola* upto 1.5 per cent CO_2 (optimal concentration). Growth was nearly 1.9 folds higher at this concentration. Although further increase in CO_2 suppressed the growth of the Cyanobacterium but growth was higher than control upto 7.5 per cent. Cyanobacterium tolerated upto 10 per cent CO_2 when growth was reduced to almost half. The Chl *a* and carotenoid content followed a similar trend as was recorded for protein content. The PC/Chl *a* ratio was 5.0 in contol condition, which gradually increased with increasing concentration of CO_2. The ratic of Chl *a*/carotenoid was not so much altered, as the value ranged between 0.5 to 0.6.

Table 9.1: Photosynthetic Pigment and Protein Content of *Nostoc calcicola* Grown Diazotrophically in Varying CO_2 Concentrations. The protein and pigment contents were estimated after 10 days of growth. The data presented are mean values of 4 replicates ± SD. The growth medium was buffered with HEPES (10mM).

CO_2 (%, v/v)	Protein (μg ml^{-1})	Chlorophyll a (μg ml^{-1})	Carotenoid (μg ml^{-1})	Phycocyanin (μg ml^{-1})	PC/Chl a	Chl a/ Carotenoid
Control (ambient level)	142±9.84	1.9±0.36	3.4±0.34	9.5±0.69	5.00	0.55
0.1	180±7.21	2.4±0.17	4.3±0.34	12.0±0.86	5.00	0.56
0.5	204±13.11	2.7±0.20	4.9±0.34	14.1±0.69	5.20	0.55
1.0	240±9.16	3.3±0.30	5.3±0.17	17.2±0.72	5.20	0.62
1.5	264±13.11	3.8±0.35	6.8±0.87	21.0±1.11	5.52	0.56
2.0	168±5.29	2.3±0.12	4.6±0.52	12.8 ±1.7	5.56	0.50
2.5	168±10.06	2.2±0.20	4.1±0.65	12.5±0.45	5.68	0.54
5.0	156±10.06	2.1±0.17	4.0±0.52	12.2±0.72	5.80	0.53
7.5	144±5.29	1.9±0.23	3.1±0.65	11.2±0.52	5.89	0.61
10.0	98±8.0	1.2±0.20	2.3±0.40	7.4±0.34	6.16	0.52

The results of the experiment conducted on the effect of different concentrations of CO_2 on growth of HCO_3^{-R} mutant have been presented in Table 9.2. It is evident from the table that the mutant showed a concentration dependent growth in varying CO_2 concentrations till 5 per cent CO_2. Growth at this concentration was 2.5 folds higher compared to control (protein content). The growth of HCO_3^{-R} mutant was better than control till 10 per cent CO_2 (38 per cent higher). The mutant has the capacity tolerate 20 per cent CO_2. Chl *a* and carotenoid content followed a similar trend as was observed for protein content. Phycocyanin/Chl *a* ratio increased gradually with increasing concentration of CO_2. Carotenoid/Chl *a* ratio did not follow any trend and the values ranged between 0.30-

0.39. Thus it can be said that the mutant has the capacity to tolerate higher concentrations of $NaHCO_3$ and CO_2 in comparison to wild type but when $KHCO_3$ was used, as bicarbonate source growth of wild type strain was better than the mutant strain.

Table 9.2: Photosynthetic Pigment and Protein Content of Bicarbonate Resistant Mutant of *Nostoc calcicola* Grown Diazotrophically in Varying CO_2 Concentrations. The protein and pigment contents were estimated after 10 days of growth. The data presented are mean values of 4 replicates (±SD). The growth medium was buffered with 10mM HEPES buffer.

CO_2 (%, v/v)	*Protein (µg ml^{-1})*	*Chlorophyll a (µg ml^{-1})*	*Carote-noid (µg ml^{-1})*	*Phycocyanin (µg ml^{-1})*	*PC/Chl a*	*Chl a/ Carote-noid*
Control (ambient CO_2)	120 ± 7.21	0.73 ± 0.34	2.0 ± 0.07	3.5 ± 0.20	4.8	0.36
1	185 ± 17.77	0.84 ± 0.04	2.3 ± 0.09	4.0 ± 0.20	5	0.35
1.5	217 ± 8.0	1.00 ± 0.05	2.8 ± 0.09	5.0 ± 0.34	5	0.35
2.5	272 ± 6.92	1.35 ± 0.03	3.5 ± 0.11	6.9 ± 0.35	5.1	0.37
5	300 ± 11.13	1.80 ± 0.07	4.9 ± 0.09	9.0 ± 0.40	5	0.37
7.5	285 ± 13.11	1.40 ± 0.09	3.8 ± 0.11	7.3 ± 0.72	5.2	0.36
10	159 ±4.35	0.80 ± 0.04	2.0 ± 0.09	4.5 ± 0.35	5.6	0.38
20	65 ± 3.46	0.30 ± 0.02	0.79 ± 0.04	1.7 ± 0.2	5.6	0.37

3.4 SDS-PAGE

During the course of present investigation it was observed that the mutant strain was able to tolerate even 1 M $NaHCO_3$. Though at this concentration growth is very slow. It was thought that in the initial stage of transfer to high $NaHCO_3$ it may show stress response and synthesize stress proteins to overcome the initial shock and gradually it gets adapted. In order to check this possibility the HCO_3^{-R} was first grown in basal medium (-$NaHCO_3$), then transferred to 1M $NaHCO_3$ and after1h incubation, pulsed the population with ^{35}S (methionine) for 3h (along with a control). The proteins synthesized during this period were then analyzed by SDS-PAGE followed by autoradiography. The comparison between 2 protein profiles, (control and treated population) in the fluorogram

(Figure 9.2) revealed that during these 4h of incubation in high $NaHCO_3$, the cells normal protein synthesis has stopped (lane B) and only the synthesis of a few proteins of control (lane A) were maintained and a protein of about 35 KD is probably being synthesized *de novo* during this period. But since our HCO_3^{-R} mutant survives the initial shock we also analyzed the shocked cell's proteins after 24 h, last 3h being with ^{35}S (methionine) (lane C). We have found that almost all the bands present in the control have reappeared indicating that these proteins (lane B) might help the mutant to overcome the initial stress condition and gradually it gets adapted to the condition.

4. Discussion

Cyanobacteria constitute a large and diverse group of photosynthetic prokaryotic organisms that are capable of oxygenic photosynthesis. They are found in wide range of habitats and are frequently major contributors to primary productivity. In a general sense the photosynthetic apparatus of these organisms resembles that of C_3 higher plants. Growth of both the strains of Cyanobacteria increased with increasing concentrations of C_i compounds either supplemented in the medium or in gas incubation phase, which could be the outcome of increase in photosynthesis. Since the concentration of inorganic carbon present in the natural environment (both in aquatic and atmosphere) is always limiting for their photosynthesis, the increase in concentration of inorganic carbon would enhance the photosynthesis and hence the growth. Growth and tolerance level of *N. calcicola* and its mutant differed in different carbon source *viz.* $NaHCO_3$, $KHCO_3$ and CO_2. $NaHCO_3$ promoted maximum growth in *N. calcicola* indicating its inherent capability of tolerating higher concentrations of salt in its natural habitat *i.e.* 'usar' land. Besides Na^+ also known to play an essential role in Cyanobacterial metabolism (Kratz and Myers, 1955; Fernandez and Avendano, 1993; Apte and Thomas, 1983). Requirement of Na^+ has also been directly related with requirement of Na^+ in CO_2 and HCO_3^-transport and photosynthesis (Miller and Canvin, 1987). Moroney and Somanchi (1999) reported three types of transporter involved in transport of inorganic carbon in Cyanobacteria- (i) Na^+-independent HCO_3^-transporter, (ii) Na^+-dependent HCO_3^- transporter and (iii) a CO_2 transporter. So increased growth of *N. calcicola* in $NaHCO_3$ supplemented medium as compared to $KHCO_3$

or CO_2 supplementation might be correlated to Na^+ dependent HCO_3^- transport mechanism, which may be preferred in this case as compared to Na^+ independent HCO_3^-transport or CO_2 transport. Growth and tolerance of bicarbonate resistant mutant of *N. calcicola* in HCO_3^-medium was strongly Na^+ dependent and the difference in growth in three carbon sources ($NaHCO_3$, $KHCO_3$ and CO_2) was even more pronounced. It is interesting to note that though the mutant can tolerate 4 times higher concentration of $NaHCO_3$ as compared to wild type but its tolerance capacity was much lower (75mM) than wild type (100mM) when $KHCO_3$ was used as carbon source.

Cyanobacteria in general, follow the universal pattern of stress response in synthesizing a set of specific proteins, the heat shock proteins (hsp)/stress proteins when they are subjected to environmental stresses like sudden upshift of ambient temperature, high salinity, nutrient scarcity, anoxia and so on (Borbely *et al.*, 1985; Bhagwat and Apte, 1989; Hageman *et al.*, 1991). Most of these stress proteins function as chaperones, which enable the cells to get rid of stress, induced abnormal proteins and thus overcome stress (Vierling, 1991; Parsell and Lindquist, 1993; Hendric and Franz-Ulrich, 1993). Osmotic and ionic stress responses in Cyanobacteria have been reported by several workers (Bhagwat and Apte, 1989; Apte *et al.*, 1998; Fulda *et al.*, 1999). In our experiment with the HCO_3^{-R} mutant we presumed that the immediate effect of transfer of the organism to1M $NaHCO_3$ from control condition (ambient level CO_2) was an ionic stress caused by high concentration of Na^+ resulting from dissociation of $NaHCO_3$ in water. When $HCO_3^-$is transported into the cell, Na^+ is either symported with it or exchanged with H^+ (antiport) energizing the HCO_3^-transport system (Price *et al.*, 1998; Moroney and Somanchi, 1999). This results nito elevation of Na^+ inside the cell. During a study on salt stress in a salt tolerant species of *Synechococcusl.* Fulda *et al.*, (1999) observed that a sudden increase in NaCl concentration in the medium (*i.e.* Na^+ ion) had a stressor effect on these species that are otherwise tolerant to the concentration used. These species synthesized stress proteins in the initial phase to overcome the stress effect and ultimately get adapted to the condition by efficient extrusion of Na^+ by accumulating osmolytes. Our experiment with ^{35}S-methionine–pulsed population also showed that though the mutant strain HCO_3–^{R} tolerated a concenteration of 1M $NaHCO_3$ (as against the parent which could tolerate only upto 250mM $NaHCO_{3)}$, a sudden shift to 1M $NaHCO_3$

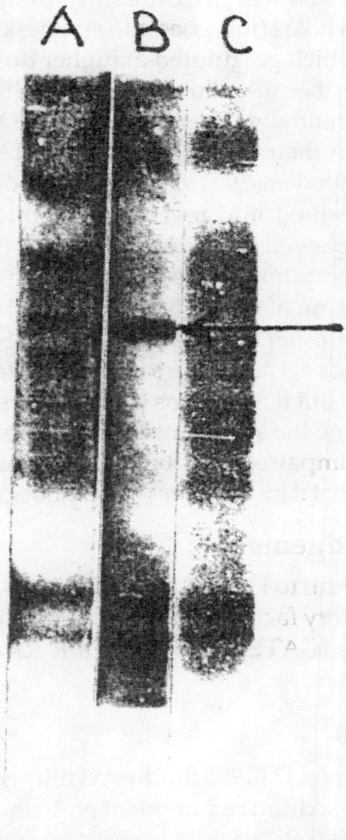

Figure 9.5: Fluogram of Gel Showing Protein Synthesized by Bicarbonate Resistant (HCO_3^{-R}) Mutant of *Nostoc calcicola* Grown in Ambient Level CO_2 (A) and in Medium Supplemented with 1M $NaHCO_3$ for 4h(B) and 24 h (C). The cells incubated under above conditions were pulsed with 35S (methionine) for last 3h. The arrow indicates the stress protein synthesized *de novo* (35KDa).

proved to be a stress condition for the strain which showed a typical stress response with loss of most of the housekeeping proteins from the protein profile after 4h in 1M $NaHCO_3$, but continued synthesis of some proteins–the stress proteins, of which atleast one was synthesized de novo (Figure 9.5). As the stress becomes more and more severe the cell lose its capability of synthesizing normal proteins and its whole effort goes into synthesizing stress proteins/ HSPs, many of which get diluted at higher dose of stress (Fulda *et al.*, 1999). This is because during stress, cell's protein synthesis machinery preferentially translates stress mRNA proteins though the cell still retain their normal pool of mRNA (Joshi–Barve *et al.*, 1992; Roy and Raghavan, 1994). However, a prolonged period under the condition resulted into reversal of normal protein synthesis showing stress-relieved condition, ultimately (Fig.5). Little growth in the high concentration of the salt is probably due to the energy dependent extrusion of excess Na^+ ions, which the strain has to do at the expense of other metabolic activities. So we conclude that high $NaHCO_3$ does act as a severe ionic stress for HCO_3^{-R} mutant in the initial period. But it maintains synthesis of some protein, which help it to overcome the stress until it adapts itself to the condition. But with highly impaired metabolism (especially N_2 fixation and CO_2 fixation), so that it just survives but cannot grow in 1 M $NaHCO_3$.

5. Acknowledgement

We are thankful to Head, Department of Botany for providing necessary laboratory facilities. One of the authors (PJ) is thankful to funding agencies (GATE; JRF/SRF; CSIR RAship) for providing financial assistance.

6. References

Allen M B and Arnon D I 1955 Studies on nitrogen fixing blue-green algae. II. The sodium requirement of *Anabaena cylindrical. Plant Physiol* 8: 653-660.

Apte S K and Thomas J 1983 Impairement of photosynthesis by sodium deficiency and its relationship to nitrogen fixation in Cyanobacterium *Anabaena torulosa. FEMS Microbiol Lett* 16: 153-157.

Apte S K, Fernandes T, Badran H and Ballal A 1998 Expression and possible roles of stress responsive proteins in *Anabaena. J Biosci* 23: 399-406.

Bhagwat A A and Apte S K 1989 Comparative analysis of proteins induced by heat shock, salinity and osmotic stress in nitrogen

G and Whitton B A (eds) *The Biology of Cyanobacteria*. Blackwell Scientific Publication, Oxford.

Shapiro B M and Stadtman E R 1970 Glutamine synthetase (*Escherichia coli*). pp910-922 *In*: Tabor H and Tabor C W (eds) *Methods in Enzymology* Vol XVII, Academic Press, New York.

Summons R E, Jahnke L L, Hope J M and Logan G A 1999 2-methylhalopanoids as biomarkers for Cyanobacterial oxygenic photosynthesis. *Nature*, 400: 554-557.

Vierling E 1991 The role of heat shock proteins in plants. *Ann Rev Plant Physiol Plant Mole Biol* 42: 579-620.

Kashyap A K, Gupta S L and Johar G 1982 pH dependent uptake and reduction of nitrate by *Nostoc calcicola*. *Zpflanzenphysiol* 106: 81-84.

Kratz W A and Myers J 1955 Nutrition and growth of several blue green algae. *Amm J Bot* 42: 282-287.

Laemmli U K 1970 Cleavage of structural proteins during the assembly of the head of bacteriophage T_4. *Nature* 227: 680-685.

Lowry O H, Rosebrough N J, Farr A L and Randall R J 1951 Protein measurement with Folin phenol reagent. *J Biol Chem* 193: 265-275.

Miller A G and Canvin D T 1987 Na^+ stimulation of photosynthesis in Cyanobacterium *Synechococcus* UTEX 625 grown in high levels of inorganic carbon. *Plant Physiol* 84: 118-124.

Moroney J V and Somanchi A 1999. How do algae concentrate CO_2 to increase the efficiency of photosynthetic carbon fixation? *Plant Physiol.* 119: 9-16.

Myers J and Kratz W A 1955 Relations between pigment content and photosynthetic characteristics in blue-green algae. *J Gen Physiol* 39: 11-92.

Parsell D A and Lindquist S 1993 The function of heat shock proteins in stress tolerance: Degradation and reactivation of damaged proteins. *Annu Rev Genet* 27: 437-496.

Price G D, Sultmeyer D, Klughammer B, Ludwig M and Badger M R 1998 The functioning of CO_2 concentrating mechanism in several Cyanobacterial strains: a review of general physiological characteristics, gene, protein and recent advances. *Can J Bot* 76: 973-1002.

Rai A K (Ed) 1997 *Cyanobacterial Nitrogen Metabolism and Environmental Biotechnology*. Narosa Publishing House, New Delhi.

Roy S and Raghavan N 1994 Control of heat shock responses on young gametophytes of sensitive fern is at translation level. *J Exptl Bot* 45: 477-483.

Schopf J W and Walter M R 1982 Origin and early evolution of Cyanobacteria: The geological evidence. pp 543-564 *In*: Carr N

G and Whitton B A (eds) *The Biology of Cyanobacteria*. Blackwell Scientific Publication, Oxford.

Shapiro B M and Stadtman E R 1970 Glutamine synthetase (*Escherichia coli*). pp910-922 *In*: Tabor H and Tabor C W (eds) *Methods in Enzymology* Vol XVII, Academic Press, New York.

Summons R E, Jahnke L L, Hope J M and Logan G A 1999 2-methylhalopanoids as biomarkers for Cyanobacterial oxygenic photosynthesis. *Nature*, 400: 554-557.

Vierling E 1991 The role of heat shock proteins in plants. *Ann Rev Plant Physiol Plant Mole Biol* 42: 579-620.

Kashyap A K, Gupta S L and Johar G 1982 pH dependent uptake and reduction of nitrate by *Nostoc calcicola*. *Zpflanzenphysiol* 106: 81-84.

Kratz W A and Myers J 1955 Nutrition and growth of several blue green algae. *Amm J Bot* 42: 282-287.

Laemmli U K 1970 Cleavage of structural proteins during the assembly of the head of bacteriophage T_4. *Nature* 227: 680-685.

Lowry O H, Rosebrough N J, Farr A L and Randall R J 1951 Protein measurement with Folin phenol reagent. *J Biol Chem* 193: 265-275.

Miller A G and Canvin D T 1987 Na^+ stimulation of photosynthesis in Cyanobacterium *Synechococcus* UTEX 625 grown in high levels of inorganic carbon. *Plant Physiol* 84: 118-124.

Moroney J V and Somanchi A 1999. How do algae concentrate CO_2 to increase the efficiency of photosynthetic carbon fixation? *Plant Physiol.* 119: 9-16.

Myers J and Kratz W A 1955 Relations between pigment content and photosynthetic characteristics in blue-green algae. *J Gen Physiol* 39: 11-92.

Parsell D A and Lindquist S 1993 The function of heat shock proteins in stress tolerance: Degradation and reactivation of damaged proteins. *Annu Rev Genet* 27: 437-496.

Price G D, Sultmeyer D, Klughammer B, Ludwig M and Badger M R 1998 The functioning of CO_2 concentrating mechanism in several Cyanobacterial strains: a review of general physiological characteristics, gene, protein and recent advances. *Can J Bot* 76: 973-1002.

Rai A K (Ed) 1997 *Cyanobacterial Nitrogen Metabolism and Environmental Biotechnology*. Narosa Publishing House, New Delhi.

Roy S and Raghavan N 1994 Control of heat shock responses on young gametophytes of sensitive fern is at translation level. *J Exptl Bot* 45: 477-483.

Schopf J W and Walter M R 1982 Origin and early evolution of Cyanobacteria: The geological evidence. pp 543-564 *In*: Carr N freshwater and is being extensively used for various purposes in medicine, in cosmetics, specialized animal feed and as a general food supplement. *Spirulina* is of commercial importance due to its nutritional qualities.

A large number of formulations are available for the preparation of novel food using *Spirulina* powder. It is a great use in fermented food like 'pastalina' (whole wheat noodle) *Spirulina* protein powder– 10 per cent alga, soya, egg). The biliprotein present in *Spirulina* are used in immuno diagnostics (Venkataraman 1989). Clinical and biochemical evaluation of *Spirulina* for successful treatment of obesity has been reported by Becker *et al.* (1986). Indian workers have evaluated clinically by a Japaness group and in rats cholesterol lowering effect. Successful treatment of malignant or acute anaemia and improvement of hemoglobin in humans have been reported following administration of *Spirulina* 2 to 5 gm/day. *Spirulina* has also been claimed to have curative effects on patients suffering from pancreatitis, cirrhosis, and hepatitis and also acts as a prophylactic against cancer (Venkataraman and Becker 1986) *Spirulina* can also be used for chaemoprevention of oral cancer. Polysaccharides of *Spirulina* can improve the body immune functions. Phycocyanin of *Spirulina* plays a major role in renal failure caused by mercury and pharmaceutical drugs.

The United Nations world food conference in 1974 declared that *Spirulina* was the best food of tomorrow. Today it is a part of diet for space travelers and the U.S Olympic team, used for treating obesity in Germany, being used in cosmetics formulations and given to nursing mothers in many South East Asian countries. Much progress has been made in the last 20 years in developing the biotechnology for algal mass culture. So, far three algal species have attracted major commercial interest chlorella, *Spirulina* and Dunaliella. Multiplication and utilization of *Spirulina* is much easier as compared to two other algae. It needs less energy input and hardware. *Spirulina* can substitute fishmeal or oilseed meal, vitamins and minerals on isoprotenic basis for both layers and broilers. Even supplementation of commercial poultry feed at 5 per cent level improves feed conversion efficiency and egg qualities. Specialized feed formulations for fish, Prawn and Shrimp are now available containing *Spirulina* (Venkataraman and Becker, 1986).

In spite of wonderful properties, it is indeed unfortunate that its cost of production is very high, making it beyond the reach of

Spirulina, a blue-green alga derives its name probably from its spiral shaped chloroplast arrangement. Its 65-70 per cent protein, 10 times beta carotenoid content over carrots, two fold B_{12} content over liver, 20 fold iron content over any natural food, compliment of fat/water soluble vitamins, consortium of trace minerals and PUFA (licosapentaenoic acid), Cumulatively impart it hematinic, B.P normalizing and stress relieving near miracle health supplement characteristics (Table 10.1). *Spirulina*, a non heterocystous, a non nitrogen fixing but filamentous Cyanobacterium belongs to family oscillatoriaceae is the present leader in commercial ventures globally and has overtaken the lucrative chlorella production which was very well exploited by Japanese in the seventies. The alga grows in

Table 10.1: Nutritional Value of Some Common Food and *Spirulina*

Food	*Protein Content (gm/10 gm)*
Egg	1.10
Milk	0.32
Soyabean	3.7
Pulses	1.8
Spirulina	6.6
Iron Content (mg/100 gm)	
Spinach	1.14
Soyabean	10.4
Amaranth	27.3
Egg	2.1
Milk	0.2
Spirulina	58.0
Vit. B_{12} (mcg/g)	
Mutton	0.14
Egg	0.163
Milk	0.437
Spirulina	3.0
β-carotene content (mcg/100gm)	
Carrot	1890
Spinach	5580
Papaya	666
Mango	2743
Spirulina	190000

Source: Sunova Biocare Centre, Laxmi Nagar, Vikas Marg, New Delhi.

better results with continuous bubbling with commercial grade Co_2 for optimizing the photosynthesis. The quantitatively optimized medium initiated *Spirulina* propagation in tubes, scaled up to shake flasks of various sizes and ultimately the inoculum transcended in 25 and 50 litre capacity glass carboys. At each stage of scale up, the initial optical density (OD) of the inoculum must be about 0.35-0.40 and the final OD prior to transfer to the next scale up stage should 1.0 or above. The light intensity in the laboratory must be monitored to provide diffused light. The ideal illumination is 2-3K lux and pH of the medium should be around 9.0 under laboratory conditions the alga is grown in the CFTRI medium prescribed earlier.

6. Production of *Spirulina*

Today *Spirulina* is commercially cultivated in several countries with a total annual production of few hundred tones. Third world society is in search of a rich source of protein that can be produced under local conditions using local resources, that is marginal land and saline water not suitable for conventional agriculture as well as animal and human wastes. Initially, *Spirulina* was considered as an unconventional source of protein for feed and food uses. Currently, there is an enormous interest to use them in natural and health foods and as an unconventional source of energy. There is considerable scope to use them for production of oxygen and recycling wastes for space travel.

Production of *Spirulina* in synthetic culture medium is a costly affair because of usually high mineral requirements of alga. As compared to clean water cultivation of *Spirulina* its production in sewage is more economical as it combines recycling of waste nutrients with synthesis of algal protein and water reclamation. By utilizing sewage and organic waste for large-scale production of *Spirulina*, the nutrient inputs can be minimized, there by effectively lowering the overall cultivation cost. However, the nutrient media containing commercial grade constituent has considerable reduced the cost of cultivation.

The large-scale production of this alga is done in cemented tanks or raceways lodged in a covered space for round the year production. The roof of the enclosure is covered with transparent polythene sheets to permit natural sunlight for algal growth during the daytime. This will lower the energy consumption on artificial illumination. A raceway is a shallow pond of about 40 cm depth liquid medium can be used for routine work. For fully-grown culture on agar slant, the mouth of the tube may be covered from polythene secured with rubber band to reduce the evaporation.

Constituents	*g/l*
Sodiumbicarobate	4.5
diPotassium hydrogen phosphate	0.5
Sodium nitrate	1.5
Potassium sulphate	1.0
Sodium chloride	1.0
Magnesium sulphate	1.2
Calcium chloride	0.4
Ferrous sulphate	0.01
A_5 Micronutrient solution	1.0 ml
pH of the medium	9.0
A_5 Micronutrient solution	
Boric Acid	2.86
Maganese chloride	1.81
Zinc sulphate	0.222
Sodium molybdate	0.017
Copper sulphate	0.079

5. Preparation of Inoculum

Spirulina strain is genetically stable, rapidly multiplying and preferably locally adapted is highly desirable. Although *Spirulina* is ubiquitous for production purposes, strains adapted to local environment have afforded sustained/higher productivity. The media for inoculum preparation and production, to say the least differed from strain to strain. In general the medium was fortified with nutrients which afforded optimized osmotic balance, ionic balance, intracellular and extracellular pH mass and energy transfer so that an integrated strategy of inoculum preparation and its further propagation for production could be easily monitored, rectified and controlled. The inoculum medium comprised of demineralized water fortified with Na, K, Mg salts, besides ammonium salts, phosphate sulphate, chloride, carbonate and trace elements. The medium gives

Venkataraman L V and Becker E W 1982 *Biotechnology and Utilization of Algae: The Indian experience*, Department of Science and Technology, New Delhi.

Venkataraman L V 1989 *Spirulina*: State of the art and emerging prospects. *Phykos* 28: 231-250.

Vincent W A 1969 Algae for food and feed. *Process Biochem* 4: 45-47.

necessitating addition of carbon to the cultures. There may be regulation by adding appropriate amount of bicarbonate or supplying the compressed Co_2 to the medium. Nutrients supply should be maintained as some of the nutrients become short in medium due to the continuous culturing of the organism in the pond. Some trace elements and nitrogen and ammonia may be lost due to high pH and precipitation as high pH is usually maintained by Cyanobacterial growth.

6.4 Maintenance of Temperature and Other Stresses

The climatic conditions in India are fluctuating which are harmful for its cultivation round the year. The efforts should be made to maintain optimum temperature (30°C). Arid and semi-arid zones are more suitable for its growth as there is plenty of light and optimum temperature. High temperature during summer accelerates the rate of water loss from open tanks results in the enhancement of salinity in open ponds. Under such conditions it would be appropriate to develop and exploit the salt and temperature tolerant strains. Using artificial illumination should minimize the dark period, though it increases the cost of production but checks the dark respiration.

By taking some measures *Spirulina* cultivation may prove beneficial and the use of this wonder alga as health food for human may be helpful to eradicate the mal nutrition problem from the under nourished population but the quality control measures must be adopted before using the alga for human consumption. Specific quality varies from importer to importer and country to country. Contents and assay on vitamin B_{12} as well as β-carotene is a common requirement. The uniform law must be followed in this regard.

7. References

Becker E W 1994 *Biotechnology and Microbiology.* University Press, Cambridge, UK.

Borowitzska M A and Borowitszka L J 1988 *Microalgal Biotechnology.* University Press, Cambridge, UK.

Goyal S K and Goyal Dinesh 1996 *Spirulina*: Promise and potential for commercialization. *The Botanica* 46: 201-204.

Seshadri C V and Jeeji Bai (eds) 1992 *Spirulina*. Proc ETTA National symposium Murugappa Chettiar Research Center, Chennai, India.

Venkataraman L V and Becker E W 1982 *Biotechnology and Utilization of Algae: The Indian experience*, Department of Science and Technology, New Delhi.

Venkataraman L V 1989 *Spirulina*: State of the art and emerging prospects. *Phykos* 28: 231-250.

Vincent W A 1969 Algae for food and feed. *Process Biochem* 4: 45-47.

necessitating addition of carbon to the cultures. There may be regulation by adding appropriate amount of bicarbonate or supplying the compressed Co_2 to the medium. Nutrients supply should be maintained as some of the nutrients become short in medium due to the continuous culturing of the organism in the pond. Some trace elements and nitrogen and ammonia may be lost due to high pH and precipitation as high pH is usually maintained by Cyanobacterial growth.

6.4 Maintenance of Temperature and Other Stresses

The climatic conditions in India are fluctuating which are harmful for its cultivation round the year. The efforts should be made to maintain optimum temperature (30°C). Arid and semi-arid zones are more suitable for its growth as there is plenty of light and optimum temperature. High temperature during summer accelerates the rate of water loss from open tanks results in the enhancement of salinity in open ponds. Under such conditions it would be appropriate to develop and exploit the salt and temperature tolerant strains. Using artificial illumination should minimize the dark period, though it increases the cost of production but checks the dark respiration.

By taking some measures *Spirulina* cultivation may prove beneficial and the use of this wonder alga as health food for human may be helpful to eradicate the mal nutrition problem from the under nourished population but the quality control measures must be adopted before using the alga for human consumption. Specific quality varies from importer to importer and country to country. Contents and assay on vitamin B_{12} as well as β-carotene is a common requirement. The uniform law must be followed in this regard.

7. References

Becker E W 1994 *Biotechnology and Microbiology.* University Press, Cambridge, UK.

Borowitzska M A and Borowitszka L J 1988 *Microalgal Biotechnology.* University Press, Cambridge, UK.

Goyal S K and Goyal Dinesh 1996 *Spirulina*: Promise and potential for commercialization. *The Botanica* 46: 201-204.

Seshadri C V and Jeeji Bai (eds) 1992 *Spirulina*. Proc ETTA National symposium Murugappa Chettiar Research Center, Chennai, India.

cell biomass was maximum. The Cyanobacterium followed a life cycle similar to 'commune type' aseriate stage, hormogonia, filament with differentiated heterocyst, colony formation. At temperature close to 10º C or in low light intensity, the hormogonium formation was arrested and the cyanobacterium remained at aseriate stage. Nitrognease activity (C_2H_2-reduction) of *N. commune* was compared with *Nostoc muscorum*, *N. calcicola*, *Anabaena doliolum* and *Gloeocapsa* sp. The temperature profile of acetylene reduction (5-30ºC) for *N. commune* revealed (a) that the highest rate of nitrogenase activity was at 25 ºC. (b) that it was low (69 per cent) in comparison to *N. muscorum* and (c) that the nitrogenase activity continued at lower temperatures, which was not evident for other Cyanobacteria. The result suggested that *N. commune* is adapted to lower temperature in terms of nitrogen fixation.

1. Introduction

Field and laboratory investigation have suggested that the cyanobacterium *Nostoc commune* grows as free living thalli (or mats) as well as epiphyte on mosses, and, therefore, is of ecological significance as a contributor of new nitrogen through nitrogen fixation (Davey and Marchant 1983, Nakatsubo and Ino 1987). Also their widespread occurrence in lakes and soil indicates that the ecosystems outside the influence of marine animals are rather infertile due to low nitrogen reservoir.

Most studies on nitrogen fixation by Antarctic cyanobacteria are based on natural samples which are heterogeneous assemblage of different organisms, and, therefore, do not reflect the nitrogen fixation potential of a cyanobacterial species. Temperature seems to be the main factor governing the rate of acetylene reduction by cyanobacterial mats in nature (Jones 1977 à, b).

Besides the temperature stress conditions in Antarctica, physiological availability of water is crucial to the metabolic activity of cyanobacteria. Desiccation is postulated to play a major role in limiting nitrogen-fixation by cyanobacteria in natural habitats (Stewart 1974). It is predictable that during austral summers the cyanobacterial communities absorb water and nutrients and hence their metabolic activities are expressed at their optimal level. As the temperature decreases, reaching below freezing point, water is transformed into ice and due to high aridity it sublimes. This leads to dehydration and the thalli remain in the form of dried flakes until water available in the next austral summer. As yet we do not know as to how quickly water is taken up or lost by these thalli and how much time is taken to initiate or cease the metabolic activity.

The present work was undertaken to investigate some of the biological properties of *Nostoc commune* under culture conditions. In order to compare some of the field observations, especially in relation to water loss and gain by field material, similar experiments were conducted with the laboratory grown *Nostoc commune*.

2. Materials and Methods

2.1 Collection, Isolation and Cultivation of *Nostoc commune*

Soil samples with macroscopically visible surface crusts of cyanobacterium were removed from the ground near Maitri station, Schirmacher Oasis (Figure 11.1), Antarctica. The samples was carefully freed from adhering sand and gravel and was lodged in hard glass culture tube containing sterile glass beads (100-200 µm). The tube was manually agitated till the filaments were broken into small pieces. A small quantity of water (5 ml) was added in the tube, shaken and allowed to stand. One ml supernatant was removed and plated on agar plates containing modified Chu No 10 medium. The plates were incubated in light (50 mE/m^2/s^{-1}) for about 10 days until visible colonies were evident. The colonies were picked up with the help of sterile glass capillaries and transferred into culture

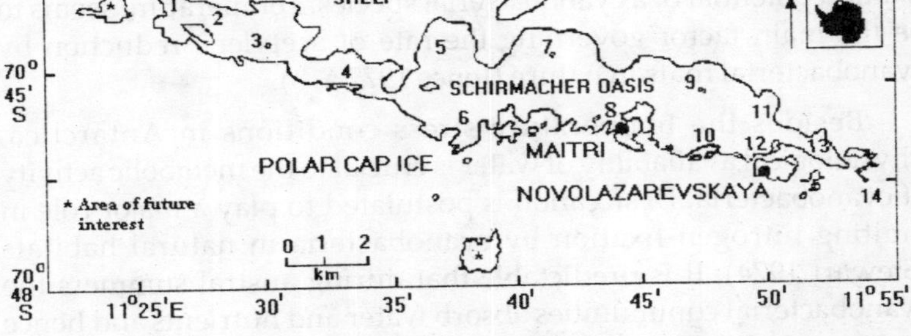

Figure 11.1: Map of Schirmacher Oasis (SO) Showing the Location of 14 Different Sampling Sites. The location of the Indian Antarctic Station, Maitri and the Russian Antarctic Station, Novolazarevskaya are also shown.

tubes containing 10 ml medium. The tubes were incubated as above for additional fortnight. In order to purify the algal filaments, the suspension was centrifuged, washed two times with water, crushed with glass beads and streaked on agar plates. Following incubation, the colonies were picked up as above and culture were maintained in culture room.

2.2 Morphological Observations

The morphological observations made are based on clonal culture of the cyanobacterium raised from a hormogonium consisting of 5-6 cells. The cultures were routinely cultivated in BG-11 medium without supplementation with an inorganic nitrogen sources, designated as BG-N medium. If the medium was supplemented with nitrate or ammonium, they have been designated as BG-NO_3 and BG-NH_4 medium. Colony morphology and life cycle pattern was monitored on medium solidified with agar (1.0 per cent, w/v) in petridishes. The cultures were incubated in light (50 µE/ m^2/s^{-1}) at 25+1°C, and have been referred as standard condition.

In some of the experiments where soil conditions were simulated, the cyanobacterium was inoculated on quartz sand layered in petridishes. The silica sand was moistened with liquid BG-N medium periodically and incubated under standard condition. The cyanobacterium was allowed to form a thick mat (2-3 mm) on the surface of sand.

2.3 Other Cyanobacterial Strains

For some of the experiments where a comparison was desired, some other cyanobacterial strains were employed. They are *Nostoc muscorum, N.clacicola, Gloeocapsa* sp, *Plectonema boryanum* and *Phormidium* sp (Antarctic strain). *N. muscorum*. *Plectonema boryanum* and *Phormidium* sp were obtained through the courtesy of Dr. R. Haselkorn (University of Chicago, USA), Dr R.S. Safferman (US Department of Health and Education, USA) and Dr P.A. Broady (University of Canterbury, New Zealand), respectively. *Nostoc calcicola* was isolated from 'user'soils (pH 10.5) while rest of the strains were isolated from rice fields. All these strains were maintained in the laboratory.

2.4 Developmental Behavior

A loopful of washed and homogenized cyanobacterial suspension was inoculated in the center of the nutrient agar plate and incubated under photoautrophic growth conditions. Developmental sequence and morphology were followed up microscopically. All the photomicrographs were taken with bright field illumination using Nikon microscope fitted with 35 mm camera and photographed on Fujicolor film.

2.5 Extraction of Total Pigments and Absorption Spectra

Method of Colbey and Miranda (1983) was followed for extraction of total pigments. Cyanobacterial samples, after centrifugation (5000 g, 5 min), were suspended in buffer (15mM KH_2PO_4, 1mM mercaptoethanol and 1 per cent Triton X-100, pH adjusted to 7.8 with KOH). The cell suspension was sonicated (Vibronix, India) for 20 min with constant cooling (4°C). Sonicated material was centrifuged (2000 g, 15 min) and adsorption spectra of the supernatant was recorded with a Backmen spectrophotometer (Model 35, USA). The total amount of chlorophyll *a*, carotenoid and phycocyanin were calculated using adsorption coefficients of 82.04, 200 and 7.5 for chl *a* (Myers and Kratz 1955) carotenoids (Allen 1968) and Phycocyanin (Brody and Brody 1961), respectively.

2.6 Nitrogen Fixation at Different Temperatures and Determination of Activation Energy

Nitrogenase activity was estimated according to Schollhorn and Burris (1967). The organism was incubated at different temperatures either in BOD incubator or in a circulating water bath set a different temperature (5-30°C), and illuminated with day light fluorescent tubes (50 μE $m^{-2}s^{-1}$). For each temperature value, kinetic constant VT was determined and from Arrhenius plot of Log VT against 1/T, activation energy Ea was calculated by the following relationship:

$$Ea = \frac{-2.303R \log VT}{1/T}$$

where,

Ea = Activation Energy

R = Gas constant

T = Absolute temperature, and

VT= Rate of acetylene reduction

2.7 Hydration and Dehydration Experiments

Cyanobacterial mats grown on sand in laboratory or occurring in field were harvested, the adhering sand particles were removed and mats were air dried. For water uptake experiments, completely air dried piece of algal mat was submerged in BG-N medium and increase in weight was estimated by weighing the thalli at regular intervals of time. While weighing residual water adhering to the mats was removed by placing the thalli between two sheets of blotting paper. For water loss experiments, completely hydrated thalli were cut in to pieces with the help of cork-borer, placed in open petridishes and exposed to air (temperature 20-25°C). The thalli were periodically weighed until constant weight was obtained. For each experiment, 45 pieces were used and the data presented in Result are average of three independent experiments.

2.8 Chlorophyll a Estimation

The thalli were paced in culture tube and 10 ml acetone (80 per cent) was added. The tubes were plugged with rubber stopper and stored refrigerated (4°C) for 24 h in dark. The supernatant was removed and optical density (=663 nm) was measured with the help of spectrocolorimeter (Bosch and Lomb, USA). Quantitative estimate was obtained by using the equation suggested by Myers and Kratz (1955).

3. Result

3.1 Morphological Observation

3.1.1 Field Specimens

Nostoc commune occurred in soils of Schirmacher Oasis as thalli which were blue-green to yellowish brown in colour. Microscopic observation revealed that the cyanobacterium remained predominantly at the aseriate stage (Figure 11.2). The cells were either barrel shaped or spherical, 3-5 μm in diameter. The colonies were covered with hyaline mucilage sheath, usually 1.0-1.5 μm thick. Individual colonies bore spherical to elliptical heterocysts at the ends, measuring 4.0-5.5 μm. The number of cells within each microcolony was variable which ranged between 12-40 cells. The colonies were 30-75 μm long and 8-25 μm wide. Representative colonies of *N. commune* at different stages of development has been represented in Figure 11.2. The samples which were hydrated

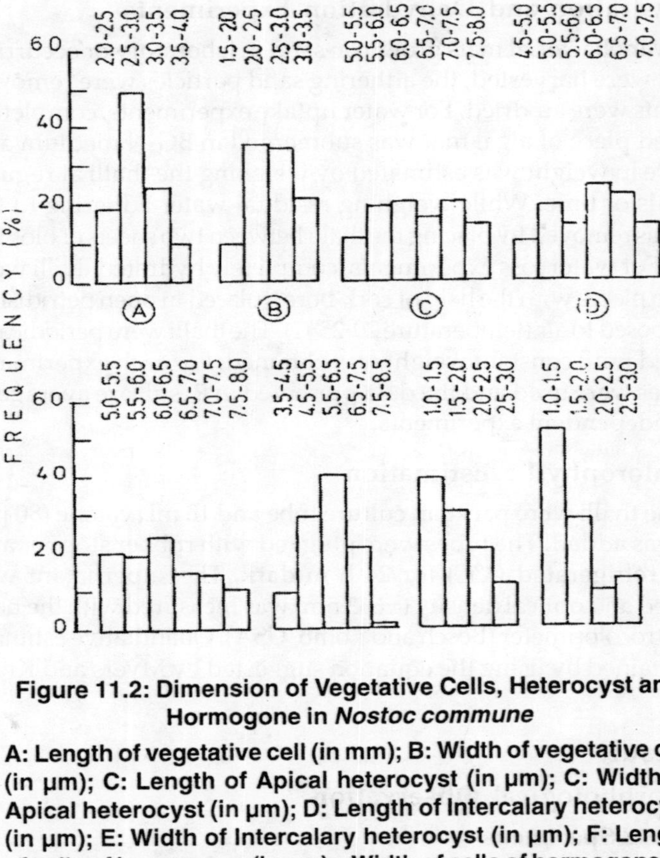

Figure 11.2: Dimension of Vegetative Cells, Heterocyst and Hormogone in *Nostoc commune*

A: Length of vegetative cell (in mm); B: Width of vegetative cell (in μm); C: Length of Apical heterocyst (in μm); C: Width of Apical heterocyst (in μm); D: Length of Intercalary heterocyst (in μm); E: Width of Intercalary heterocyst (in μm); F: Length of cells of hormogone (in μm); : Width of cells of hormogone (in μm)

showed formation of filaments with swollen sheath (ca 2-6 mm). Hormogonia formation was not detected in nature with certainty.

3.1.2 Morphological Features in Culture

When clonal population of *N. commune* was raised in nitrogen-free liquid medium, the cyanobacterium formed mucilaginous mass and the cultures were olive to blue-green in colour. The cyanobacterium did not form homogenous suspension in liquid cultures. Microscopic observation (Figure 11.2) revealed that the cyanobacterium was predominantly in filamentous stage. A trichome consists of vegetative cells and heterocysts at regular intervals. Formation of akinetes was never observed even in three months old cultures. The vegetative cells were elliptical in shape. They were 2.0-4.0 mm long, however, 48 per cent of the population exhibited a length of 2.5-3.0 mm (Figure 11.2a). The width of cells ranged between 1.5-3.5 mm but 35 per cent cells measured 2.0-2.5 mm and another 34 per cent had a length of 2.5-3.0 mm (Figure 11.2b).

The cyanobacterium formed both apical as well as intercalary heterocysts which were elliptical to spherical in shape. The length of apical heterocysts ranged between 6.0-8.0 mm. When they were classified at intervals of 0.5 mm, two major fractions of population were constituted by 6.0-6.5 mm (20 per cent) and 7.0-7.5 mm (20 per cent) (Figure 11.2c). Similarly, width of heterocysts ranged between 4.5-7.5 mm, the major fraction constituted by heterocysts having a length of 6.0-6.5 (23.5 per cent) and 6.5-7.0 mm (20.5 per cent) (Figure 11.2.d).

The measurement of intercalary heterocysts revealed that they could be classified into different class intervals. The length ranged between 5.0-8.0 mm where 24 per cent population had a dimension of 6.0-6.5 mm (Figure 11.2e). The width of heterocysts ranged between 3.5-8.5 mm, where 40 per cent population measured 5.5-6.5 mm (Figure 11.2f).

When the trichomes of liquid cultures were plated on solid medium, hormogones were formed within 48 h of plating. The number of cell in individual hormogones ranged between 5-22 cells, where 70 per cent of population had 8-9 cells. The length of individual cells ranged between 1.0-3.0 mm, the major fraction of population (38 per cent) had a length of 1-1.5 mm (Figure 11.2g). Similarly the width of cells ranged between 1.0-3 mm, the major fraction of population (51.5 per cent) measured 1.0-1.5 m m (Figure 11.2h).

3.2 Life Cycle Pattern

The result on experiment on life cycle pattern is schematically presented in Figure 11.3. When liquid suspension of one month old culture was plated on solid medium, the inoculum contained aseriate stage of the cyanobacterium. Hormogonia were formed after about 48 of light incubation under standard growth conditions. These hormogonia were not formed if the plates were incubated at 10°C or

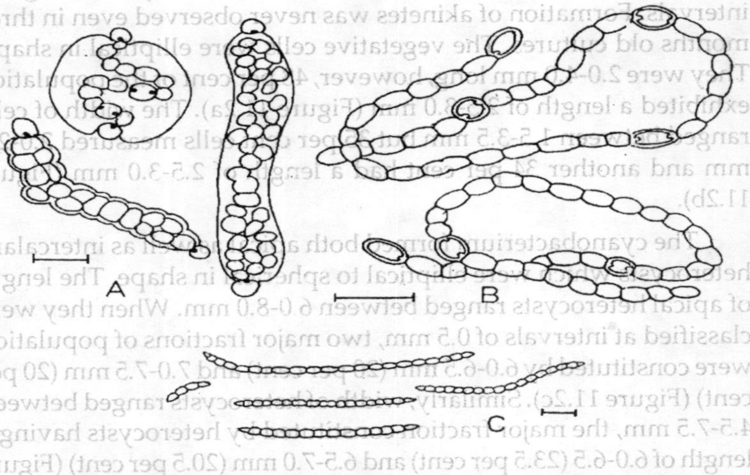

Figure 11.3: Life Cycle Pattern in *Nostoc commune.*
A. Aseriate stage B. Filamentous stage C. Hormogonial stage.
The Bar represents 10 mm.

in low light intensity (10 $mE^{-2}s^{-1}$). These hormogonia were devoid of heterocysts and were motile. After motility of approximately 24 h, the motility ceased and heterocysts differentiated following growth and increase in filament length. Colony development took place, after differentiation of intercalary heterocysts. Filament anastomosis was also observed which led to development of aseriate stage. The colonies developed gave rise to secondary hormogonia which were similar to those arising from the aseriate stage. Such hormogonia, after irregular cell division in horizontal and vertical plane, gave rise to compact cell mass ultimately leading to aseriate stage.

3.3 Growth Characteristics

The growth for *N. commune* in media supplemented with different inorganic nitrogen sources was determined. In preliminary experiment, the growth of cyanobacterium in graded concentrations of nitrogen sources was tested. The concentration supporting maximum growth was further selected for study.

Choice of these two parameters are based on the fact that the cyanobacterium forms copious mucilage in liquid medium and therefore, dry weight estimations are not reliable unless supplemented with additional parameters. Nevertheless, chlorophyll estimations may not be reliable, because the amount of pigments may change substantially on per cell basis, especially when the media are supplemented with nitrogen sources. When dry weight estimations are taken into consideration, it was observed that highest cell mass was obtained when the cyanobacterium was cultivated in Chu-N medium (0.75 mg dry weight per 5 ml culture on 16th day) followed by nitrate and ammonium. The results were apparently similar when chlorophyll *a* was taken in to consideration. The result, therefore, suggested that *N. commune* grew best in BG-N medium without supplementation with inorganic nitrogen sources although estimation of specific growth rate, calculated during exponential phase (2–4 d in terms of log 10 d^{-1}) suggested that it was BG-N medium where maximum specific growth was observed (1.31). This was followed by ammonium where the value was 1.01 followed by nitrate supplemented medium (0.66). It was, therefore evident that BG-N medium supported best growth either in terms of cell mass on the basis of specific growth rate.

3.4 Nitrogenase Activity

Influence of temperature on nitrogenase activity (C_2H_2–reduction) of *N. commune* and other cyanobacteria is presented in Table 11.1. Rate of acetylene reduction by *N. commune* was lower in comparison to *N. mucorum* (*Anabena* sp PCC 7120, a widely used cyanobacterium for the study of diazotrophic physiology, biochemistry and genetics), *N. calcicola, Anabena doliolum* and *Gloeocapsa* sp. The reason for low nitrogen fixation by *N. commune* could be due to heterocyst frequency which was variable during different stages of the life cycle. The heterocyst frequency in filaments was 5.9±0.8 per cent whereas it was less than 2 per cent in aggregated cell mass (aseriate stage). In contrast, hormogonia were devoid of heterocysts. All the stages were encountered in the batch culture and were so well mixed that the contribution of each stage was rather difficult to assess.

Further experiments were conducted to determine the temperature at which *N. commune* exhibited maximum nitrogenase activity. Parallel experiments were conducted with *N. muscorum* for the sake of comparison. The results in Figure 11.4 indicate that optimal acetylene reduction by *N. commune* occurred at 25°C, and a further increase in temperature caused inhibition of nitrogenase activity. Although *N. muscorum* showed higher rate of C_2H_2–reduction at 25°C than *N. commune* and slight increase at 30°C, it showed only 5.3 per cent of its maximum activity at 15°C. Subsequent decrease in temperature had inhibitory effects on acetylene reduction. Unlike *N. muscorum*, the cyanobacterium *N. commune* was capable of C_2H_2–reduction at 5°C or 10°C, the rates being 24 and 38 per cent, respectively, of its optimal rate at 25 °C. Analysis of the data in Figure 11.4 revealed that variation in nitrogenase activity of *N. muscorum* ($F1, 2 = 9.8$) and *N. commune* ($F1, 2 = 11.5$) due to temperature, was statistically significant at 10 per cent probability level. Also, the two strains differed significantly ($p<0.0025$) in their nitrogenase activity with respect to the temperature profile ($F2, 15 = 9.4$).

Table 11.1: Nitrogen Fixation (C_2H_2-reduction) by *N. commune* (Antarctic isolate) and Other Cyanobacteria

Species	*Source*	*Heterocyst Frequency*	*Nitrogenase Activity**
Nostoc muscorum	Pro.R.Haselkorn Univ. of Chicago	5.6±0.2	9.6±0.3
Nostoc calcicola	User land soil (pH 10.5)	6.2±0.3	7.2±0.1
Nostoc commune	Schirmacher Oasis, Antarctica	4.8±0.8	6.6±0.4
Anabena doliolum	Rice field, B.H.U., Varanasi	5.3±0.5	8.6±0.6
Gloeocapsa	Rice field, B.H.U., Varanasi	–	8.2±0.2

* Nitrogenase activity (n mole C_2H_2/µg Chl a per h) at 24°C, light intensity 50 $\mu Em^2 s^1$.

The temperature profile of acetylene reduction permitted to estimate the energy of activation for *N. muscorum* and *N. commune*, respectively. The Arrhenius plot for the former cyanobacterium revealed that the energy of activation was 108 KJ/mol (temperature range 15-25°C). On the other hand, *N. commune* exhibited two levels of energy activation: one at a temperature range of 5–15°C (59 KJ/mole), the other at 15-25°C (33 KJ/mole).

3.5 Pigment Composition

Following the method of Cobley and Miranda (1983), a comparison of absorption spectra of *N. commune* and *N. calcicola*

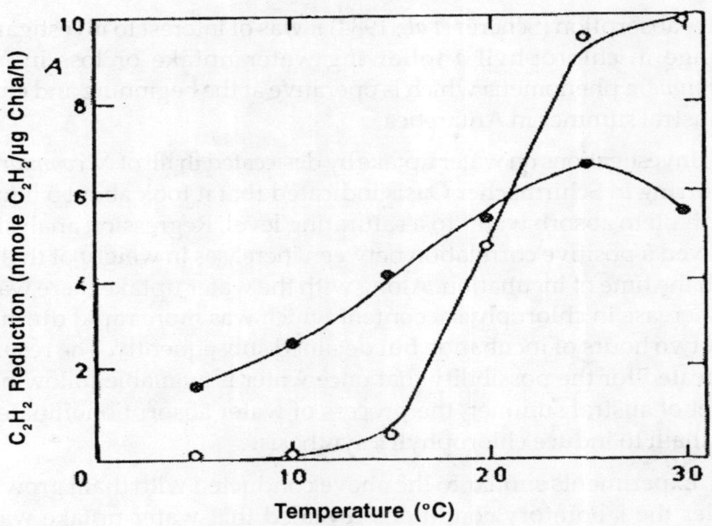

Figure 11.4: Effect of Temperature on Nitrogenase Activity (C_2H_2–reduction) of *Nostoc commune* (dark box) and *Nostoc muscorum* (hollow box) at light intensity 25mE/m²/s⁻¹

was made. Apparently there was no difference in absorption spectra of two species. Further experiments were conducted to quantify the pigments. Considering the ratio of Chl a/Carotenoids it may be suggested that all the cyanobacteria had more or less similar ratio which did not differ among themselves significantly. When Chl a/PC ratio are considered, it was observed that *N. commune* and *Phormidium* sp. did not differ significantly among themselves. On the other hand, *N. calcicola* and *P. boryanum* differed significantly, the latter show highest PC content. This could be due to the possible differences in habitats from which these two species were isolated. A similarity in Chl a/PC ratio of Antarctic species reflects that they occupied similar habitat.

3.6 Water Uptake and Loss by Field and Laboratory Thalli

It has been demonstrated that nitrogen fixation is more sensitive to water stress (Jones 1977) and that dessicated thalli exhibit a series of integrated physiological responses following absorption (Scherer *et al.*, 1984). Since photosynthesis precedes nitrogen fixation after water absorption (Scherer *et al.*, 1984) it was of interest to investigate change in chlorophyll *a* following water uptake or loss in *N. commune*, a phenomena which is operative at the beginning and end of austral summer in Antarctica.

Investigations on water uptake by desiccated thalli of *N. commune* occurring in Schirmacher Oasis indicated that it took about 5 h for the thalli to absorb water to a saturating level. Regression analysis showed a positive correlation between increases in weight of thalli with the time of incubation. Along with the water uptake there was an increase in chlorophyll *a* content which was more rapid during first two hours of incubation but declined subsequently. The result indicated for the possibility that once water is available following onset of austral summer, the process of water absorption initiates the thalli to induce chlorophyll *a* synthesis.

Experiments similar to the above conducted with thalli grown under the laboratory conditions revealed that water uptake was more rapid than the field specimens because it took only 25 min to regain water. Regression analysis showed a positive significant correlation between weight of thalli and incubation time. Although pattern of water uptakes curve was essentially similar to those observed for field specimens, the Chl *a* synthesis followed a biphasic pattern. During the first ten minutes of incubation there was no significant increase in Chl *a*. However, there was a rapid increase during the next ten minutes of incubation. The reason for such an apparent differences between the field and laboratory grown thalli are not yet clear but it could be due to growth conditions of cyanobacterium, since under former condition *N. commune* predominantly exists in aseriate stage (Lazroff 1973) while in the latter as filament and hormogonial stage only.

Experiments conducted with fully hydrated thalli revealed that water loss was complete within a period of 6 h and there was a significant negative correlation between weight of thalli and period of incubation. However, loss of Chl *a* followed a biphasic pattern, the initial phase lasted for three hours where only 60 per cent loss was evident. Subsequently 35 per cent loss occurred during next 3 h. Results on water loss experiments conducted with laboratory grown thalli revealed that there was strong parallelism with those of field experiment. It took about 6 h for complete dehydration of thalli. Nevertheless, differences in Chl *a* loss was apparent in the laboratory grown thalli where biphasic pattern was not evident. At the end of 6 h of dehydration only 65 per cent loss was observed. In contrast, the results obtained with field material showed that during initial 65 per cent dehydration of natural thalli there was a rapid loss of Chl *a* (56 per cent). When the thalli lost additional 20 per cent water, the Chl *a* loss was resisted (5 per cent loss only). Additional 20 per cent loss of water occurred during 120-300min, which was associated with 35 per cent loss of Chl *a*. In contrast the loss of Chl *a* in laboratory thalli occurred in only one step and no loss beyond 65 per cent occurred).

4. Discussion

Cyanobacteria are world wide in distribution, colonizing even extreme environments, including Antarctica. The cyanobacterium *N. commune* is a terrestrial species which is abundant in temperate, tropical and polar regions, on isolated island as well as continent (Geitler 1932; Fogg *et al.*, 1973). Such world wide distribution is presumably related both to the antiquity of the group and to the high degree of resistance which most of the members show towards adverse conditions.' Although the distribution of *N. commune* has been described from various habitats, we still do not know about its life cycle from isolates of various regions under culture conditions.

During the culture of cyanobacteria, reproductive bodies are usually produced at a definite stage of growth. This indicates that some cyanobacteria have a distinct life cycle, although its precise nature varies considerably according to growth conditions (Fogg *et al.*, 1973). Little information is available regarding the distinct alternations of the morphological stages making up the developmental cycle of cyanobacteria, and it refers mainly to three species: *N. muscorum* (Lazaroff and Vishniac 1961, 1962), *Chlorogloea frtischii* (Fay *et al.*, 1964) and *N. commune* (Robinson and Miller 1970). Many of the *Nostoc* strain reported in literature do not display a developmental cycle, probably due to their inability to form hormogonia (Lazaroff 1973). Kantz and Bold (1969) extended their observations on 25 isolates of *Nostoc* and *Anabaena* and concluded that the life cycle of *Nostoc* sp did not show the complexity described for *Nostoc muscorum*. On the basis of their life cycle, *Nostoc* species can be divided into two groups: the 'piscinale type' and the 'commune type'. The authors also demonstrated that in *Nostoc* sp.

the motile (reproductive) and nonmotile (vegetative) phases would alternate regularly. No such alternations of phases could be observed in *Anabaena* sp. The observation made in the Antarctic isolate of *N. commune* showed that the cyanobacterium followed a similar pattern as described for 'commune type' but with certain deletions, truncation and/or prolongation of certain stages Forest (1968).

The laboratory observations that the hormogonia liberation from the aseriate stage did not occur when the incubation temperature was lowered to 10°C indicate that if temperature is low, the organism would tend to remain at the aseriate stage. This explains, partly if not completely, the occurrence of *N. commune* at the aseriate stage in Schirmacher Oasis and that as long as the microenvironment has a low temperature, the aseriate stage would not produce hormogonia for further growth and development of the organism. Possibly, it also explains limited growth of the cyanobacterium in the region.

Although we do not know how far availability of water would regulate the life cycle of *N. commune* in nature, but the observations that hydrations of dry thalli leads to a rapid synthesis of Chl *a* in the cells indicates that it would require at least six hours of wetting before any metabolic activity may be initiated. Similar is the case when water is withdrawn from the wet thalli, *i.e.*, it takes at least six hours after water withdrawn that metabolic activity is ceased. Using three *Nostoc* sp. dried for two years, Scherer *et al.* (1984) demonstrated that the sequence of reactivation upon rewetting was respiration, photosynthesis, then nitrogen fixation. Potts and Morrison (1986) further showed that upon rewetting of cells the rise in ATP pool size was instantaneous and was shown to be due to ATP synthesis. Further, Sherer and Potts (1989) have shown that *N. commune* accumulates a novel group of acidic proteins when colonies are subjected to repeated cycles of drying and rehydration. Although, experimental proof for such a protein has not been demonstrated in Antarctic cyanobacterium, it is tempting to speculate that such proteins should exit in such cyanobacteria. The observation that optimal acetylene reduction by *N. commune* occurred above 10°C has also been reported by various other workers (Alexander *et al.*, 1978, Basilier *et al.*, 1978, Devey 1983, Smith 1984, Nakatsubo 1987). Devey (1983) reported 20°C as the optimal temperature for acetylene reduction by natural samples of *N. commune* collected from Vestfold Hills. Similarly, a higher temperature was found to be more favorable for acetylene reduction by cyanobacteria in moss association on East Ongul Island (Nakatsubo and Ino 1987). The apparent differences in temperature optima exhibited by *N. commune* under field conditions could be reflection of microclimatic conditions experienced by the cyanobacterium (Friedmann *et al.*, 1987).

5. Acknowledgement

The authors thank the Head, Department of Botany, Centre of Advanced Study in Botany, Banaras Hindu University, Varanasi, Secretary, Department of Ocean Development (Government of India, New Delhi) and Principal, Government Post-Graduate College, Rishikesh (Dehradun) for their help.

6. References

Alexander V Billington M and Schell D M 1978 Nitrogen fixation in Arctic and Alpine tundra. pp 539-558 *In* 'Vegetation and Production Ecology of an Alaskan Arctic Tundra: *Ecological Studies* 29 (Ed. LL Tieszen). Springer, Berlin.

Allen M M 1968 Simple condition for growth of unicellular blue-green algae on plates. *J Phycol* 4: 1-4.

Basilier K, Granhall U and Stenstrom T 1978 Nitrogen fixation in wet microtrophic moss community of a sub-arctic mire. *Oikos* 31: 236-246.

Brody S S and Brody M 1961 A quantitative assay for the number of chromophores on a chromoprotein, its application to phycoerythrin and phycocyanin. *Biochem Biophys Acta* 50: 348-352.

Cobley J G and Miranda R D 1983 Mutations affecting chromatic adaptation in the cyanobacterium *Fremyella diplosiphon*. *J Bacteriol* 153: 1486-1492.

Davey A 1983 Effect of abiotic factors on nitrogen fixation by blue-green algae in Antarctica. Polar Biol 2: 95–100.

Davey A and Merchant HT 1983 Seasonal variation in nitrogen fixation by *Nostoc commune* Vaucher at the Vestfold Hills, Antarctica. Phycologia 22: 377-385.

Fay P, Kumar H D and Fogg GE 1964 Cellular factors affecting nitrogen fixation in the blue-green alga *Chlorogloea fritschii*. *J Gen Microbiol* 35. 351-360.

Fogg G E, Stewart W D P Fay P and Watsby A E 1973 *The Blue-green Algae*. Academic Press, London.

Forest H 1968 The approach of a modern taxonomist. pp 185-199 *In: Algae, Man and Environment'* (Ed D F Jackson). University Press New York.

Friedman, El, Mckay CP and Nienow JA 1987 Cryptoendolithic environment in Ross Desert of Antarctica: Satellite transmitted continuous nanoclimate data. 1984-1986. *Polar biology* 7: 273-287.

Gaitler L 1932 Cyanopohyceae. Kademische Verlagsgesellschaft, MbH, Leipzig.

Jones K 1977 The effect of moisture on acetylene reduction by mats of blue-green algae in subtropical grassland. *Ann Bot* 41: 801-806.

Jones K 1977a Acetylene reduction by mats of blue-green algae in subtropical grassland. *New Phytol* 78, 421-426.

Jones K 1977 b The effects of light intensity on acetylene reduction by blue green algal mats in subtropical grassland. *New Phytol* 78: 427-431.

Kantz T and Bold HC 1969 Phycological studies. IX. Morphological and taxonomic investigation of *Nostoc* and *Anabaena* in culture, University Publication No. 6924, Texas.

Lazaroff N 1973 Photomorphogenesis and Nostocacean development. pp 279-319 *In: Biology of Blue-Green Algae* (eds) NG Carr and BA Whitton University of Califonia Press, Berkely and Los Angeles.

Lazaroff N and Vishniac W 1961 The effect of light on the developmental cycle of *Nostoc muscorum*, a filamentous blue-green alga. *J Gen Microbiol* 25: 365-374.

Lazaroff N and Vishniac W 1962 The participation of filament anastomsis in the developmental cycle of *Nostoc muscorum*, a bluegreen alga. *J Gen Microbiol.* 28: 203-210.

Myers J and Kratz WA 1995 Relation between pigment content and photosythetic characteristics in blue-green algae. *J Gen Physiol* 39: 11-21.

Nakatsubo T and Ino Y 1987 Nitrogen cycling in an Antarctic ecosystem, 2. Estimation of the amount of nitrogen fixation in a moss community on East Ongul Island. *Ecol Res* 2: 31-40.

Pandey K D and Kashyap A K 1989 Genetic control of hormogonium production and developmental morphogenesis in *Anabaena doliolum*. *Israel J Bot* 38: 9-24.

Potts M. Friedman RO, Bowman MA and Tozum B 1983 *Chroococcus* N24 and *Chroococcus* N41 (Cyanobacteria): Morphological. biochemical and genetic characterization and effect of water stress on ultrastructure. *Arch Microbiol*, 135: 81-90.

Potts M and Morrison NE 1986 Shifts in the intracellular ATP pools of immobilized *Nostoc* cells (Cyanobacteria) induced by water stress. *Plant and Soil* 90: 211-221.

Robinson, BL and Miller JH 1970 Photomorphogenesis in the blue-green alga *Nostoc commune* 584. *Physiol Plant* 23: 461-472.

Scherer S, Ernst A, Chen TW and Boeger P 1984 Rewetting of drought resistant blue-green algae. Time course of water uptake and reappearance of respiration, photosynthesis and nitrogen fixation. *Oecologia* 62: 418-423.

Scherer S and Potts M 1989 Novel water stress protein from a desiccation tolerant cyanobacterium. *J Biol Chem* 264: 12546-12553.

Schollhorn R and Burris RH 1967 Acetylene as a competitive inhibitor of N_2-fixation. *Proc Natl Acad Sci* USA 58: 213-216.

Smith VR 1984 Effects of abiotic factors on acetylene reduction by cyanobacteria epiphyte on moss at a subantarctic island. *Appl Environ Microbiol*, 48: 594-600.

Stewart WDP 1974 Blue-green algae. pp 202-237 *In: The Biology of Nitrogen Fixation* A Quispel (Ed), North Holland Publications, Amsterdam.

Chapter 12

Cyanobacteria in Relation to Human and Animal Health

A. Khare and M. Kumar

Contents

Glimpses of Cyanobacteria, 2006 182–200
Daya Publishing House, New Delhi

ABSTRACT

Cyanobacteria are also known as blue-green algae, so named because these organisms have characteristics of both algae (ability to photosynthesize like plants) and bacteria (prokaryotic cell organization). They have the ability to grow in most diverse ecological conditions and are very beneficial to mankind in many aspects; besides, they produce some toxins, which cause deleterious effects on human and animal health.

Cyanobacterial blooms can produce health and environmental hazards in water and affect potability of water and recreational purposes. How, why and when the blooms are produced and how to deal with them are questions whose answers are necessary for the safety of human and animal health in regions where these organisms occur in abundant. The blooms are linked to eutrophication (*i.e.* the biological response of water to over enrichment by plant nutrients) of water and this chapter discusses the growth of cynobacterial blooms their nature, distribution, and effect on human and animal health.

1. Introduction

Cyanobacteria is the name given to blue-green algae comprised under the class Cyanophyceae. Blue-green algae possess simple prokaryotic organization and occur naturally in various habitats. They are a group of chlorophyll containing thalloid plants of the simplest type having no true root, stem, leaves, or leaf like organs. They are very minute and polymorphic organisms, which are visible under the microscope as a single cell, colonies, or trichomes. Blue-green algae are also called as Cyanobacteria in order to emphasize their close relationship with bacteria. Blue-green algae are a heterogeneous assemblage of organisms, which includes unicellular organisms as well as branched filamentous forms. It constitutes cellular organization.

Cyanobacteria appeared on our planet about three billion years ago. Based on the international code of botanical nomenclatures the class cyanophyceae contains about 150 genera and 2000 species (van den Hoek *et al.*, 1995). Due to their ability to get adapted to a wide range of environmental factors including extreme ones, they colonize most aquatic and terrestrial ecosystems. This ability of cyanobacteria to withstand adverse ecological conditions, their capacity to thrive well in hostile environments and to respond to the onset of dry conditions, have distinguished this group as pioneers of plant successions.

Cyanobacteria are aerobic phototrophs. They require only water, carbon dioxide inorganic substances and light for the growth. They are the first plants to colonize bare areas of rock and soil. Morphologically, the group includes simple unicellular organisms, unbranched filamentous forms, and strains with more elaborate filamentous structures. The cell wall is made up of peptidoglycan. Basically, the prokaryotic cellular organization in cyanobacteria is characterized by presence of thylakoid membranes, which incorporate in the photosynthetic pigments and therefore constitute the photosynthetic apparatus of the cell. The cytoplasm contains a variety of granular inclusions of different composition and function. These structures provide buoyancy to the cells permitting the Cyanobacteria to occupy an optimum position within a water body. Under specific environmental conditions, many of these Cyanobacteria initiate the formation of two types of cells called akinetes and heterocysts, from vegetative cells. Akinetes are viewed as resting cells with a capacity for germination. The heterocyst is characterized by the deposition of additional envelope layers external to the vegetative cell wall, by its relatively pale colour and by the presence of a conspicuous nodule at each junction with the vegetative cell. Heterocyst is the main site of nitrogen fixation performed by the members of Cyanophyceae.

Many scientists gave the classification of this group but the classification given by Fritsch (1945) is followed by most of the workers of Cyanobacteria. According to his classification, the group Cyanophyceae is divided in to five orders designated as Chroococcales, Chaemosiphonales, Pleurocapsales, Nostocales, and Stigonematales. The classical taxonomy of Cyanobacteria is based on the morphological features and their nomenclature is ruled by the Botanical code.

Cyanobacteria have tremendous potential in environment management; as soil conditioners; biofertilizers; biomonitors of soil fertility; water quality; amelioratory agents; feed for animals; protein supplements and rehabilitation of degraded ecosystems through bioabsorption of metals (Whitton and Potts, 2000). They are also important in the formulation of the strata containing petroleum rich deposits (Radwan and Al-Hassan, 2000). The use of Cyanobacteria for soil fertility and as biofertilizers is most important. A number of Cyanobacteria, which may be heterocystous (Wolk *et al.*, 1999) or non- heterocystous (Stal, 2000) are known to fix atmospheric nitrogen. The free-living diazotrophic Cyanobacteria are the largest contributors of the process of biological nitrogen fixation, the second most important biological process on this planet.

Besides their vast beneficial aspects there are some harmful aspects of BGA. Most common harmful aspect of BGA is formation of algal blooms, which cause several harms to the human and animal life. A bloom of Cyanobacteria (toxic BGA) is a common term used to describe an increase in the number of algal cells to a point where they can seriously reduce the water quality. They can discolour water, form surface scum, produce unpleasant tastes and odours, and create problems for aquatic life. Excessive algal growth in drinking water sources is responsible for toxin generation. Blooms of Cyanobacteria (toxic BGA) can produce health and environmental hazards in water, including water used for drinking or recreation purposes. The blooms are linked to eutrophication of water. As the bloom ages and begins to die, concentrations of toxins may increase. Some toxins may persist for more than three months before they are degraded by sunlight and microbial activity.

2. Cyanobacterial Distribution

Cyanobacteria are truly ubiquitous being distributed throughout the globe under widely varying environmental conditions. Cyanobacteria have a worldwide distribution, living in the sea, in brackish and freshwater areas, in damp soil, glaciers, and deserts as well as in hot springs (van den Hoek *et al.*, 1993). They belong to the most archaic organisms on the earth. The Proteozoic era is some times called the 'Age of Cyanobacteria' because, at that time, between 2.5 and 0.5 billion years ago, BGA dominated the biota (van den Hoek *et al.*, 1993). The great antiquity of cyanobacteria is well documented. The earlier unicellular and filamentous forms of cyanobacteria were reported in sedimentary rocks formed 3500 million years ago. Endolithic forms were observed in rocks formed 500 million years ago. Heterocystous forms and forms with true branching were observed more recently after the Precambrian era.

Cyanobacteria are prokaryotes with oxygen evolving photosynthetic system and are found to in habit a wide range of saline environments. Cyanobacteria, a dweller of diverse ecological habitats, constitute larger and fascinating group of photosynthetic nitrogen fixing prokaryotes. Their ability of adaptations such as UV absorbing sheath pigments, increase their fitness in the relatively exposed land environment. Many species are capable of living in the soil and other terrestrial habitats. Most organisms show structural, physiological, biochemical, and genetic adaptations to their habitat.

2.1 Factors Favouring their Distribution

The most important factor, which favours their distribution widely through out the world is their ability to survive in the extreme conditions of the environment. Along with this some environmental factors also play an important role in the growth of BGA as light, temperature, pH etc. the distribution of edaphic BGA in rice fields is regulated by many environmental factors. Amongst them hydrogen ion concentration is an important factor followed by temperature, light, soil type and available nutrients.

2.1.1 Light

Light plays an important regulatory role in the occurrence and abundance of Cyanobacteria while very little or sometimes no growth is obtained in dark. Many Cyanobacteria are obligate photoautotrophs *i.e.* they grow in the presence of light. It has been reported that the heterocystous forms were three times more abundant at high incident irradiation.

2.1.2 Temperature

High temperature suits the growth of BGA. The maximum temperature at which a Cyanobacterium has been recorded is about 73°C (Ward and Castenholtz, 2000). *Synechococcus* is the single Cyanobacterium, which grows at this temperature in many thermal springs of North America.

2.1.3 pH

Cyanobacteria have generally been reported to prefer neutral to alkaline medium. Numerous records for Cyanobacteria in freshwater and soils indicate that their diversity and abundance are greatest at higher pH. Good Cyanobacterial growth needs an optimum pH about 7.0 and sometimes as high as 10. Many BGA like *Nostoc, Anabaena, Aulosira* and *Calothrix* have been reported from acidic habitats with pH as low as 3.5. Algal inoculations in such habitats require proper soil amendment for pH to have desired effect on the crop yield.

2.1.4 Other Factors

Some species of Cyanobacteria are highly tolerant to UV B and C radiations (Castenholtz and Garcia-Pichel, 2000). This feature was highly responsible for early evolution of BGA. Along with this, the blue-green algae can survive prolonged periods of drought. Many forms are able to survive at very high temperature. Both in xeroseres and hydroseres, these appear as the first plant community leading to the establishment of higher plant associations.

2.2 Population Dynamics

Cyanobacteria belong to most diverse and widely distributed group of prokaryotes that enjoy worldwide distribution. They occur in various habitats *viz.* freshwater, marine, epilithic, epipelic, epiphytic, endophytic endozoic, etc. They are common in a great variety of natural habitats and are often abundant in freshwater and seawater, as well as in the treatment environments. These algae occur as phytoplankton or as attached algae in standing or running water. They predominate in the flora of arid deserts and are abundant on rocks and soils exposed to the full intensity of tropical sunlight. Different species show massive growth in temperate freshwater lakes. Cyanobacteria are the dominant plant life in the frigid rivers of the Garhwal Himalayas where they form extensive benthic mats, scum under or over to water, ponds, streams and over the surfaces of moist soils.

2.3 Seasonal Variation

Extensive growth of *Microcoleus, Anabaena, Tolypothrix* and *Fischerella* develop from mid Sept. to mid Dec. when the soils dry out. It has been suggested by several workers that in equatorial Senegal the seasonal pattern is different. In these regions, the microbial flora becomes dominated after rain and during early part of rice cultivation cycle. Highest biomass have been recorded at the time of development of panicles. Singh (1961) noted that extensive growth of Cyanobacteria takes place during early part of rainy season (From June until mid Oct.). This flora is dominated by species of *Aphanothece, Microcoleus, Aulosira, and Scytonema*. It has also been demonstrated by Roger and Reynaud (1976) that in Senegal soils, the dry period spores of heterocystous Cyanobacteria constituted more than 95 per cent flora.

3. Cyanobacterial Toxins

Blue-green algae occur naturally in many ponds. There are a number of different species: some are harmless, some form toxins that affect the nervous system and other produce toxins that affect mainly the liver. Cyanobacteria are particularly troublesome organisms because apart from producing large amount of biomass, they secrete very toxic substances to water that are known as Cyanobacterial toxins. Any release of toxin into surrounding water can present a significant hazard to livestock and humans drinking the same water.

Cyanobacterial toxins are the group of organic compounds with very diverse chemical structure. They are divided into two groups– Cytotoxins and Biotoxins. Cytotoxins are not lethal for people and animals. They are enzymes, antibiotics, and anticarcinogenic factors with very complicated chemical structure. Biotoxins are very toxic for people and can even cause lethal effect. They are divided into neurotoxins (affecting nervous system), hepatotoxins (affecting liver) and dermatotoxins (affecting skin). Hepatotoxins occur more frequently than neurotoxins. They inhibit the activity of protein phosphatase, which leads to the contraction of hepatocytes, the liver cells (El Saddi and Cameron, 1993; Falconer, 1996). The cells start to separate, and blood which retains between them leads to local hepatocellular damage and a shock lethal dose leads to Chronic disorder of digestive system and liver (Osiecka, 1995).

High pH of water, scarcity of CO_2 dissolved in water and high amount of phosphorus is conductive to the production of the biomass of Cyanobacteria (Carmichael, 1994; Gajdek, 2000). Cyanobacteria survive the period of unfavourable conditions in the form of spores and then, in favourable conditions quickly achieve quantitative mass development. In this time they give the water a distinct colour, referred to as bloom. The authors are not unanimous about the number of algal or Cyanobacterial cell population, which results in water bloom (Fawell *et al.*, 1993).

During the bloom, the number of cells can reach the number of 10 million per one milliliter (10 mln/ml). Blooms can discolour water, form surface scums, produce unpleasant tastes and odours and create problems for aquatic life. They may vary in colour from green to blue, red, brown, dark green or black. Species of BGA may dominate and increase excessively in water when:

1. Nutrient levels; particularly phosphorus and nitrogen are sufficient to support growth.
2. Water is still or turbulence is minimal.
3. Weather pattern is stable for a long time.
4. Weather is warm.

The longer the period of calm weather conditions, the greater the 'bloom-forming potential'. Such potential exists most in flowing rivers or in lakes, dams, weirs and reservoirs.

Several hypothesis have been put forward to explain algal toxin production. Some of them are described as follows: Turner and Tester (1997), Kirk and Gilbert (1992), Carlsson *et al.*, (1995) suggested that Grazer repellence is probably the most studied hypothesis. Other hypothesis suggests that toxins are responsible for allelopathic activity (Windust *et al.*, 1996; Pushparaj *et al.*, 1999) or functioning as a nitrogen source (Dale and Yentsch, 1998) or those toxins are the results of different metabolic processes (Turner and Tester, 1997).

3.1 Distribution of Cyanotoxins

Cyanobacterial toxins can grow rapidly in favourable conditions such as calm nutrient rich freshwaters in warm climates or during the late summer months in cooler parts of the world. Cyanobacteria tend to occur repeatedly in the same water, posing repeated risk to human populations. Cyanobacterial toxins in lakes, ponds and dugouts in various parts have long been known to cause poisoning in animals and humans; one of the earliest reported toxic effect was in China 1000 years ago (Chorus and Bartram, 1999).

3.2 Influence of Cyanotoxins on Various Body Organs

Cyanobacterial toxins are classified by how they affect the human body. Hepatotoxins are produced by some strains of Cyanobacteria as *Microcystis, Anabaena, Oscillatoria, Nostoc,* and *Cylindrospermum.* Neurotoxins are produced by some strains of

Aphanizomenon and *Oscillatoria.* Cyanobacteria from the species *Cylindrospermum raciborskii* may also produce toxic alkaloids, causing gastrointestinal symptoms or kidney diseases in human. Not all Cyanobacteria of these species form toxins and it is likely that there are unrecognized toxins.

People are mainly exposed to Cyanobacterial toxins by drinking or bathing in contaminated sources including algal food tablets. Some species form a scum on the water, but high concentration may also be present throughout the affected water. Surface scums, where they occur, represent hazard to human health because of their particularly high toxin contact.

3.3 World Scenario of the Problem

Cyanobacteria have been linked to illness in various regions throughout the world, including South America, Africa, Australia, Europe, Scandinavia, and China. There is no reliable fixed number of people affected worldwide. The only documented and scientifically substantial deaths due to Cyanobacterial toxins have been due to exposure during dialysis.

Since the first publication on toxicity from Cyanobacteria given by Francis (1878), many similar incidents have been reported from all parts of the world. Gale *et al.* (1987) and Reynolds (1980) reported it in Cattle while Main *et al.*, (1977) reported in sheep. It was also reported in swine (Beasley *et al.*, 1983), dogs (Edwards *et al.*, 1992; Nehring, 1993). Tencalla *et al.*, (1994) reported Cyanobacterial toxicity in fish and Pybus and Hobson (1985) in bats. An extensive work has been carried out to identify the poisonous strains of Cyanobacteria, to isolate the toxins they produce and to study their effects upon living organisms.

4. Effect of Cyanotoxins on Human and Animal Health

4.1 Cyanotoxins and Human Health

Many Cyanobacterial Species produce toxins that affect animals and humans. Consumption of contaminated water by people in any form may cause Cyanobacterial toxicity. The most frequently adverse health affects are caused by drinking water containing the toxins or by recreational water contact.

4.1.1 Disease Caused by Cyanotoxins In Human Beings

Disease due to cyanobacterial toxins varies according to the type of toxin and the type of water related exposure (drinking, skin contact etc.). Humans are affected with a range including skin irritation, stomach cramps, vomiting, nausea, diarrhoea, fever, sore throat, muscle, and joint pain, blisters of the mouth and liver damage. Swimmers in water, which contains cyanobacterial toxins, may suffer allergic reactions such as asthma, eye irritation, rashes around the mouth and nose. Water contaminated by BGA constitutes a health risk to water users in the following ways:

Drinking Water

This causes a number of diseases as nausea, headache, vomiting, abdominal pain, diarrhoea, gastroenteritis, muscle weakness, pneumonia, and paralysis. It is said that boiling the water destroys toxins but it is not true. Boiling will kill algae but in doing so will release toxins into the water. In fact, some toxins become more dangerous after boiling the water.

Skin Contact

Contact with BGA (toxic and non-toxic forms) through water based activities can cause problems such as skin rashes, swollen lips, eye irritation and redness, ear ache and itchiness, sore throat, hay fever symptoms, asthma and possibly promotion of skin tumors. The risk increases as contact time increases (Bell and Codd, 1994). Wearing wet suits may result in greater risk of skin irritation because algal material trapped inside the wet suit may be in close contact with the skin for long periods. Table 12. 1 indicates the level of risk for some water activities.

Table 12.1: Frequency of Risks Caused by Different Activities of Human Beings

Level of Risk	*Human Activities*
High	Swimming, diving, sail boarding, water skiing
Medium	Canoeing, sailing, rowing
Low	Fishing, pleasure cruising, passive, shoreline recreation

Some Other Effects

In addition to the above cases of acute poisoning there is

evidence of more long-term chronic human health hazards produced by Cyanobacterial toxins. It is found that the algal toxins produced more chromosome damage than done by Benzene. All the results suggest that toxins produced by Cyanobacteria may be a more serious environmental hazard than generally recognized (Repavic *et al.*, 1990). The impact of toxins on public health has also been discussed by Piotois *et al.* (2000) and Falconer (1999).

4.1.2 Precautions

1. Reducing nutrient build up on lakes and reservoirs, especially by management of wastewater disposal systems and control of pollution by fertilizer from agriculture.
2. Educating the staff in the health and water supply sectors, as well as the public aware of drinking, bathing or water sports in water likely to contain high densities of Cyanobacteria.
3. Water treatment to remove the organisms and their toxins from drinking water should be appropriate.

4.2 Cyanotoxins and Animal Health

Nearly all animals–including cattle, sheep, horses, pigs, dogs, ducks, fish and wild animals can be poisoned by BGA. They are also implicated in the poisoning of wild animals including rodents, amphibians, fishes, pelicans, waterfowl, bats and zooplankton.

Cyanobacteria have been reported to affect animals in a wide range. Stock deaths have occurred after drinking contaminated water. Pets can also be affected; particularly dogs are susceptible, as they tend to lick their coats after swimming. Sometimes blooms wash onto the shore of a water body where a crust of dead algae can form and dry. In the dry crust, toxins may be active for months. Consumption of such material by animals may cause illness or even death.

4.2.1 Cyanotoxins Poisoning in Animals

Livestock affected with the nervous system toxins may show signs including muscle tremors, decreased movement, and difficult breathing. In the field, many cases show no signs except sudden collapse and death. Animals affected with the liver toxins may show weakness, pale coloured mucous membranes, mental derangement, bloody diarrhoea, and ultimately death.

If they survive blue-green poisoning, they may lose weight and become chronic poor doors. A very few may develop photosensitization–sunburn that affect lighter areas of the skin.

4.2.2 Effect of Cyanobacteria on Zooplanktons and Fishes

Cyanobacteria have commonly been considered as low quality food for Zooplankton, due to their morphology, low nutritional value, and toxin content. Many Cyanobacteria have a filamentous or colonial morphology and form aggregates, which could reduce feeding rates. Unpotability of a food species *i.e.* bad taste or bad odour (van den Hoek *et al.*, 1993) is also considered to be a characteristic of poor food quality.

The low nutritional content of Cyanobacteria may be due to lack of essential compounds (Brett and Muller–Navarra, 1997; Muller–Navarra *et al.*, 2000). Demott *et al.* (1991) reported that Zooplankton mortality is increased due to algal toxins or other secondary metabolites.

The hepatotoxins of Cyanobacteria are considered extremely harmful to vertebrates (Sivonen and Jones, 1999). Keshavanath *et al.*, (1994) demonstrated that grazing studies with herbivorous fish have shown that feeding decreases when the percentage of toxic Cyanobacteria cells increases. Cyanobacterial toxins have caused fish kills (Penaloza *et al.*, 1990). Cyanotoxins have been detected in fish livers and have been found to accumulate in fish tissues (Sipia *et al.*, 2001 a, b). The mechanisms by which toxic Cyanobacterial mass occurrences affect fish behaviour and recruitment are still largely known.

4.3 Diagnosis

Animals affected with blue-green algae poisoning may show signs given above. Most times, however, animals are found dead with no signs observed by the producer, if you check the edges of the pond, you may find some carcasses along the shoreline. They may be clues to the presence of the algae. A complete set of tissue samples including liver, brain and stomach contents from recently dead animals can assist making a diagnosis.

A water sample from the suspected area may be obtained and examined to determine the presence of BGA species. The samples must be taken as soon as possible after discovery of deaths, since winds may shift and disperse the algae. It is recommended that 1 gallon of water be collected and frozen, and a small amount (5ml) taken and refrigerated and another small amount taken and diluted 50:50 with 10 per cent formalin. These samples can be sent to the South Sakota Animal Disease Research and Diagnostic Laboratory, North Campus Drive, SDSU for further investigations.

5. Prevention and Control of Cyanobacteria

Blue-green algae poisoning is unpredictable and sporadic. Poisoning is more likely in stagnant natural waters following hot dry weather. We can:

1. Fence off downwind drinking areas and force animals to drink from areas where concentration of blue-green organisms is unlikely.
2. Pump water from several yards below the surface of a fenced off pond to a nearly livestock tank.
3. Add copper sulfate to the water as an algicide.

5.1 Composite Remedies for Control of Cyanobacteria

The objectives of control and treatment of Cyanobacteria are to:

1. Minimize nutrient concentrations in streams and storages.
2. Encourage water conservation measures.
3. Develop or improve techniques to treat or detoxify contaminated water.

5.2 Classical Methods of Removing Toxins Secreted by Cyanobacteria

Cyanobacterial toxins can be removed in process of coagulation, adsorption, biodegradation and microfiltration (Nawrocki *et al.*, 2000). The degradation of hepatotoxins under the influence of chlorination with chlorine and chloramines was researched by Nicholson, Rositano and Burch (1994). We can also use chlorination, ozonation, irradiation with UV rays, coagulation, biodegradation, microfiltration. The influence of UV radiation on Cyanobacteria *Microcystis aeruginosa* was researched by Alam *et al.* (2001).

5.3 Photocatalytic Oxidation Process

The photocatalytic degradation of toxins secreted to water by Cyanobacteria are an interesting alternative as the supplement of the methods applied so far.

5.4 Immediate Action for Control of Cyanobacteria

If you suspect that blue-green algae have infested a water source you use for drinking, stock water in or domestic purposes, you should refrain from using water from that source until a suitably trained person has identified the algae.

5.5 Short Term Action for Control of Cyanobacteria

If the presence of blue-green algae is suspected, alternative water sources should be used. These might include bottled, carted or tank water or good quality ground water. Treatment of the water with herbicides, copper, or other algicides should not be carried out, as such treatment break open algal cells and release toxins into the water. It has been demonstrated by recent studies that some blue-green algal toxins when released to water from more toxic products that may persist for many months. In case of absence of any other alternative source, filtration of the water through activated carbon may be used for domestic application.

5.6 Long Term Action for Control of Cyanobacteria

Currently a number of strategies are being used to control blue-green algae. A few of these are:

1. Integrated catchments management that is aimed at reducing a managing nutrient flows into waterways from natural and artificial sources.
2. Promoting natural algal grazers through selective fish stocking, aquatic plant restoration, and reduction in the contamination of waterways with pesticides.
3. Supporting the maintenance and restoration of riparian vegetation (*i.e.* the plants near or at the edges of water bodies).
4. Community involvement in tackling catchments management issues.
5. Research programs focused on toxin identification, rapid testing procedures for field use, toxin stability, and degradation under different environmental conditions.

6. Conclusion

It has been clearly explained in the chapter that along with the vast beneficial aspects of Cyanobacteria, they, may also be harmful to human and animal health by producing some toxins. Applying certain measures can stop their growth. To conclude, prevention of blooms in the primary aim but as they form part of the natural ecosystem, this may not always be possible. It is, therefore, important to be aware of the potential dangers associated with blooms. By adhering to the procedures as outlined above in point 5, dangers and risks can be minimized and we along with our pets can live a healthy life.

7. Acknowledgement

Authors are thankful to the UGC, New Delhi for providing financial support in the form of a Major Research Project on Cyanobacteria.

8. References

Alam M Z B, Otaki M. Furumai H, and Ohgaki S 2001 Direct, and indirect inactivation of *Microcystics aeruginosa* by U V-radiaton. *Water Res* 35: 1008-1014.

Beasley V R Coppock R W, Simonj Ely R, Buck W B and Corley R A 1983 Apparent blue-green algae poisoning in swine subsequent to ingestion of a bloom dominated by *Anabaena spiroides. J American Vet Med Assoc* 182: 413-415.

Bell S G and Codd G A 1994 *Cyanobacterial toxins and human health.* Reviews in Medical Microbiol 5: 256-264.

Carmichael W W 1994 The toxins of Cyanobacteria. *Scienti Ameri* 270: 78-86.

Castenholtz R W and Garcia–Pichel F 2000 Cyanobacterial response to UV radiation. 591-611 *In*: B A Whitton and M. Potts (eds.). *The Ecology of Cyanobacteria*. Kluwer Academic Publ., Dodrecht.

Carlsson P, Graneli E, Finenko G and Maestrine S Y 1995 Copepod grazing on a phytoplankton community containing the toxic dinoflagellate *Dinophysis acuminate. J Plankton Res* 17: 1925-1938.

Chorus I and Bartram J 1999 *Toxic Cyanobacteria in Water: A Guide to Their Public Health Consequences, Monitoring all Management*, World Health Organization, Geneva.

Dale B and Yentsch M 1978 Red tide and paralytic shellfish poisoning. *Oceanus* 21: 41-49.

DeMott W R and Moxter F 1991 Foraging on Cyanobacteria by copepods responses to chemical defenses and resource abundance. *Ecology* 72: 1820-1834.

Edwards C, Beattie K A, Scrimgeour C M and Codd G A 1992 Identification of anatoxin–a in benthic cyanobacteria and in associated dog poisonings at Loch Insh, Scotland. *Toxicon* 30: 1165-1675.

El Saddi O and Cameron A S 1993 Illness associated with blue-green algae. *The Medical Journal of Australia* 158: 792-793.

Fawell J K, Hart J, James H A and Parr W 1993 Blue-green algae and their toxins: analysis, toxicity treatment and environmental control. *Water Supply* 11: 109-121.

Falconer I R 1996 Potential impact on human health of toxic Cyanobacteria. *Phycologia* 335: 6-11.

Falconer I R 1999 An overview of problems caused by toxic blue-green algae (Cyanobacteria) in drinking and recreational water. *Environmental Toxicology* 14: 5-12.

Frances G 1878 Poisonous Australian lakes *Nature* 18: 11-12.

Fritsch F E 1945 *The Structure and Reproduction of the Algae.* Vol-II. Cambridge Univ Press London.

Gajdek P 2000 Mikrocystyny sinic W zbiornikach wodnych. *Wiadomollci Chemiczne* 54: 637-650..

Keshavanath P, Beveridge M C M, Baird D J, Lawton L A, Nimmo A and Codd G A 1994 The functional grazing response of a phytoplanktivorous fish Oreochromis niloticus to mixtures of toxic and non-toxic of the Cyanobacterium *Microcystis aeruginosa. J Fish Biol* 45: 123-129.

Kirk K L and Gilbert J J 1992 Variation in herbivore response to chemical defences. Zooplankton foraging on toxic Cyanobacteria. *Ecology* 73: 2208-2217.

Main D C, Berry P H, Peet R L and Robertson J P 1977 Sheep mortalities associated with the blue-green alga *Nodularia spumigena. Australian Vet J* 53: 578-581.

Muller-Navarra D C 1995 Biochemical versus mineral limitation in Dapnia. *Limnol. Oceanogr* 40: 1209-1214.

Muller-Navarra D C, Brett M T, Liston A M and Goldman C R 2000 A highly unsaturated fatty acid predicts carbon transfer between primary producers and consumers. *Nature* 403: 74-77.

Nawrocki J and Bilozor S 2000 Uzdanianie wdy. Procesy chemiczne I biologiczne. Warszawa.

Nehring S 1993 Mortality of dogs associated with a mass development of Bodularia spumigena in a brackish lake at the German North Sea coast. *Journal of Plankton Research* 15: 867-887.

Nicholson B C, Rositano J and Burch M D 1994 Destruction Cyanobacterial peptide hepatotoxins by chlorine and chloramine *Water Res* 28: 1247-1303.

Osiecka R 1995 Mutageniczne I cytoksyczne dzialanie toksyn sinicowych. 111-124 *In*: Zelewski M (Ed.) Procesy biologiczne W. ochronie I rekultywacji nizinnych Zbiornikow Zaporowych.

Penaloza R, Rajas M, Vila I and Zambrano F 1990 Toxicity of a soluble peptide from *Microcystis* sp to zooplankton and fish. *Freshwater Biol* 24: 233-240.

Piotois S, Jackson M H and Wood B J B 2000 Problems associated with the presence of Cyanobacteria in recreational and drinking waters. *Internati J of Environ Health Res* 10: 203-218.

Pushparaj B, Pelosi E and Juttner F 1999 Toxicological analysis of the marine Cyanobacterium *Nodularia harveyana. J Appl Phycol* 10: 527-530.

Pybus M J and Hobson D P 1985 Mass Mortality of bats due to probable blue-green algal toxicity. *J Wildlife Diseases* 22: 449-450.

Radwan S S and Al Hassan R 2000 Oil pollution and Cyanobacteria pp304-319 *In*: Whitton BA and Potts M (eds) *The Ecology of Cyanobacteria*, Kluwer Acad Publ, Dodrecht.

Reynolds C S 1980 Cattle death and blue-green algae: A possible instance from Cheshire, England. *J of the Institute of Water Engineers and Scientists* 34: 74-76.

Repavich W W, Sanzogni W C, Standridge J H, Wedepohl R E and Meisner L F 1990 Cyanobacteria in Wisconsin Waters: Acute and Chronic toxicity. *Water Research* 24: 225-231.

Singh R N 1961 *Role of Blue Green Algae in Nitrogen Economy of Indian Agriculture.* Indian Council of Agricultural Research, New Delhi.

Sipia V 2001 Accumulation of Cyanobacterial hepatotoxins and okadaic acid in mussel and fish tissues from the Baltic Sea. *Finnish Inst. Mar. Res.*, Contributions 200: 1-51.

Sipia V O, Kankaanpaa H T, Flinkman J, Lahti K and Meriluoto J A 2001 Time dependent accumulation of cyanobacterial hepatotoxins in flounders *(Platichthys flesus)* and mussels *(Mytilus edulis)* from the northern Baltic Sea. *Environ Toxicol* 16: 330-336.

Sipia V, Kankaanpaa H, Lahti K, Carmichael W W and Meriluoto J 2001 Detection of nodularin in flounders and cod from the Baltic Sea. *Environ Toxicol* 16: 121-126.

Sivonen K and Jones G 1999 *Cyanobacterial toxins; pp*41-111 *In: Chorus I and Bartram J (eds) Toxic Cyanobacteria in water. A guide to their public health consequences, monitoring and management*, E and F N Spon London.

Stal L J 2000 Cyanobacterial mats and stromatolites; pp61-120 In: BA Whitton and M. Potts (eds.). *The Ecology of Cyanobacteria.* Kluwer Acad Publ, Dodrecht.

Tencalla F G, Detrich D R and Schlatter C 1994 Toxicity of *Microcystis aeruginosa* peptide toxin to yearling rainbow trout *(Oncorhynchus mykiss). Aquatic toxicology* 30: 215-224.

Turner J T and Tester P A 1997 Toxic marine phytoplankton, Zooplankton grazers and elagic food webs. *Limnol. Oceangr* 42: 1203-1214.

Van den C, Mann D G and Jahns H M 1993 Algae–An introduction to Phycology University Press, Cambridge.

Van den C, Mann D G and Jahns H M 1995 Algae–An introduction to Phycology University Press, Cambridge.

Ward D M and Castenholtz R W 2000 Cyanobacteria in geothermal habitats; pp37-59. In: BA Whitton and M. Potts (eds) *The Ecology of Cyanobacteria.* Kluwer Acad Publ, Dodrecht.

Whitton B A and Potts M 2000 *The Ecology of Cyanobacteria.* Kluwer Acad Publ Dodrecht 233-255.

Windust A J, Wright J L C and McLachlan J L 1996 The effect of the diarrhetic shellfish poisoning toxins, Okadaic acid and dinophysistoxin-1, on the growth of microalgae *Mar Biol* 126: 19-25.

Wolk C P, Ernest A and Elhai J 1999 Heterocyst metabolism and Development; pp769-823 *In*: Bryant DA (eds.) *The Molecular Biology of Cyanobacteria*. Kluwer Acad Publ, Dodrecht.

Chapter 13

Impact of UV-B (312nm) Radiation on Nitrate, Ammonium and Phosphate Uptake in *Spirulina fusiformis*

V. Prasad, A. Kumar and H.D. Kumar

Contents

ABSTRACT

Due to ozone depletion and increasing incidence of UV-B on earth's surface, the nitrogen uptake mechanism was severely affected in highly protein–rich cyanobacterium, *S. fusiformis*.

Glimpses of Cyanobacteria, 2006 201–212
Daya Publishing House, New Delhi

UV-B exposure of 312nm wavelength and 2. $5Wm^{-2}$ dose, showed 1h duration for 50 per cent survival and 2h for complete killing the *Spirulina* filaments. Nitrate and Phosphate uptakes were stimulated by short exposure of UV-B but inhibited by longer exposure. Ammonium uptake showed opposite results.

Higher Nitrate uptake occurred by deficient filaments, which was still stimulated by UV-B irradiation. Darkness inhibited the uptakes whereas fluorescent light favours the uptakes in UV-B exposed filaments for longer period.

1. Introduction

Cyanobacteria, the chief primary producer of organic matter and/or first organism to release O_2 into the earth's primitive atmosphere. They are responsible for a major evolutionary change leading to the development of aerobic metabolism (Holm–Hansen, 1968). Thus Cyanobacteria are called the architects of the earth's atmosphere.

The growing global population, expanding human activities and Industrial Revolution especially since the middle of the 20th century by developed countries, have resulted threat of climatic change and the thinning of the ozone layer (Kumar, 2002).

Due to ozone depletion and increasing incidence of UV-B on earth surface (Lubin and Jensen, 1995), one of the architects *Spirulina* that is highly protein-rich (70 per cent) is severely damaged. Protein synthesis is dependent on nutrient uptakes and therefore this present paper is concentrated on the impact of UV-B on uptake mechanism, in *Spirulina*.

A large number of inorganic and organic nitrogenous compounds can serve for the growth of various prokaryotic micro algae. Most of the Cyanobacteria prefer nitrate or ammonium as a source of nitrogen (Boussiba 1989). Besides nitrogen, phosphorus is also one of the major nutrients required for normal growth of micro algae. It is responsible for the regulation of Cyanobacterial blooms (Pearl, 1988).

The increasing incidence of UV-B on the earth surface due to stratospheric ozone depletion is highly deleterious to many physiological process such nitrogen metabolism of several micro algae and marine phytoplankton (Behrenfelds *et al.*, 1995; Häder *et al.*. 1998).

Nitrate uptake in nutrient limited medium was studied in Microcystis and Hapalosiphon (Dwivedi, 1992). UV-B effects on nitrate uptake in marine and freshwater algae were little studied by Cullen and Lesser, 1991; and Braune and Döhler, 1994. The UV-B effect on UV-B excised root of *Phaseolus aureus* is inhibitory on phosphate uptake (Tanada, 1955).

The ammonium uptake was reduced by 10 per cent in culture exposed to solar radiation (Sinha *et al.*, 1996). The ammonium uptake in diatoms is usually more affected by UV-B radiation than that of nitrate and it decreased with increasing level of UV-B (Döhler and Stolter, 1986).

The impact of UV-B and light on nitrate in normal and nitrate deficient medium in *S. fusiformis* was not yet investigated. The effect of UV-B on ammonium uptake immediately and role of fluorescent light to UV-B treated culture were also discussed in this present paper. This present paper for the first time deals with the UV-B effect on phosphate uptake in Cyanobacteria.

2. Materials and Methods

2.1 Test Organisms and Growth Condition

Axenic strain of *S. fusiformis* collected from Sambhar Salt Lake, Sambhar, Rajasthan, India, was routinely grown in Zarrouk's medium at PH 9.5 (Zarrouk, 1966) in 250 ml Erlenmeyer flasks containing 100 ml medium at standard condition 25 ± 1°c, and continuously irradiated with 14. 4 Wm^{-2} fluorescent light (flr.) for $14hd^{-1}$.

2.2 Source of UV-B Irradiation

The UV-B radiation (Fotodyne, Inc. USA lamp cat no. 3-4408, main output 312 nm) fluorescent light (14. 4 Wm^{-2}) was supplied at 18 cm distance from the sample. An instrument, j-21 Longwave length Radiometer UV-P Inc. Cultures were stirred continuously with the help of magnetic stirrer, determined the intensity of artificial UV-B irridiation.

2.3 Survival Test During UV-B Exposure

For determining the per cent survival at known interval of time, 1.0 ml irradiated sample was withdrawn and diluted ten times. 0. 1 ml diluted sample was plated on agar plates, incubated in dark for

40h and then kept in fluorescent light. Per cent survival was scored by colony counts before 15 days with colony counter and plotted semi logarithmically. Similarly, samples were withdrawn at desired intervals for measuring chlorophyll estimation.

2.4 UV-B Treatments and Nitrate Uptake

Exponentially growing filaments were harvested by centrifugation, washed thrice with double glass distilled water, then thrice with N^-medium and resuspended in medium containing 1mM NH_4cl for 40h. The fresh inoculum was also transferred into deficient medium having only 0. 25 gl^{-1} $NaNO_3$ (2. 5 gl^{-1} $NaNO_3$ in normal medium). The filaments were again harvested, washed thrice with double distilled water and then thrice with N^-medium and lastly resuspended in N^-medium. The cell number was adjusted to about $2x10^4$ filaments per millilitre, with the help of haemocytometer.

In the first set, during UV-B treatment 20 ml aliquots having starved filaments were supplemented with 100 mM $NaNO_3$ and distributed into a 75 mm sterile Borocil petri-dish with removed lid. The cultures were treated with UV-B/flr. light directly and uptake was estimated immediately. In the second set, the cultures were exposed to fluorescent light immediately after UV-B treatment for different periods. At desired intervals the triplicate aliquots were withdrawn, diluted, centrifuged immediately and concentration of residual nutrients were measured by using the supernant. Nitrate uptake was determined calorimetrically by Brucine-Sulphuric acid method (Nicholas and Nason, 1957).

2.5 UV-B Treatments and Ammonium Uptake

Exponentially nitrate grown filaments were suspended in N^- medium, after several washing with distilled water as well as N^- medium and supplemented with 400 mM NH_4cl during UV-B treatments. Ammonium uptake was also estimated calorimetrically with Nessler's reagent (APHA, 1985) in immediate treated cultures and fluorescent light exposed cultures to UV-B treated culture,

2.6 UV-B Treatments and Phosphate Uptake

Exponentially growing filaments were harvested by centrifugation (2000 x g, 5 min), washed thrice with double distilled water, and then again thrice with p^-median and resuspended in medium for 72h to make phosphate starved. The phosphate starved filaments were exposed to UV-B/flr. light and uptakes were estimated immediately. Then, some of the UV-B exposed cultures were exposed to fluorescent light and uptakes were estimated and determined calorimetrically by stannous chloride immediately (APHA, 1985).

2.7 Chlorophyll Estimation

Chlorophyll content of treated cultures was estimated calorimetrically at 663 nm with 80 per cent acetone and calculated by Myers and Kratz method (1955).

3. Results

3.1 Survival of the Organism

Figure 13.1 shows the survival curve based on the loose colony counts after UV-B treatment for different durations. There was 50 per cent killing at 60 min. whereas complete killing occurred after 120 min. exposure.

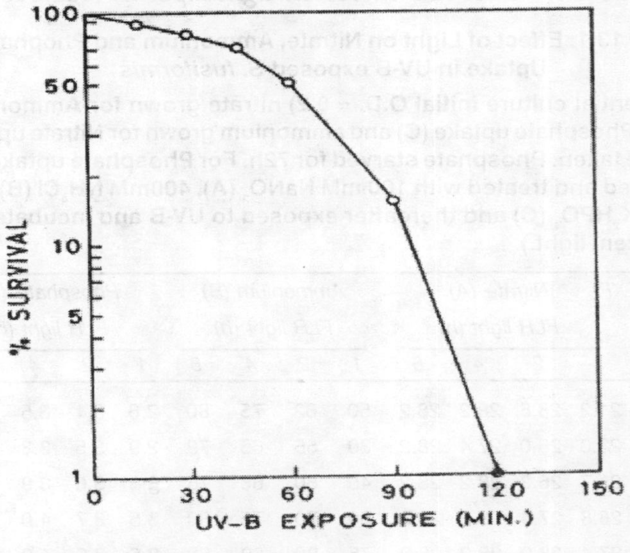

Figure 13.1: Effect of UV-B on Survival of *S. fusiformis* (Exponential culture, initial O.D. = 0. 2), was exposed to UV-B, 312nm, 2. 5Wm^{-2}, 18 cm distance. Plates were incubated in dark for 41h after plating and thereafter incubated in flr. light, 14. 4Wm^{-2}. Colony counts were made after 15 days)

3.2 NO_3^-, NH_4^+ and PO_3^-Uptake During UV-B Irradiation

The nitrate uptake was fast under UV-B than in fluorescent light upto 2h. Although uptake was enhanced upto 90 min and declined with increased duration of UV-B exposure (Figure 13.2a). Figure 13.2b showed that ammonium uptake was slowly increasing under UV-B exposure which was inhibited for 1h and then sharply stimulated after 120m exposure in comparison to control. *S. fusiformis* preferably has higher ammonium uptake than nitrate. Phosphate uptake was similar to nitrate uptake under UV-B exposure (Figure 13.2c).

In *S. fusiformis* (Table 13.1) when UV-B treated filaments were incubated in fluorescent light there was linear increase in nitrate and phosphate uptake with increasing duration of light. Higher nitrate and phosphate uptake were shown by the UV-B exposed filament (Tables 13.2a and c), but ammonium uptake was inhibited in 60 min UV-B treated filaments whereas enhanced in 120 min UV-B treated filaments for 6 hr fluorescent light exposure (Table 13.2b).

Table 13.1: Effect of Light on Nitrate, Ammonium and Phophate Uptake in UV-B exposed *S. fusiformis*

(Exponential culture initial O.D. = 0.2) nitrate grown for Ammonium (B) and Phosphate uptake (C) and ammonium grown for Nitrate uptake (A) were taken. Phosphate starved for 72h. For Phosphate uptake (C) Harvested and treated with 100 mM $NaNO_3$ (A), 400mM NH_4Cl (B) and 50 mM K_2HPO_4 (C) and thereafter exposed to UV-B and incubated in fluorescent light.)

UV-B EXP. (Min.)	*Nitrate (A) FLR light (h)*				*Ammonium (B) FLR light (h)*				*Phosphate (C) FLR light (h)*			
	1	*2*	*4*	*6*	*1*	*2*	*4*	*6*	*1*	*2*	*4*	*6*
Control	21.2	28.5	28.2	28.2	50	62	75	80	2.6	3.1	3.5	4.1
30	23.0	25.0	27.4	28.2	30	55	65	72	2.9	3.5	3.8	4.1
60	25.7	26.3	28.2	28.2	45	60	68	75	3.3	3.6	3.9	4.1
90	26.8	27.4	28.2	28.2	60	70	75	80	3.5	3.7	4.0	4.1
120	27.4	28.2	28.2	28.2	75	80	80	80	3.6	3.8	4.0	4.1

3.3 Nitrate Uptake in Nitrate Deficient Media

Figure 13.3a showed nitrate uptake in both normal and deficient media but the uptake was faster in deficient medium both under

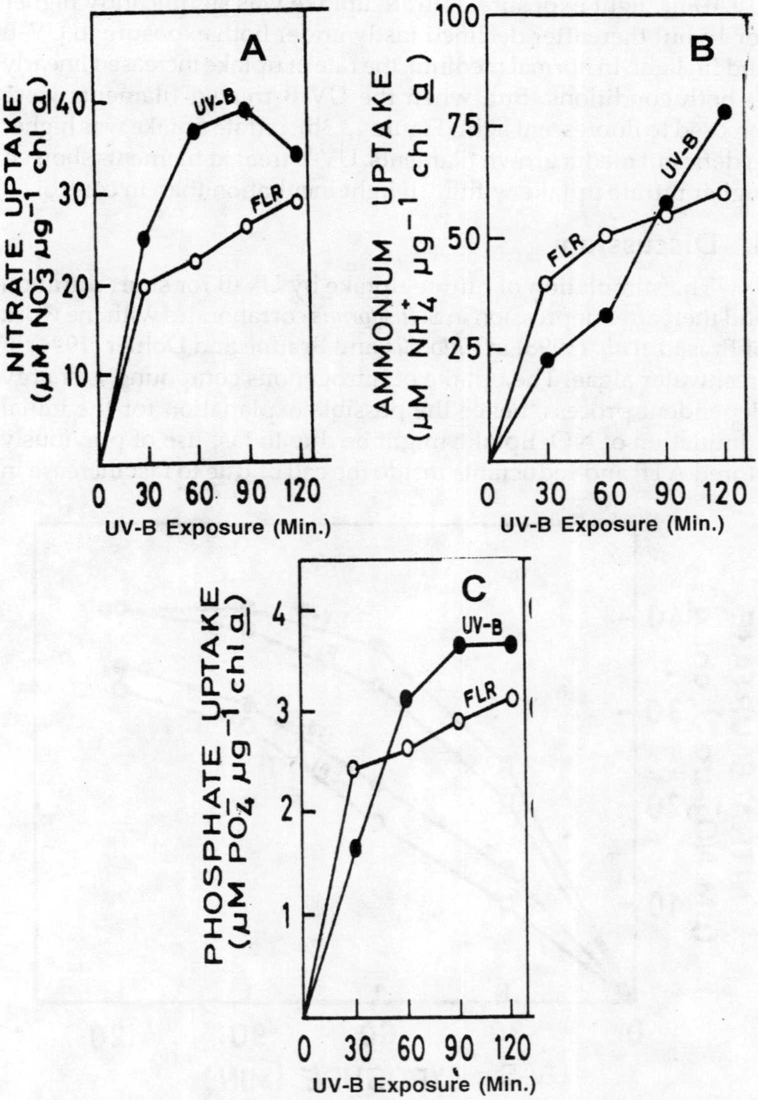

Figure 13.2: Effect of UV-B on Nitrate (A), Ammonium (B) and Phosphate (C) uptakes in *S. fusiformis* (4 days 1mM NH_4Cl/ Normal $NaNO_3$ grown culture having $2x10_4$ filaments respectively treated with 100 mM $NaNO_3$ (A), 400 mM NH_4cl and 50 mM K_2HPO_4 and thereafter exposed to UV-B/flr. light)

UV-B/flr. light exposure. Nitrate uptake was significantly higher for 1h but thereafter declined fastly under both exposure to UV-B and flr. light. In normal medium, the rate of uptake increased linearly in both conditions. But, when the UV-B treated filaments were exposed to fluorescent light (Figure 13.3b), nitrate uptake was higher in deficient media grown filaments. UV-B treated filaments showed higher nitrate uptake within 1h light incubation than in control.

4. Discussion

The stimulation of nitrate uptake by UV-B for short duration and thereafter depression in *S. fusiformis* corraborates with the work of Prasad *et al.*. (1998) in *Chlorella* and Braune and Döhler (1994) in freshwater algae. The uptake of nitrogenous compound is energy dependent process, hence the possible explanation for the initial stimulation of NO_3 uptake might be due to fast use of previously stored ATP and reductants inside the cell or due to fast increase in

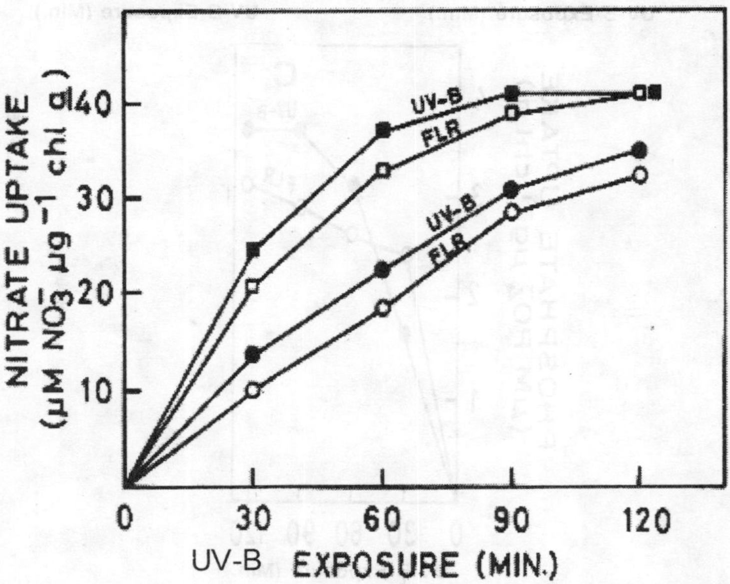

Figure 13.3A: Effect of UV-B on Nitrate Uptake in Normal (o ●) and Deficient Medium (▯ ◆) in *S. fusiformis* (Normal medium, 2.5 gl^{-1} and deficient medium, 0.25 gl^{-1}, 5 days grown culture initial O.D. = 0.2, haivested and treated with 100 M $NaNO_3$ and thereafter exposed to UV-B/flr. light.

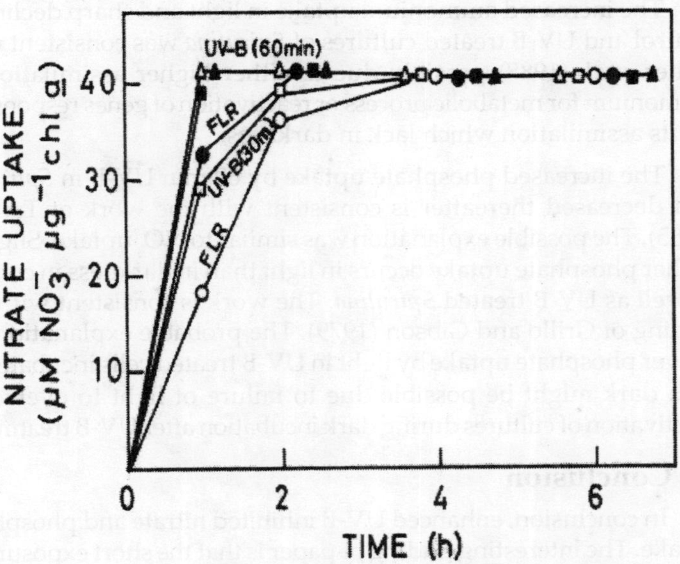

Figure 13.3B: Effect of Light on Nitrate Uptake in Normal (open) Deficient Medium (closed) Grown and 30 and 60 min. UV-B exposed filaments of *S. fusiformis* (conditions same as in Figure 13.3A)

membrane permeability by UV and later declination might be due to depletion of ATP and ferredoxin supply due to inactivation of cell as well as ATPase (Murphy, 1983). In *S. fusiformis*, fast nitrate uptake in light and slower in darkness might be due to assimilation of nitrate during nitrogen metabolism and photosynthesis in light. (Döhler, 1997).

Faster nitrate uptake in deficient medium in *Spirulina* was still stimulated by UV-B irradiation for 1h which coincides with the work of Dwivedi (1992) but contradict with Cullen and Lesser (1991) possibly might be due to greater deficiency pool by fast conversion of NO_3 to NO_2 due to stimulation of NR activity under UV-B exposure.

The sensitivity of ammonium uptake by UV-B in *Spirulina* for early 45 min was enhanced but lower down fastly which coincides with the work of Döhler and Hagmeier (1997). It is possible due to ATP and reductant deficiency during UV-B irradiation but latter faster uptake was presumably due to decrease in cytoplasmic pH by UV-B resulting from increased K^+ efflux and appearance of HCO_3 inside the cell.

The increased ammonium uptake in light and sharp decline in control and UV-B treated cultures of *Spirulina* was consistent with Fisher *et al.* (1988) possibly due to either higher assimilation of ammonium for metabolic process or reactivation of genes responsible for its assimilation which lack in darkness.

The increased phosphate uptake by 60 min UV-B in *Spirulina* and decreased thereafter is consistent with the work of Tanada (1955). The possible explanation was similar to NO_3 uptake. Slightly higher phosphate uptake occurs in light than in darkness in control as well as UV-B treated *Spirulina.* The work is consistent with the finding of Grillo and Gibson (1979). The probable explanation for slower phosphate uptake by light in UV-B treated cells incubated in 24 h dark might be possible due to failure of light to overcome inactivation of cultures during dark incubation after UV-B treatment.

5. Conclusion

In conclusion, enhanced UV-B inhibited nitrate and phosphate uptake. The interesting finding of paper is that the short exposure of UV-B stimulated nitrate and phosphate uptakes, but inhibited the ammonium uptake in *Spirulina.* Nitrate uptake declined later on when ammonium uptake was enhanced and faster than nitrate and phosphate.

Nitrate uptakes were stimulated by light but dark incubation decreased the rate of uptakes both in control and UV-B treated filaments of *Spirulina.* Nitrate uptake was higher in deficient medium, which was more enhanced by UV-B. Filaments exposed to UV-B for longer period takes short duration for maximum uptakes.

6. Acknowledgement

Dr. V. Prasad is grateful to the authorities of Tribhuvan University for granting leave for research and to the colloquies of the department, T. R. M. Campus, Birgunj, for valuable suggestions.

6. References

A P H A 1985 *Standard Method for the Examination of Water and Wastewater*, APHA, Washington D C.

Behrenfeld M J 1995 Ultraviolet–B radiation effects on inorganic nitrogen uptake by natural assemblages of Oceanic plankton. *J Phycol* 31:25-36.

Boussiba S 1989 Ammonia uptake in the Alkalophilic Cyanobacterium *Spirulina plantensis*. *Plant Cell Physiol* 30:303-308.

Braune W and Döhler G 1994 Impact of UV-B radiation on ^{15}N-ammonia and ^{15}N-nitrate by *Haematococcus lacustris* (Volvocales). 1 Differential response of flagellates and aplanospores. *J Plant Physiol* 144: 8-44.

Cullen J J and Lesser M P 1991 Inhibition of phytoplankton photosynthesis by ultraviolet radiation as a function of dose and dosage rate. *Mar Biol* 111:183-190.

Döhler G 1997 impact of UV radiation of different wavebands on pigments and assimilation of ^{15}N- ammonium and ^{15}N-nitrate by natural phytoplankton and ice algae in Antarctica. *J Plant Physiol* 151: 550-555.

Döhler G and Hagmeier E 1997 UV- effects on pigments and assimilation of ^{15}N- ammonia and ^{15}N- nitrate by natural marine Phytoplankton of the North Sea. *Botanica Acta* 110: 481-488.

Döhler G and Stolter H 1986 Impact of UV-B (290-320 nm) radiation on photosynthesis mediated uptake of ^{15}N- ammonia and ^{15}N–nitrate in several marine diatoms. *Biochem Physiol Pflanzen* 18: 533-539.

Dwivedi A 1992 Ecophysiological Studies on Alkalophilic Cyanobacteria with special reference to the Effect of pH on *Microcystis aeruginosa* and *Hapalosiphon welwitschii*, PhD Thesis Banaras Hindu University Varanasi India.

Fisher T R, Morrissey K M, Carlson P R, Alves L F and Melack J M 1988 Nitrate and ammonium uptake by phytoplankton in an Amazon River flood plain Lake. *J Plankton Res* 10: 7-29.

Grillo J F and Gibson J 1979 Regulation of phosphate accumulation in the unicellular cyanobacterium *Synechococcus*. *J Bacteriol* 140: 508-517.

Hader D P, Worrest R C, Kumar H D and Smith R C 1998 Effects on aquatic ecosystems. *J Photochem Photobiol B Biol* 46: 53-68.

Holm–Hansen O 1968 Ecology, Physiology and Biochemistry of Blue-Green Algae. *Ann Rev Microbiol* 22: 47-70.

Kumar H D 2002 *Environmental Technology and Biosphere Management*. Science Publishers, Inc Enfield, USA.

Lubin D and Jensen E H 1995 Effects of clouds and stratospheric ozone depletion on ultraviolet radiation trends. *Nature* 377: 710-713.

Murphy T M 1983 Membranes as targets of ultraviolet radiation. *Physiol Plant* 58:381-388.

Myers J and Kratz W A 1955 Relationship between pigment content and photosynthetic characteristics in a blue-green alga. *J Gen Physiol* 39:11-21.

Nicholas D J D and Nason A 1957 Determination of Nitrate and Nitrite. pp 982-987 *In*: Colowick S P and Kaplan N O (eds) *Methods in Enzymology* Vol III Academic Press, New York.

Pearl H W 1988 Nuisance phytoplankton blooms in coastal, esturine and inland waters. *Limnol Oceanogr* 33: 823-847.

Prasad V, Kumar A and Kumar H D 1998 Effects of UV-B on certain metabolic processes of the green alga *Chlorella vulgaris*. *Intern J Environmental Studies* 55: 129-140.

Sinha R P, Singh N, Kumar A, Kumar H D, Hader M and Hader D P 1996 Effects of UV- radiation on certain physiological and biochemical processes in Cyanobacteria. *J Photochem Photobiol B* 32: 107-113.

Tanada T 1955 Effect of ultraviolet radiation and calcium and their interaction in salt absorption by excised mung bean roots. *Plant Physiol* 30: 221-225.

Zarrouk C 1966 *Contribution a l'etude d'une Cyanophyceae Influence de divers facteurs physiques et chimiques sur la croissance et la photosynthese de Spirulina maxima* (setch et Gardner) Geitler. PhD Thesis University of Paris, France.

Chapter 14

Various Roles of Blue-Greens in Human Life

M. Masroor A. Khan

Contents

Glimpses of Cyanobacteria, 2006 213–230
Daya Publishing House, New Delhi

11. **Food and Food Additives**
12. **Biofertilizers**
13. **Pharmaceutical Uses of Blue-Greens**
14. **Measures to Control the Growth of Blue-Greens**
15. **References**

ABSTRACT

Blue-green algae (blue-greens, cyanobacteria) are prokaryotic and the simplest group of algae. Blue-greens have a considerable and ever-increasing impact on social economical and environmental aspects of human life and society. They have both harmful and beneficial effects on human life. Among the harmful effects of blue greens is production of toxic compounds in human and livestock-water supplies. These toxins also make the swimming and recreation in lakes and pools hazardous, and blue-green blooms are unaesthetic *i.e.* unpleasant to view. Poisoning can cause not only the death of animals but humans too. Recent research shows that among the neurotoxins, β-*N*-methylamino-L-alanine is commonly produced by all types of blue-greens (PNAS, USA 102: 5074-5078, 2005). Nuisance and undesirable conditions are also developed in water-bodies due to these excess algal blooms which overwhelm the water surface. On the other hand, there are many beneficial effects, including nitrogen-fixation. Several of them are known to be capable of reducing the extraordinarily inert nitrogen gas into ammonia. *Spirulina*, containing more than 60 per cent protein, B vitamins and beta-carotene, shows great promise in helping to overcome malnutrition. Certain blue-green species are used as biofertilizers. Many blue-greens have been proved to be sources of pharmaceutical compounds with antibiotic, antiviral and antineoplastic (anticancer) properties.

1. Introduction

The algae are (with numerous exceptions) aquatic organisms that (with frequent exceptions) are photosynthetic, oxygenic autotrophs that are (except for the kelps) typically smaller and less structurally complex than land plants (Graham and Wilcox, 2000).

However, the prokaryotic blue-green algae are the simplest of the algae. Blue-greens have a considerable and increasing economic importance; having both beneficial and harmful effects on human life. Blue-greens have no nucleus, the structure that encloses the DNA, and no chloroplast, the structure that encloses the photosynthetic membranes, structures that are present in photosynthetic eukaryotic algae. In fact blue-greens are more akin to bacteria, which have similar biochemical and structural characteristics. The process of nitrogen fixation and the occurrence of gas vesicles are especially important to the success of nuisance species of blue-greens. The blue-greens are widely distributed over land and water, often in environments where no other vegetation can exist. They were probably the chief primary producers of organic matter and the first organisms to release elemental oxygen, O_2, into the primitive atmosphere, which was until then free from O_2. Thus blue-greens were most probably responsible for a major evolutionary transformation leading to the development of aerobic metabolism and to the subsequent rise of higher plant and animal forms. They are referred to in literature by various names, *viz.* Cyanophyta, Myxophyta, Cyanochloronta, Cyanobacteria, Chloroxybacteria, blue-green algae, blue-green bacteria and blue-greens. Their fossils have been identified as over three billion years old.

Blue-greens can grow in full sunlight and in almost complete darkness. However, the majority are aerobic photoautotrophs: their life processes require only oxygen, light and inorganic substances. A species of *Oscillatoria* found in mud at the bottom of the Thames- is able to live anaerobically. Blue greens can live in extremes of temperatures -60°C to 85°C. A number of species of blue-greens can tolerate temperatures as high as 72°C in the thermal pools of Yellow Stone National Park (USA). A few species are halophilic or salt tolerant. Some blue greens tolerate salt concentrations as high as 27 per cent (for comparison, the concentration of salt in seawater is 3 per cent). Blue greens are often the first plants to colonize bare areas of rock and soil, such as those resulting from cataclysmic volcanic explosion at Krakatoa, Indonesia in 1883. Unlike many more advanced organisms, blue-greens need no organic nutrients. At the onset of nitrogen limitation during bloom conditions, certain cells in *Anabaena* and *Aphanizomenon* develop into heterocysts, which convert nitrogen gas into ammonium, which is then distributed to the neighboring cells of a filament. In addition, blue-greens that form mutually beneficial symbiotic relationships with a wide range of other life forms, can convert nitrogen gas into ammonium (Chapman, 1968; Lee 1999; Graham and Wilcox, 2000; Rai et al. 2002). Finally, at the onset of adverse environmental conditions, some blue-greens can produce modified cells, called akinetes. Akinetes contain large reserves of carbohydrates, and owing to their density and lack of gas vesicles, eventually settle to the bottom of waterbodies. Akinetes can tolerate adverse conditions such as the complete drying of a pond or the cold winter temperatures, and, as a consequence, akinetes serve as "seeds" for the growth of juvenile filaments when favorable conditions return. Heterocysts and akinetes are unique to the blue-greens.

2. Terrestrial and Freshwater Planktonic Blue-greens

Certain species, *Nostoc commune*, are often conspicuous on the soil surface. In temperate regions blue-greens are especially common in calcareous and alkaline soils. Acid soils, however, lack blue-greens and are usually dominated by diatoms and green algae. Unicellular and filamentous blue-greens are almost invariably present in freshwater lakes, frequently forming dense planktonic populations or water blooms in eutrophic (nutrient rich) waters. In temperate lakes there is a characteristic seasonal succession of the bloom-forming species, due apparently to their differing responses to the physico-chemical conditions created by thermal stratification. Usually filamentous forms (such as *Anabaena* species, *Aphanizomenon flos-aquae* and *Gloeotrichia echinulata*) develop first soon after the onset of stratification in late spring or early summer, while the unicellular-colonial forms (like *Microcystis* species) typically bloom in mid-summer or in autumn. The main factors which influence the development of planktonic populations are light, temperature, pH, nutrient concentrations and the presence of organic solutes.

3. Attached and Benthic Populations in Lakes

Many blue-greens grow attached on the surface of rocks and stones (epilithic forms), on submerged plants (epiphytic forms) or on the bottom sediments (epipelic forms, or the benthos) of lakes. The epilithic community displays a clearly discernable zonation in lakes. Members of the genera *Pleurocapsa, Gloeocapsa* and *Phormidium* often dominate the dark blue-black community of the spray zone. *Scytonema* and *Nostoc* species form olive-green coatings and are more frequent about the water line, whilst the brownish *Tolypothrix* and *Calothrix* species are more typical components of the subsurface littoral community. The epiphytic flora of lakes is usually dominated by diatoms and green algae, and blue-greens are of less importance in this community. However, species of the genera *Nostoc, Lyngbya, Chamaesiphon* and *Gloeotrichia* occasionally encrust submerged plants. The epipelic community commonly includes blue-greens such as *Aphanothece* and *Nostoc,* particularly in the more eutrophic lakes. Benthic blue-greens growing over the littoral sediments and on submerged plants may be responsible for the occasional high rates of N_2-fixation measured in oligotrophic lakes (Palmer, 1980, Carpenter et al. 1992).

It is an established fact that blue-green algal blooms have considerable social, economic and environmental impacts. Blue greens influence the operation of water supplies, human health, agriculture (livestock), fish, native flora and fauna, recreation and tourism (Schwimmer and Schwimmer, 1955, 1968). Examples of these impacts are described below.

4. Water Supplies

In addition to the production of toxins, blue-green algal blooms alter the physical and chemical characteristics of water, which may influence water supply operations. Dense algal blooms cause increased suspended solid loads which increases turbidity, thereby reducing the efficiency of disinfection, and can also block filters and sprinklers. In addition, increased water pH due to photosynthesis can further reduce the acceptability of water for drinking and industrial uses. It also increases maintenance costs for treatment and distribution systems. A wide range of blue-green algae produce the chemicals geosmin and methyl-isoborneol which impart a disagreeable earthy taste and odour to the water. While these compounds may not pose health risks, they may make the water unpalatable. However, these compounds can alter the form and toxicity of heavy metals and ammonia and change the chemical reactions in water treatment. As the algae decay, they increase the dissolved organic carbon load and the biochemical oxygen demand. This increases the demand for flocculants in treatment. It also increases the chlorine required for treatment and the potential for trihalomethane production, both of which have possible health effects (Australian Water Resources Council, 1991). There are numerous reports from both the mainland and within Tasmania of stock deaths associated with blue-green algal blooms. The most common genera involved in cases of stock death have been *Microcystis, Anabaena, Nodularia* and *Aphanizomenon.* All cases of stock deaths relate to dense blooms with accumulation of scum on the water surface and shore. However there is no detailed information on the concentration of cells in the water which cause stock deaths (Palmer, 1980).

5. Swimming and Recreation

The scale of a hazard, or risk of injury, is likely to increase with the duration of contact. Ingestion of concentrated algal cells in the form of scum constitutes the greatest hazard. The potential risk from various water based activities depends on the likelihood of that activity bringing the individual into contact with the scum. Pleasure cruising, for example would constitute a low risk activity, and swimming particularly in near shore blue-green blooms, a high risk activity (Hardin, 1961). The use of wetsuits for water sports may result in the greater risk of rashes, because algal material in the water trapped inside the wetsuit will be in close contact with the skin for long periods of time (Beer *et al.*, 1968). Severe illness may occur to the people who come in contact with water containing high concentrations of blue-green algae. Effects can include influenza like symptoms, sore throats, blistering in and around the mouth, vomiting, diarrhoea and fever (Asai *et al.*, 1982). Conspicuous blue-green algae growths will have an obvious impact on local tourism and recreation and particularly on the economy of communities adjacent to dams and storages used for these purposes (Green Algal Task Force, 1995). Economic costs can include losses associated with activities supporting recreation and tourism (accommodation, transport, food) and losses incurred by commercial recreation facilities such as caravan parks and tourist parks. (ARMACANZ, 2000).

6. Irrigation

There is little evidence that toxins from blue-green algae affect irrigation use and plant productivity, and it is much more likely that agricultural produce from irrigation areas using affected water will be subject to scrutiny from the human health perspective. Although storages that are primarily established for agricultural purposes may require minimal management, it is often the case that the rural community uses this water for domestic needs and this should be kept in mind when dealing with a potentially toxic bloom (AWRC, 1991). Furthermore, according to recent reports, blue-greens produce certain neurotoxins and hepatotoxins. The bioaccumulation of these toxins in additional food-chains and ultimate destination in our body cells may prove a health hazard. One of the most detrimental toxins, β-*N*-methylamino-L-alanine has been reported in brain tissues of Canadian Alzheimer patients (Murch *et al.*, 2004).

7. Mussels (Shellfish)

Research into the toxicity of edible mussels from Peel Harvey Estuary in Western Australia during a *Nodularia* bloom has clearly shown a consumer hazard (Falconer and Choice, 1992). The mussels ingest toxic blue-green algae and accumulate toxins in the digestive system of the mussel. The toxins are retained even after freezing and boiling the mussels and their detrimental effects can be demonstrated by death of mice receiving extracts of the mussels. It is unclear what effects freshwater algal blooms may have particularly in terms of downstream transport on estuarine aquaculture activities (Schwartz and Tabershaw, 1945; Bowszyc, 1966).

8. Nuisance and Undesirable Conditions

The formation of water blooms results from the redistribution and often rapid accumulation of buoyant planktonic populations. When such populations are subjected to suboptimal conditions, they respond by increasing their buoyancy and move upward nearer to the water surface. Water turbulence usually prevents them reaching the surface. If, however, turbulence suddenly weakens on a calm summer day, the buoyant population may become unable to submerge, becoming lodged at the water surface. There the cells are exposed to most unfavourable conditions, such as O_2 supersaturation, rapidly diminishing CO_2 concentrations and intense solar radiation, which are inhibitory to photosynthesis and N_2-fixation, These conditions cause photo-oxidation of pigments, inflicting irreversible damage to the genetic constitution of cells. A frequent outcome of surface bloom formation is massive cell lysis and rapid disintegration of large planktonic populations. This is closely followed by an equally rapid increase in bacterial numbers, and in turn by a fast deoxygenation of surface waters which could be detrimental to animal populations within the lake (Hallegareff, 1993). Water blooms are objectionable for recreational activities, and more importantly, create great nuisance in the management of water reservoirs (Fresh Water Algal Blooms in Queensland, 1992). Most of these conditions are produced by *Anabaena flos-aquae, Aphanizomenon flos-aquae, Microcystis aeroginos* and particularly *Cylindrospermopsis* (which is becoming increasingly common). An oversupply of nutrients, especially phosphorus and possibly nitrogen, will often result in excessive growth of blue-greens because they possess certain adaptations that enable them to outcompete eukaryotic algae. Perhaps the most important adaptation is their positive buoyancy, which is regulated by their gas vesicles which are absent in eukaryotic algae.

9. Poisonous Conditions and Cyanotoxins

Poisonous blue-greens occur in ponds and lakes throughout the world (Wild, 1961; Kingsbury, 1964: Chorus and Bartram, 1999). In Canada, they primarily occur in the prairie provinces. Poisoning has caused the death of cows, dogs, and other animals. Although humans ordinarily avoid drinking water that displays a blue-green bloom or scum, they may be affected by toxic strains when they swim in recreational water bodies during a bloom. Typical symptoms include redness of the skin and itching around the eyes; sore, red throat; headache; diarrhoea; vomiting; and nausea (Grant, 1962; Fraser and Lyell, 1963, Carmichael, 1994). Some cases of 'swimmers' itch' has been attributed to contact with *Lyngbya majuscula, Schizothrix calcicola* and *Oscillatoria nigroviridis*, which are commonly found in tropical and subtropical seawaters (Cohen and Reif, 1953; Kersten and Zschunke, 1955; Moikeha and Chu, 1971; Moikeha et al.,1971; Krauskop, 1973; Carmichael and Falconer, 1993). The toxins responsible for skin irritation are lipid-soluble phenolic compounds (Chu, 1969). The species of cyanobacteria that have been established to produce toxins include some species or strains of *Anabaena, Anabaenopsis, Hapalosiphon, Aphanizomenon, Cylindrospermopsis, Microcystis, Nodularia, Nostoc and Planktothrix* etc. The structure and biosynthesis of cyanobacterial toxins has been reviewed by Rinehart, *et al.* (1994). Cyanobacteria can generate toxins hazardous to human health including hepatotoxic cyanotoxins, also called hepatotoxins (damaging liver) and neurotoxins (damaging brain). Hepatotoxins are of two types microcystins (produced by *Microcystis, Anabaena, Planktothrix, Hapalosiphon, Anabaenopsis,* and *Nostoc)* and nodularins (produced by *Nodularia*). Human death is also possible due to cyanotoxins. Recently, a teenager died in Madison, Wisconsin, USA from exposure to anatoxin (a type of cyanotoxin) present in cyanobacterial contaminated water (Figure 14.1). Cases also reported for dialysis patients in Mexico (Personal communication with Prof. Linda Graham of Wisconsin University, USA). Reported cases affecting humans list *Anabaena* as the main culprit. Most of the recorded toxic blooms are caused by *Microcystis aeruginosa* which manufactures "microcystin", which yields 7 (or 14) amino acids upon hydrolysis. It causes enlargement and congestion of the liver followed by necrosis and haemorrhage, and may also exhibit neurotoxic activity. But many toxic blooms are also produced by either *Anabaena flos-aquae* (manufactures "anatoxins") or *Aphanizomenon flos-aquae* (manufactures "aphantoxins"). Alkaloid toxins (anatoxins, aphantoxins) act on the nervous system, leading to paralysis of muscles needed for breathing (McElhenny *et al.*, 1962). Two other genera, *Oscillatoria* and *Nodularia* are also known to produce toxic populations. Whether the animal survives the poisoning depends primarily upon the concentration of toxin ingested. Blue-green toxins may act on zooplankton and might be an effective mechanism of protection against grazing pressures. Little is known about the per cent of blooms that are toxic (upto 25 per cent quoted in literature), and also why a toxic population is produced. A complicating factor is that part of a bloom can be toxic and another part nontoxic within the same lake. It has been suggested that toxic strains may develop only under a particular set of environmental conditions, or that toxin production may be associated with plasmid-mediated gene transfer.

Lipopolysaccharide endotoxins, hepatotoxin tumor promoters, and neurotoxins are three general classes of toxins produced by many cyanobacteria (Rantala *et al.*, 2004). These affect not only zooplankton and other herbivores, but also humans and aqua-cultured food organisms. Neurotoxins produced by cyanobacteria in drinking water can poison dogs and farm animals and even humans. Death by respiratory arrest may occur within minutes to hours after toxin ingestion. Incidents of animal poisonings are associated with large populations or blooms of cyanobacteria. Some 50-75 per cent of cyanobacterial blooms generate toxins, and some

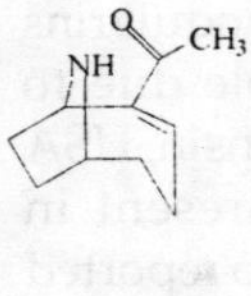

Anatoxin-A

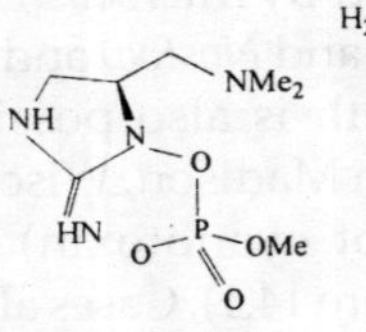

Anatoxin-A(S)

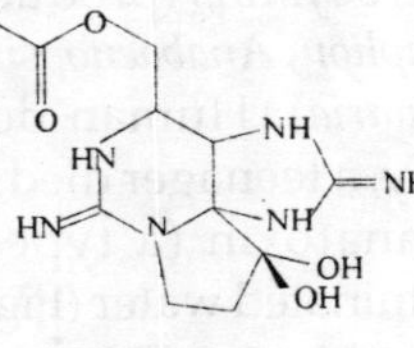

Saxitoxin

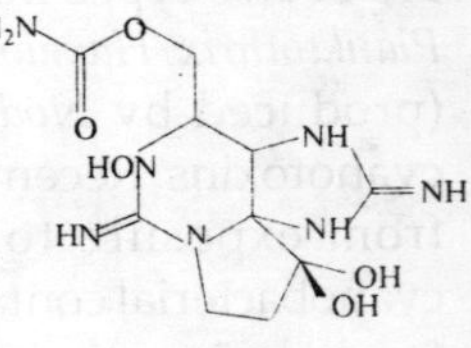

Neosaxitoxin

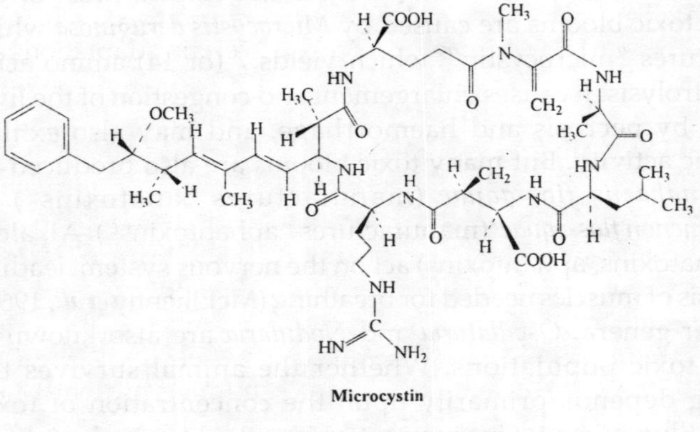

Microcystin

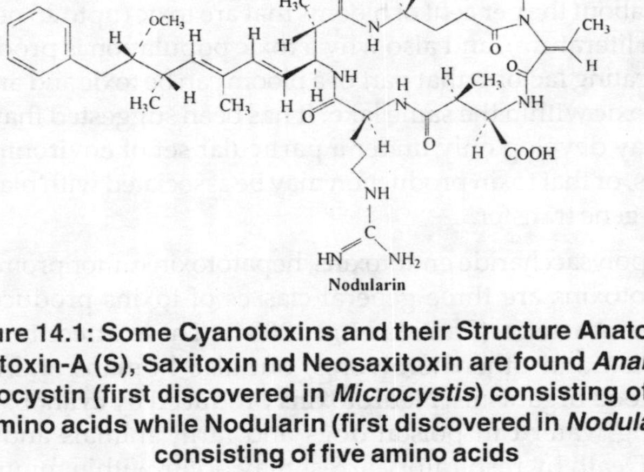

Nodularin

Figure 14.1: Some Cyanotoxins and their Structure Anatoxin-A, Anatoxin-A (S), Saxitoxin nd Neosaxitoxin are found *Anabaena*. Microcystin (first discovered in *Microcystis*) consisting of seven amino acids while Nodularin (first discovered in *Nodularia*) consisting of five amino acids

experts believe that all cyanobacterial blooms should be regarded as toxic until disproved. Although cyanobacterial toxins are generally regarded as defense-related, it has been proposed that they might serve additional cellular functions, such as nitrogen storage (Graham and Wilcox, 2000). It is estimated that more than forty cyanobacterial species are either demonstrated or suspected toxin producers. Toxins can be identified by high performance liquid chromatography and / or mass spectrometry.

Recently, many toxic compounds from the blue-greens that cause not only the death of animals but humans too as discussed above (Carmichael, 1994, Cox et al. 2003, Cox et al. 2005). Among the neurotoxins produced by blue-greens, ß-*N*-methylamino-L-alanine (BMAA) is commonly produced by all types. The ubiquity of cyanobacteria in diverse terrestrial and aquatic environments suggests that ingestion of BMAA may occur through even less esoteric routes, including direct consumption of cyanobacteria or cyanobacterial hosts, bioaccumulation in additional food chains, or exposure to cyanobacterial contaminated water supplies. The recent hypothesis that BMAA accumulates in proteins, which collectively function as an endogenous neurotoxic reservoir within the human body, and then slowly released time to time as these proteins are metabolized.

Advantages of Blue-Greens

Though blooms are considered a nuisance in recreational lakes and water supply reservoirs in some regions of the world. The near continuous blooms of blue-greens in some tropical lakes can be a valuable source of food for humans. Some blue-green species make major contributions to the world food supply by naturally fertilizing soils and rice paddies. R.N. Singh of the Banares Hindu University in India has shown that the introduction of blue-green algae to saline and alkaline soils in the state of Uttar Pradesh increases the soils' content of nitrogen and organic matter and also their capacity for holding water. This treatment has enabled formerly barren soils to grow crops. Astushi Watanabe of the University of Tokyo found the introduction of *Tolypothrix tenuis* resulted in a 20 per cent increase of rice crop. W.E. Booth of the University of Kansas showed through experiments in Kansas, Oklahoma and Texas, that a coating of blue-greens on prairie soil binds the particles of the soil to their mucilage coating, maintains a high water content and reduces erosion.

Humans also consume *Spirulina*. It contains all of the amino acids essential for humans, and its protein content is high (± 60 per cent). It is a staple food in parts of Africa and Mexico. In China, Taiwan and Japan, several blue-greens are served as a side dish and a delicacy. Several areas in North America culture and commercially process certain blue-greens for various food and medicinal products such as vitamins, drug compounds, and growth factors.

Heterocystous blue-greens possess the unique ability to simultaneously evolve O_2 in photosynthesis (in vegetative cells) and H_2 by nitrogenase catalyzed electron transfer to H_+-ions (in heterocysts), in the absence of N_2 or other substrates of nitrogenase. This is the basis for the attempts of several workers to exploit the potential through the development of a 'biophotolytic system' for solar energy conversion, even though to date the thermodynamic efficiency has been disappointingly low. Nevertheless, the utilization of blue-greens in food production and in solar energy conversion may hold immense potential for the future, and could be exploited for human economy. Progress in the study of the genetics of blue-greens may enable us to manipulate the N_2-fixation (*nif*) and associated genes, and produce strains which fix N_2, evolve H_2 or release ammonia with great efficiency.

10. Hydrogen Production

Strains of *Synecococcus, Microcystis* and *Oscillatoria* have received much attention as a clean type of fuel. They produce hydrogen at a significant rate. The strains of MiamiBG7 of *Oscillatoria* is particularly suitable as it lacks the uptake hydrogenase which catalyses the uptake of hydrogen by the cells. This enzyme has an adverse effect on the net yield of hydrogen gas. Philips and Mitsui observed that immobilization of the strain Miami BG7 with agar significantly improves the rates and duration of hydrogen production as compared to that by free cell suspensions. The rates of hydrogen production of about 12 microlitres H_2 per mg dry weight per hour were recorded and hydrogen production could be sustained for three weeks (Kumar, 1990; Kumar and Singh 1995).

11. Food and Food Additives

The cyanobacterium *Spirulina* has traditionally been harvested from African lakes and consumed in sauces at the rate of some 9-13 grams per meal. The protein level of *Spirulina* can be as high as nuts, grains and soybeans, ranging form 50-70 per cent of algal dry weight, and shows promise for use as a protein supplement for malnourished populations. Because the cell walls are composed of mucopolysaccharides rather than cellulose, *Spirulina* is more digestible than some other microalgae. *Spirulina* is also naturally high in B vitamins and essential unsaturated fatty acids as well as high levels of β-carotene, which is converted to vitamin A during digestion. Vitamin A can prevent xerothalmia, a form of blindness that arises in malnourished children. *Spirulina* is commercially cultivated in shallow raceways, 500-5000 m^2 in area, while being stirred with paddles.

12. Biofertilizers

The blue-green algae, which are rich in nitrogen and phosphorus, are better fertilizers than seaweeds. In tropical countries, the bottom mud of dried up ponds is regularly used as manure in crop fields. The mud of this as a manure (or soil amendment) is mostly due to the high content of blue-green algae. A suitably blended mixture of seaweed and cyanophycean manures (*e.g.*, bloom of *Microcystis*) may serve as an ideal fertilizer and this can relieve the acute shortage of fertilizers in developing countries.

Regarding the nitrogen status of natural habitats, the nitrogen-fixing blue-green algae deserve special mention. These algae grow luxuriantly in tropical habitats, *e.g.*, rice fields. Researches conducted with *Tolypothrix tenuis* in Japan and with *Aulosira fertilissima* in India have shown that the yield of paddy is substantially increased following the inoculation of fields with these algae. Phycologists at the Central Rice Research Institute, Cuttack (India), inoculated rice fields with four species of nitrogen-fixing blue-green algae; the grain yield increased by nearly 30 per cent. The nitrogen-fixing blue-green algae have been successfully used as a potent biofertilizer in several Indian states and are now being widely applied in rice fields. *Azolla-Anabaena* system has also been found to contribute significantly to the nitrogen status of rice field. Further work is needed to isolate ammonia-excreting strains of *Azolla* and also those strains of nitrogen-fixing blue-green algae that can grow and fix nitrogen even in those fields which have received chemical fertilizer. In the Sambhar salt Lake in Rajasthan, cartloads of algal consisting of A*nabaenopsis* and *Spirulina* are produced annually, from September to December, and are employed by local farmers as manure to paddy and other crops. (Kumar and Singh, 1995; Bilgirami and Saha, 1992).

13. Pharmaceutical Uses of Blue-Greens

Many of algae have been proved to be sources of compounds with antibiotic and antineoplastic (anticancer) activity as well as other pharmacologically useful products. Patterson, *et al.* (1994) found that a wide variety of cyanobacteria produce compounds, including sulfolipids, that were active against a herpes virus, a pneumonia virus, and HIV. Also certain cyanobacteria produce potential antitumor compounds. A potent antifungal antibiotic, tolytoxin, which has low toxicity to humans, is produced by the cyanobacteria *Tolypothrix* and *Scytonema*. Tolytoxin, a macrycylin lactone that depolymerizes actin and thereby is able to disrupt eukaryotic cell division, is produced by *Scytomnema ocellatum*. A review of the antibiotics produced by cyanobacteria can be found in Patterson, et al. (1994). Once algal sources of useful medicinal compounds are identified, the algae can be cultivated on a large scale in bioreactors, harvested, and the active materials extracted, purified, and marketed (Akatsuka, 1994; Graham and Wilcox, 2000).

14. Measures to Control the Growth of Blue-Greens

A variety of algicides are widely used to prevent the growth of nuisance algae however, copper sulphate solution (1 per cent) is extraordinarily effective. A number of other algicides are phenolic compounds, amide derivatives, quaternary ammonium compounds and quinone derivatives. Dichloron aphthoquinone is selectively toxic to blue-greens. The hazards of using toxic chemicals indiscriminately in the natural environment are well documented. Micro-organisms (fungi, bacteria and viruses) appear to play an important part in regulating growth of blue-greens in freshwaters. Certain chytrids (fungal pathogens) specifically infest akinetes, and others heterocysts. Bacterial pathogens belonging to the group of Myxobacteriales can affect rapid lysis of a wide range of unicellular and filamentous blue-greens, though heterocysts and akinetes remain generally unaffected. Viral pathogens belonging to the group of cyanophages exhibit some degree of host specificity. Phage AR-1 attacks *Anabaenopsis*, phages SM-1 and AS-1 are effective against the unicellular forms, *Synechococcus* and *Microcystis*, Phage C-1 lyses *Cylindrospermum*, and the LPP-1 virus is effective against strains of *Lyngbya*, *Phormidium* and *Plectonema*.

Acknowledgement

I am highly grateful to Dr Linda E. Graham, Professor, Department of Biological Sciences, Wisconsin University, Madison, Wisconsin, USA, for very useful suggestions and critical reading of the manuscript.

References

Akatsuka I 1994 Biology of Economic Algae. S.P.B. Academic Publishing. The Hague.

ARMACANZ 2000 Agriculture and Resource Management Council of Australia and New Zealand.

Asai S, Krzanowski J J, Anderson W H, Martin D F, Polson J B, Lockey R F 1982 Effects of the toxin of red tide, *Ptychodiscus brevis*, on canine tracheal smooth muscle: a possible new asthma-triggering mechanism. Journal of Allergy and Clinical Immunology 69: 418-428.

AWRC 1991 Australian Water Resources Council, Water Resource Management Committee.

Beer W E, Jones M and Jones W E 1968 Dermatoses in lobster fishermen. Br. Med. J. 30: 807-809.

Bilgirami and Saha 1992 A Text Book of Algae. CBS Publishers. New Delhi.

Bowszyc J 1966 Redfeed dermatitis. A professional disease of workers in the fishing industry caused by Baltic herring and mackerel [Polish]. Przegl. Derm. 53: 39.

Carmichael W W and Falconer I R 1993 Diseases related to freshwater algal toxins and control measures. In: Algal Toxins in Seafood and Drinking Water. I.R. Falconer (ed). Academic Press, London.

Carmichael W W 1994 The toxins of cyanobacteria. Scientific American 270:78-86.

Carpenter E J, Capone D G and Rueter J G 1992 Marine Pelagic Cyanobacteria: *Trichodesmium* and Other Diazotrophs. Kluwer, Dordrecht, The Netherlands.

Chapman V J 1968 The Algae. Macmillan. London.

Chorus I and Bertram J 1999 Toxic Cyanobacteria in Water. E. and F. N. Spon, London, pp.113-153.

Chu G W 1969 A unique toxic lipid from a Hawaiian dermatitis-producing alga. J. Formosan Med. Ass. 68: 570.

Cohen S G and Reif C B 1953 Cutaneous sensitization to blue-green algae. J. Allergy 24: 452.

Cox P A, Banack S A and Murch S J 2003 Biomagnification of cyanobacterial neurotoxins and neurodegenerative disease among the Chamorro people of Guam. Proc. Natl. Acad. Sci.100:13380-13383.

Cox P A, Banack S A, Murch S J, Ramussen U, Tien G, Bidigare R R, Metcalf J S, Morrison L F, Codd G A and Bergman B 2005 Diverse taxa of cyanobacteria produce β-*N*-methylamino-L-alanine, a neurotoxin amino acid. Proc. Natl. Acad. Sci. 102: 5074-5078.

Falconer I R and Choice A 1992 Toxicity of edible mussels (*Mytilus edulis*) growing naturally in an estuary during a water bloom of the blue-green algae *Nodularia spumegina*. Environmental Toxicology and Water Quality 7: 119-123.

Fraser J H and Lyell A 1963 Dogger Bank itch. Lancet 1: 61.

Freshwater Algal Blooms in Queensland 1992 A report by the Queensland Water Quality Taskforce.–Volume 1 Main Report, October 1992.

Graham L E and Wilcox L W 2000 Algae. Prentice Hall, Upper Saddle River, NJ 07458, USA.

Grant W M 1962 Toxicology of the Eye. Charles C Thomas (Ed), Springfield, Ill., USA.

Green Algal Taskforce 1995 Department of Water Resources, NSW. Guidelines: For The Recreational Use of Water Potentially Containing Cyanobacteria. Occasional Paper SWR No. 1, October.

Hallegareff G M 1993 Harmful algal blooms and their apparent global increase. Phycologia 32: 79-99.

Hardin F F 1961 'Seabathers eruption'. J. Med. Ass. Georgia 50: 450.

Kersten E and Zschunke E 1955 Redfeed dermatitis, a skin affection frequent in the fishing industries. Dtsch. Gesundhwes. 10: 194.

Kingsbury J M 1964 Poisonous Plants of the United States and Canada. Englewood Cliffs, Prentice Hall, New Jersey, USA.

Krauskop F J 1973 Phytodermatitis caused by Cyanophyta and algae. Cs. Derm. 48: 195.

Kumar H D 1990 Introductory Phycology. Affiliated East-West Press. New Delhi.

Kumar H D and Singh 1995 A Text Book of Algae. Affiliated East-West Press. New Delhi.

Lee R E 1999 Phycology. Cambridge University Press. London.

McElhenny T R, Bold H C, Brown R M and McGovern J P 1962 Algae: a cause of inhalant allergy in children. Ann. Allergy 20: 739.

Moikeha S N and Chu G W 1971 Dermatitis producing alga *Lyngbya majuscula* Gomont in Hawaii. II. Biological properties of the toxic factor. J. Phycol. 7: 8.

Moikeha S N, Chu G W and Berger L R 1971 Dermatitis producing alga *Lyngbya majuscula* Gomont in Hawaii. I. Isolation and chemical characterization of the toxic factor. J. Phycol. 7: 4.

Murch S J, Cox P A and Banack S A 2004 A mechanism for slow release of cyanobacterial neurotoxins and neurodegenerative disease in Guam. Proc. Natl. Acad. Sci. 101: 12228-12231.

Palmer C M 1980 Algae and Water Pollution. Castle House Publications.

Patterson G M L, Larsen L K and Moore R E 1994 Bioactive natural products from blue-green algae. Journal of Applied Phycology 6: 151-157.

Rai A N, Bergman B and Rasmussen U 2002 Cyanobacteria in Symbiosis. Kluwer, Dordrecht, The Netherlands.

Rantala A, Fewer D P, Hisbergues M, Rouhiainen L, Vaitomaa J, Borner T and Sivonen K 2004 Proc. Natl. Acad. Sci. 101: 568-573.

Rinehart K L, Namikoshi M and Choi B W 1994 Structure and biosynthesis of toxins from blue-green algae (cyanobacteria). Journal of Applied Phycology 6: 159-176.

Schwartz L and Tabershaw I R 1945 Dermatitis in the fish industry. J. Industr. Hyg. Toxicol. 27: 27.

Schwimmer D and Schwimmer M 1968 Medical aspects of phycology. In: Jackson D.F. (Ed.) Algae, Man and the Environment. Syracuse University Press.

Schwimmer M and Schwimmer D 1955 The Role of Algae and Plankton in Medicine. Grune and Stratton. New York.

Wild H 1961 Harmful aquatic plants in Africa and Madagascar. Kirkia 2: 1-66.

Chapter 15

Algal Species Diversity of Streams of Schirmacher Oasis, Antarctica

R.K. Gupta, V.D. Pandey, K.D. Pandey and D. Vyas

Contents

ABSTRACT

Species diversity is the substitute for the biodiversity. The former can be expressed at three levels *i.e.* alpha (α), beta (β) and gamma (γ) diversity. These three diversity values has been calculated for the algal and cyanobacterial species occurring in the six freshwater streams selected in Schirmacher Oasis, Antarctica during the 11th Indian Scientific Expedition to

Glimpses of Cyanobacteria, 2006 231–241
Daya Publishing House, New Delhi

Antarctica (1991-1992). Alpha diversity comprises species richness and evenness in the distribution of individuals among species. The Simpson Index λ, Shannon-Wiener Diversity Index H' and other two indices N_1 and N_2 were calculated. In Species Richness (SR) Margalef Index (R_1) and Menhinick Index (R_2) were calculated for different streams. Different evenness indices E_1, E_2, E_3, E_4, E_5 were also calculated. For calculation of the beta (β) diversity Whittaker's measure, βw was calculated. The gamma (γ) overall diversity of all the streams can be calculated as the number of species occurring in the stream. It includes both α and β diversity. The maximum number of species was recorded in the stream SEM (24 taxa) and the minimum in stream EGF (12 taxa). Also, the species diversity was recorded to be maximum in stream SEM and minimum in stream EGF.

1. Introduction

The term biological diversity was first defined by Norse and Mc Manus in 1980 and biodiversity was given by Walter G. Rosen in 1985 (Heywood and Baste 1995). The biodiversity is the totality of genes, species and ecosystems in a region. Species diversity is considered a rough proxy for biodiversity (Ram and Singh, 1999). Whittaker (1960) explained three types of diversity: α, β and γ diversity. According to him, alpha diversity is the diversity within habitat or within community, beta diversity is the intercommunity diversity and gamma diversity is diversity of entire landscape.

Antarctica provides a harsh environment for growth and survival of organisms. The terrestrial and aquatic organisms are encountered in restricted areas, and it has been presumed that they have special physiological features enabling them to survive at low temperatures and prolonged period of darkness (Fogg and Horne, 1970). Literature shows that West and West (1911) and Fristch (1912, 1917) were the initial workers who investigated the algal flora of Victoria Land. Several investigations have also been conducted to study the patterns of distribution and taxonomy of aquatic algae in continental Antarctica (Hirano, 1965, 1979; Parker *et al.*, 1972; Karasawa and Fukushmia, 1977; Priddle and Belcher, 1981, 1982; Broady, 1982, 1989).

The investigations were undertaken by one of the authors (RKG) during XIth (1991-1992) and XIVth (1994-1995) Indian Expeditions to Antarctica. In the present article we have considered different approaches for measuring species diversity of the six freshwater streams of Schirmacher Oasis of Antarctica.

2. Methodology

2.1 Site of Study

The Schimacher Oasis has a width of 3 km and a length of about 20 km, and is oriented in east-west direction (70°46'04"-70°44'21"S; 11° 49'54"-11°26'03" E). During the polar summer, ice melts and water flows in the stream forming lakes. Two types of streams were observed: one derived from glaciers and second from snow. We examined the algae and cyanobacteria from three streams of glacier origin (WN, SEM, ST), which were about 500-1000 m in length, and the three short (ca 200-400 m) snow drift melt water streams (EGF, EM, NS).

2.2 Identification of Alga and Cyanobacteria

Algal samples were scraped, form rocks and sediments into sterile plastic bags and specimen tubes using clean implements. Algal and cyanobacterial samples collected during the expeditions were brought to the field station and examined microscopically within 24h of collection. Identification was made using Bourelly, (1966, 1968, 1970), Geitler (1932) and Desikachary (1959).

2.3 Different Diversity

2.3.1 Alpha diversity

The species diversity within a community or within a habitat is termed as alpha diversity. The alpha diversity of any location is a balance between the actions of local biotic (*e.g.* competition and predation) and abiotic elements of the environment (Whittaker, 1960).

Alpha diversity comprises tow components, species richness and evenness or equatibility. Theses indices can be calculated to the size of individuals in different species.

Many of the indices have been reported which measure both the components of species diversity (species richness and evenness) into a single index. But here we will be considering the most widely used indices *i.e.* Simpson's index (Simpson 1949), Shannon-Wiener index (Shannon and Wiener, 1949).

2.3.2 Simpson's index

Simpson (1949) proposed for the first time a widely used index which varies inversely with species heterogeneity, and in fact measures the concentration of dominance (λ)

$$9 = \sum_{i=1}^{S} pi^2 \qquad \text{or} \qquad 9^3 = \sum_{i=1}^{S} \left\{ \frac{[n_i(n_i - 1)]}{N(N-1)} \right\}$$

For representing diversity (D) Simpson's index is subtracted from its max possible value of 1 *i.e.* 1-λ

where,

λ = concentration of dominance for sample; D= diversity index; ni= number of individuals of species i in the sample; N= total number of individuals of all species in the sample; s = number of species in the sample; pi = proportion of species i in the sample.

2.3.3 Shannon Wiener index

This index of diversity is based on information theory which is a measure of the amount of uncertainity (Margalef, 1972).

Formula used for Shannon-Wiener index is:

$$H' = -\sum_{i=1}^{S} pi \ \log_2 pi$$

where,

H' = Shannon-Wiener diversity; pi= proportion of total sample belonging to ith species; $\log_2 = 3.322 \log_{10}$.

2.3.4 Indices for diversity components

(i) Species Richness

It is the most earlier concept of species diversity which is an indicator of the relative wealth of species in a community (Peet, 1974). It can be represented by the total number of species in a given community or the number of species per unit area.

$SR = S-1/\ln(n)$ (Margalef, 1958)

$SR = S/\sqrt{n}$ (Menhinick, 1964)

$SR = S(n)^{1/2}$ (Peet, 1974)

where,

S = number of species; n= number of individuals; SR = species richness index of the sample

(ii) Evenness or Equitability:

It represents the distribution of individuals among the species. It is also defined as the ratio of observed diversity to maximum diversity (Margalef, 1958). Whittaker (1972) has given an expression for equitability.

$$E = S/\log Ni - \log Ns$$

where,

S = total no. of species; N_i = no. of individual of most important species; Ns= no. of individuals of least important species; E = evenness index.

Two formulae proposed by Peet (1974) for calculating species evenness are:

$$E_1 = D - D_{min}/D_{max} - D_{min}$$

$$E_2 = D/D_{max}$$

where,

D = observed diversity, D_{max} is the value of diversity would assume if all S species were equally abundant and D_{min} is value of diversity would assume if one species was represented by N-(S+1) individuals and the other species by one individual each.

Again for evenness measurement Shannon Wiener function H' has been taken into consideration and E_3 is given as

$$E_3 = H'/H'_{max} = H'/1_n(S) = 1_n(N_1)/1_n(N_0)$$

N_0 and N_1 are Hill diversity numbers and stand for total no. of species (S) and $C^{H'}$ Sheldon (1969) gave an exponential form of E_3 as an index of evenness.

$$E_4 = e^{H'}/S = N_1/N_0$$

Hill (1973) proposed the following for representing equitability

$$E_5 = (1/\lambda)/e^{H'}$$

2.3.5 β-diversity

Inter-community or inter-habitat diversity is termed β-diversity. Routledge (1977) defined β-diversity as the inverse of the mean proportion of communities or habitats occupied by a single species β-diversity is important because it allows the comparison of habitat diversity of two different study systems and provides information about the degree of partitioning of habitats by species.

Whittaker's measure, $\beta w = S/\sqrt{\alpha}-1$

where,

βw = Whittaker's beta diversity; S= total number of species recorded in the study system which may comprise several communities; α = average no. of species found within the community samples.

2.3.6 Gamma diversity

Gamma diversity refers to overall diversity of landscape level or that of the whole region. It can be given by the number of species found in a landscape (Halffter, 1998). It include both α and β diversities (Whittaker, 1972). The following formula can be used to measure γ diversity (Schluter and Ricklefs, 1993).

$$\gamma = \alpha \times \beta \times Q$$

where,

γ = gamma diversity; α = average value of alpha diversity for a landscape unit; β = average value of beta diversity for a landscape unit and Q = total no. of habitats or sample sites.

3. Results

3.1 Occurrence of Algae and Cyanobateria

The algal and cyanobacterial species which have been encountered in different streams of Oasis have been presented in Table 15.1. It is evident that various streams situated at different locations varied in flora composition. The glacier bound stream SEM near Maitri Lake shows maximum abundance of cyanobacteria whereas the number of species encountered was minimum in Hill top snow drift bound stream EGF near German Station (extreme west of Oasis). Among the cyanobacteria more than half of species were nitrogen fixers, and *Oscillatoria limosa, P. frigidum* and *Nostoc commune* were present in all the six streams studied.

Table 15.1: List of Algae and Cyanobacteria Present in the Six Freshwater Streams of Schirmacher Oasis, Antarctica

Algae and Cyanobacteria	*Different Freshwater Streams*					
	WM	*SEM*	*ST*	*EGF*	*EM*	*NS*
Chroococcus minutus	–	+	+	–	–	–
Gloeocapsa kuetzingiana	+	+	+	–	+	–
G. ralfsiana	+	+	+	–	+	+
G.magma	+	+	–	+	–	+
Aphanothece nidulans	–	–	+	–	–	+
Chaemasiphon subglobous	–	+	–	+	+	–
Lyngbya aeusturaii	–	+	–	–	+	+
L. attenuata	+	–	+	–	–	+
Oscillatoria limosa	+	+	+	+	+	+
O. kuetlitzii	+	+	+	–	+	–
O. limnetica	+	+	–	+	–	–
O. agardhii	–	–	+	–	+	+
Phormidium fragile	–	+	+	–	+	+
P.frigidum	+	+	+	+	+	+
P. autumnale	+	–	–	+	–	+
Plectonema sp	–	+	+	–	–	–
Nostoc commune	+	+	+	+	+	+
N.sphaericum	–	+	–	–	+	+
N. punctiforme	+	–	+	+	–	–
Nostoc sp	+	+	+	–	–	+
Anabaena sp.	–	+	+	–	+	–
Calothrix gracilis	–	+	+	+	–	+
C. parietina	+	+	–	+	+	+
C. brevissima	–	+	+	–	–	–
Tolypothrix conglutinata	–	–	+	–	+	+
Uronema sp	+	+	+	+	–	+
Cosmareum laeve	–	+	+	+	+	+
Pinnularia borealis	+	+	+	–	–	–
Navicula multicopis	–	+	+	–	+	–
Nitzschia obtusa	–	+	+	–	–	+

+: Present; –: Absent.

3.2 Diversity Indices

On the basis of relative frequency and density of species in six freshwater streams, the diversity indices was calculated to determine the richness, diversity and evenness (Table 15.2). The value of Menhinck index (R2) of richness indicated that species richness of algae and cyanobacteria was highest in stream SEM followed by stream NS. The lowest species richness was recorded in stream EGF. Although the values of Margalef's index of richness (R1) are higher than R2, it also followed the same trend. The species richness in different streams was in the order of SEM>NS>ST>EM>WN>EGF> The diversity indices revealed that maximum diversity exists in stream SEM and minimum in EGF. The species diversity (NI) decreases in order of SEM>NS>ST>EM>WN>EGF whereas value of λ increases with increase in diversity index (N1). The other diversity indices (H' and N1) also followed the same trend and not much difference was recorded in the indices of ST and NS, EM and WN stream. The evenness indices E4 (Hill ratio) and E5 (Modified Hill ratio) varies between 0.9 to 0.98 the higher value seems to be related to the dominance of several species.

Table 15.2: Various Diversity Indices of Algae and Cyanobacteria

Diversity Indices		*Freshwater Streams*					
		WM	*SEM*	*ST*	*EGF*	*EM*	*NS*
Unified indices	*Simpson's index (λ)*	0.09	0.5	0.07	0.11	0.09	0.06
	Shannon Wiener index (H')	2.50	-2.98	-2.83	2.30	-2.50	-2.78
Species richness	*Menhinck index (R_1)*	3.05	5.00	4.42	2.39	3.13	3.91
	Marglef index (R_2)	1.51	2.41	2.02	1.20	1.60	1.90
Evenness indices	*Species Eveness (E_1)*	0.92	0.94	0.92	0.93	0.92	0.95
	Species Eveness (E_2)	0.81	0.82	0.77	0.83	0.81	0.85
	Species Eveness (E_2)	0.80	0.81	0.76	0.82	0.80	0.84
	Hill Ratio (E_4)	0.93	0.97	0.90	0.93	0.94	0.98
	Modified Hill Ratio (E_5)	0.93	0.97	0.90	0.92	0.93	0.98
Beta diversity	*β-diversity (βw)*	29.0	29.188	29.158	29.0	29.188	29.158
Gamma diversity	*Ψ-diversity*	435	518.5	496.88	400.2	437.82	486.355

There is not much difference in the values of β diversity of the streams studied and these ranged from 29 to 29.188 whereas the Y–diversity is highest in stream SEM and lowest in WN. Evidently, it allows the comparison of stream diversity and gives an idea about how and which stream is different with the other stream as in the present study we have recorded that minimum Y diversity is found in stream EGF and maximum in SEM. Thus we can say that both are much different in comparison to other streams.

4. Acknowlegements

The authors thank the Head, Department of Botany, Banaras Hindu University, Varanasi, Secretary, Department of Ocean Development (Government of India, New Delhi), UGC, New Delhi for financial help and the Principal, Government Post-graduate College, Rishikesh (Dehradun) for their help.

5. References

Bourrelly P 1966 Les Algues d'eau douce Intitiation a la systematique 1. *Les Algues et vertes. Boubee and Cie, Paris.*

Bourrelly, P. 1968 Les Algues d'eau douce Intiation a la systematique 2. *Les Algues et vertes. Boubee and Cie, Paris.*

Bourrelly P 1970 Les Algues d'eau douce Intiation a la systematique 3. *Les Algues et vertes. Boubee and Cie, Paris.*

Broady P A 1869 Broad scale patterns of distribution of aquatic and terrestrial vegetation at three ice free regions of Ross island, Antarctica. *Hydrobiologia* 172: 77-95.

Broady P A 1882 Taxonomy and ecology of algae in a freshwater stream in Taylor Valley. Victoria land, Antarctica. *Arch Hydrobiol Suppl* 63:331-349.

Desikachary T V 1959 Cyanophyta. *Indian Council of Agricultural Research,* New Delhi.

Fogg C H and Horne A J 1970 The physiology of Antarctic freshwater algae. pp638-732 In: R.W. Holdgate (Ed). *Antarctic Ecology.* Academic Press New York.

Fritsch F E 1912 Freshwater algae of Antarctic expedition (Under Caption Scott) 1902-1904. National Antarctic Expedition. *Nat History* 6: 1-61.

Fritsch F E 1912 Freshwater algae. British Antarctic (Terra Nova) Expedition 1910-1913. *Nat Hist Rep Bot* 1: 1-16.

Geitler L 1932 Cyanophyceae. Kademische Verlagsgesellschaft *m.b.h. Leipzing*.

Halffter G 1998 A Strategy for measuring landscape biodiversity *Bio Intl* 36: 5-17.

Heywood V H and Baste I 1995 Introduction, In: V.H. Heywood and R.T. Watson (eds) *Global Biodiversity Assessment*. University Press, Cambridge, U.K.

Hill M O 1973 Diversity and evenness: A unifying notation and its consequences. *Ecology* 54: 427-432.

Hirano M. 1965 Freshwater algae in the Antarctic regions. pp127-193 *In*: P. Van Meighem, P.Van Dye and J. Schell (eds). *Biogeography and Ecology in Antarctica*. The Hague.

Hirano M 1979 Freshwater algae from Yukidori Zava near Syoya station. Antarctica. *Mem Natl Inst Polar Res Special Issue* 11: 1-26.

Karasawa S and Fukushima H 1977 Diatom flora and environmental factors in some fresh water ponds of East Ongul Island. *Antarctica Res* 59: 46-53.

Margalef D R 1958 Information theory in ecology. *Gen Syst* 3: 36-71.

Margalef D R 1972 Homage to Evelyn Hutchinson, or why is there an upper limit to diversity. *Trans Connect Acad Arts Sci* 44: 211-235.

Menhinick E F 1964 A comparison of some species individual diversity indices applied to sample of field insects. *Ecology*. 45: 859-861.

Parker B C, Samsell G L and Prescot, G W 1972 Freshwater algae of the Antarctic Peninsula. I. Systematics and Ecology in the U.S. Palmer Station area. pp69-81 *In* G. Lian (Ed) Antarctic Terrestrial Biology. Antarctic Research Series, American Geophysical Union, Washington DC 20.

Peet R K 1974 The measurement of species diversity. *Ann. Rev. Ecol. Syst*. 285-307.

Priddle J and Belcher J H 1981 Freshwater biology at Rothra Point, Adelaide Island: II. Algae *Brit Ant Surv Bull* 55: 1-9.

Priddle J and Belcher J H 1982 An annonated list of benthic algae (excluding diatoms) from fresh water lakes on Signy Island. *Bri Ant Bull* 57: 41-53.

Routledge R D 1977 On Whittaker's components of diversity. *Ecology* 58: 1120-1127.

Sagar R and J S Singh, 1999 Species diversity and its measurement. *Botanica* 49: 9-61.

Schluter, D and R.E. Ricklefs 1993 Species diversity: an introduction to the problem. *In* R.E. Ricklefs and D. Schluter (eds) *Species Diversity in Ecological Community* (eds.) University of Chicago Press, Chicago.

Shannon C E and Weaver W 1949 *The Mathematical Theory of Communication*, University of Illinois Press, Urbana., Illinois.

Sheldon A L 1969 Equitability indices: Dependence on the species count. *Ecology* 50: 466-467.

Simpson E H 1949 Measurement of diversity. *Nature* 163-668.

West W and West G S 1911 Freshwater algae. British Antarctic Expedition, 1907-1909. Reports on Scientific investigations. 1. Part 7: 263:298.

Whittakar R H 1960 Vegetation of the Siskiyou mountains, Oregon and California. *Ecol Monogr* 30: 279-338.

Whittakar R H 1972 Evolution and measurement of species diversity. *Taxon* 21: 213-251.

Chapter 16

Azolla: A Potential Cyanobacterial Symbiont

S. Saxena and B.V Singh

Contents

ABSTRACT

Blue-green algae (Cyanobacteria) are heterogeneous assemblage of organisms, ranging from unicellular to branched

Glimpses of Cyanobacteria, 2006 242–270
Daya Publishing House, New Delhi

filamentous forms with prokaryotic cellular organization. Although they are predominantly hydrophytes but forms inha iting other habitats are not rare. Soil inhabiting forms are known as edaphic algae and those colonizing the moist soil surface are referred to as saprophytes. The latter play an important role in soil conditioning and vitality of the soil. Cyanobacteria are photosynthetic oxygen evolving and many of which can fix atmospheric nitrogen and hence help to maintain the nitrogenous fertility of the soil (Dey 1939). The ability of Blue-green algae to with stand adverse ecological conditions, their capacity to thrive well in hostile environment and to respond to the onset of dry conditions by entering into the dormant resistant state, have distinguished this group as pioneers of plant succession. Both in xeroseres and hydroseres, these appear as first plant community leading to the establishment of higher plant associations. *Anabeana azollae*, a cyanobacteria in association with *Azolla* can fix atmospheric nitrogen and serves as a potential and promising biofertilizer for rice crop. This symbiotic relationship is of interest to agriculturist since incorporation and decomposition of *Azolla* in rice field can result in an increase in available nitrogen to the companion or succeeding rice crop. Application of the symbiont improves the fertility of soil and is reported to have a residual effect on the yield of other crops succeeding rice. This symbiont of Cyanobacteria and *Azolla* is gaining importance and popularity day by day to meet the demand of nitrogen and other organic matter for the rice crop and can be exploited to produce organic rice along with other organic inputs.

The present tough task and challenge before us is to increase the crop production with minimal use of harmful chemicals. Also, there is an increasing demand of organically-grown foods among the consumers worldwide and India can harness substantial benefits from it. In this context, the use of organic manures, biofertilizers and biopesticides is gaining more and more importance. Water fern *Azolla* fixes atmospheric nitrogen in symbiotic association with the Cyanobacterium, *Anabaena Azolla*e and serves as a promising biofertilizer for rice. The positive effects of its application on growth and yield of rice are reported from India and abroad. Application of *Azolla* improves the fertility of soil and is reported to have a residual effect on the yield of other crops succeeding rice. It is also known to suppress weed growth, reduce the loss of applied chemical nitrogen fertilizer, can be used as an animal feed, human food, a medicine and a water purifier.

1. Introduction

As the world witnesses a boom in the human population, the demand to supply future food and fibre needs will be one of the greatest challenges before the agricultural community. To meet this challenge, a greet deal of efforts focusing on the soil biological system and agro ecosystem as a whole is needed to understand the complex processes and interactions governing the stability of agricultural land. Over the past half a century, agricultural production has been raised through the abundant use of inorganic fertilizers, adoption of multi cropping system and liberal application of chemical pesticides. But it is becoming clear that this agricultural production system is responsible for degradation of environment. For sustainable development a clear priority has to be given to productive agricultural practices. A healthy plant soil system is the backbone of sustainable farming without which sustainable development can never reach its maximum.

Organic farming is a production system, which favours maximum use of organic material like organic manures and discourages use of synthetically produced agro inputs for maintaining soil fertility and pest management. If we are aiming at cultivating the land and raising crops in such a way as to keep the soil alive and in good health than at the outset we need to.

Blue-green algae (BGA) and *Azolla* are the biofertilizers whose role in improving plant nutrients is very important. Algalization increases the fertilizer use efficiency of the crop plants, reduces the loss of fertilizer nitrogen and provides the benefits to the crops in a recurring manner.

Productivity of the traditional rice crop in India is poor due to adoption of this crop since centuries. Continuous mining of the nutrients by rice crop from a specific area, this crop faces serious problems in crops soil management. One of the way to overcome some of the soil fertility problems in the crop is to introduce BGA and *Azolla*, which can play an important role in improving the soil status through increasing the N fixation and organic carbon in the soil. The usar soils are one of the major problem in our country which are not suitable for agricultural practices. Cyanobacteria can check the salinity of such soil. Such fields can be inoculated with Cyanobacterial inoculum artificially. These algae while growing consume sodium and other substances and release organic matter in the form of mucilagenous secretions. Following death, they provide lot of organic matter, which brings about a change in sodium and exchangeable cation levels in the soil making them more fertile. Cyanobacterial have been reported to bring about significant changes in soil aggregation properties and also increase in dehydrogenase, urease and phosphates activities.

Cyanobacteria while growing in flooded rice soil, affects the availability of certain trace elements like Mn and Fe. The Cyanobacterial sheath also plays a very important role in regulating the availability of essential elements as well as in protecting the organism from toxic effect of metal cations. Certain chelating substances present in sheath are known to enhance the availability of Fe.

Cyanobacterial contribution to the total nitrogen fixed in paddy fields varies widely and mainly depends on physico-chemical properties of soil and many other climatic and biotic factors. A conservative estimate indicates that these BGA contribute 25-30 kg pure nitrogen per hectare/season to the rice crop and there is a progressive increase in the grain yields. Nitrogen fixed by BGA may become available to rice plants only after its release extracellularly into the surrounding either as extra cellular products or by mineralization of their intracellular contents through microbial decomposition after their death. Release of nitrogen from rapid decomposition of fresh or dry algal mass incorporates into soil has also been reported. The gradual build up of Cyanobacterial population results in soil fertility with a residual effect of succeeding crop. There is a vast scope of using water fern *Azolla* in Indian Agriculture. The biofertilizer technology is the most suited for irrigated rice areas having proper water management, which presently constitute more than 20 million hectares. There are numerous reports from India and abroad concerning the beneficial effects of *Azolla* on the growth and yield of rice.

Rice is the most important food grain crop in the world and is grown in about 40.71 million hectares in India (Mandal, 1984). The production of rice has increased considerably in recent years due to the release of improved varieties and application of better fertilization practices. Nitrogen has been found as the most limiting nutrient for increasing crop production but the production and use of inorganic N fertilizer involves not only costly investments in terms of energy and transport but also the need for complex manufacturing units/plants, which finally may lead to the environmental pollution.

The scientists are in search of suitable alternatives, which can serve as N source in rice cultivation. The use of agricultural systems/practices that include dinitrogen-fixing organisms appears to be an economically sound agricultural practice. *Azolla* is being used in rice cultivation since last several years. *Azolla* is a genus of small aquatic fern that is native to Asia, Africa and America. *Azolla* species live in lakes, swamps and streams and other small bodies of water and have been dispersed by man and natural means to various parts of the world. Some are strictly tropical or subtropical in nature while others grow and thrive in either temperate or tropical climates.

The *Azolla*-Anabaena association (symbiosis) has the unique property of being able to fix atmospheric nitrogen and retain a significant amount of nitrogenase activity in the presence of combined nitrogen, thus, making the system (association of *Azolla-Anabaena*) compatible enough with inorganic nitrogen fertilization.

The importance of *Azolla* especially as a biofertilizer for rice is well documented (Singh, 2000; Watanabe *et al.*, 1977) but its widespread use by rice farmers, is more or less confined to Vietnam and China where it has been used traditionally for centuries in rice cultivation. Even in these countries the production and utilization of *Azolla* as an integral part of organized rice production is a recent development. In many Asian countries, the use of *Azolla* in rice cultivation is still at a stage of laboratory, field experiments and trials.

2. *Azolla:* An Introduction

Azolla is small elegent heterosporous water fern, a genus belonging to monotypic family *Azolla*ceae, order Salvinales, division filicopsida of Pteridophyta group. It fixes atmospheric nitrogen in association with nitrogen fixing heterocystous Cyanobacterium *Anabaena azollae*, present in its dorsal leaf cavities; thus, these are physically separated from the environment (Moore, 1969; Peters and Mayne, 1974). This aquatic fern has potential as green manure for lowland rice due to its rapid multiplication, quick decomposition and wide adaptability (Kannaiyan *et al.*, 1982; Lumpkin and Plucknett 1980). Besides nitrogen addition to the rice crop, *Azolla* is used as a supplemental animal feed (Subhudhi and Singh 1976) and is also known to suppress weed populations in wetland rice (Moody 1981). There are seven extant and 25 fossil species of *Azolla* (Hills and Gopal 1967). The genus *Azolla* is further categorized into two subgenera *Euazolla and Rhizosperma* (Svenson 1944).

Euazolla include *A. filiculoides* (Lamarck) from Southern South America to Alaska; *A.caroliniana* (wildenow) from Eastern North America, Carribean, Mexico, and West Indies; *A.microphylla* (Kaulfuss) from tropical and subtropical America and West Indies, *A.mexicana* (Presel) from Western North America, Eastward to Illionois, Northern South America to British Columbia and another species *A.rubra* has been also reported, While *A. pinnata* (R brown) from Tropical Africa, Japan, Australia, South East Asia; *A.nilotica* (Decaisne) from upper Nile, Uganda, Congo, Tanzania, Central Africa comprise the subgenus Rhizosperma (Dunham and Fowler, 1987).

3. Cyanobacterial Symbiont

Anabaena azollae, a diazotrophic Cyanobacterium is the partner of agriculturally important symbioses (Konar and Kapoor, 1974). It is known to perpetuate through generations in the mega sporocarps of *Azolla* (Watanabe, 1978a,b; Kannaiyan, 1990) and the growth and development of both the symbiont and host is highly synchronous. Mature leaves have well-developed heterocystous *Anabaena* showing active nitrogen fixation which slows down as the leaves approach senescence. Peters (1976) found that the nitrogenase activity of the intact alga is 6-10 times higher than in free-living state. The biologically fixed nitrogen by the symbiont is absorbed by the host through branched hairs present in the cavity. Unbranched hairs transport fixed carbon from the host to the Cyanobiont (Peters *et al.*, 1978, 1980).

4. Growth Requirements

The growth curve of *Azolla* shows a typical sigmoidal growth curve and under optimum conditions, doubling time of *Azolla* falls between 2-3 days. Watanabe *et al.* (1977) reported that doubling time of *A. pinnata* was 3 days under most favourable conditions. Talley *et al.* (1977) observed a short doubling time in *A. mexicana* and about twice in *A. filiculoides.* However, Becking (1979) reported that doubling time of *Azolla* is 2-10 days for most species. Among the environmental factors, the plant density, water management, pH, availability of nutrients and occurrence of insect, pests and diseases determine the growth rate of different species of *Azolla*.

4.1 Temperature Tolerance

25-30°C is optimum temperature reported for most species of *Azolla* but under field conditions, the species are found to survive well between 15-35°C (Peters *et al.*, 1980). *A.caroliniana, A.microphylla, A.nilotica* and some varieties of *A.pinnata* are relatively tolerant to high temperatures (Tung and Watanabe, 1983; Peters *et al.*, 1980). *A.filiculoides* and *A.rubra* are tolerant to low temperatures but sensitive to high temperatures. The optimum temperature for growth and nitrogen fixation of *A. pinnata* is reported to be 20°/30° C (Moore 1969). *A. filiculoides* is reported to have an optimum temperature range of 22°-25° C although it can withstand temperature as low as– 5° C without any harm but is less tolerant to high temperature than *A. maxicana* (Talley *et al.*, 1977; Watanabe, 1978). *A. pinnata* and *A. caroliniana* are grown successfully in the field throughout the year at Cuttack where temperature during summer sometimes exceeds 40°C (Singh and Singh, 1987). The growth rate of *A. pinnata* is reduced at 38°C temperature in the fields and the fronds become chlorotic (Kannaiyan, 1984). At 20°C temperature the fronds of *Azolla* are reported to become reddish in the centre (Lumpkin, 1977). The damage to *Azolla* due to high temperature is the main problem for culture maintenance under tropical conditions particularly during summer and it can be minimized by draining the field, shading and flushing (Lumpkin, 1987).

4.2 Light Intensity

Light is another important factor and quality, intensity and optimum duration of light significantly affect the growth of *Azolla*. The intensity light required for nitrogenase activity and photosynthesis under laboratory conditions is reported between 5 and 24 K lux but growth under field conditions is maximum under 50K Lux. When *Azolla* forms a mat, significant amount of shading is caused to the individual plant due to over crowding and photosynthesis of the whole mat remains well below optimum even under full sunlight. Nitrogenase activity of the field grown *A.pinnata* is the highest during the month of April when light intensity is also almost maximum (Singh and Singh 1987). The growth limiting effect of light on *Azolla* is more pronounced when it is grown along with rice because rice plants cause varying degree of shading to *Azolla* depending upon canopy structure and growth stage. In planted fields, increased shading due to rice crop reduces growth and nitrogen fixation of *Azolla* (Singh and Singh 1986, 1987).

4.3 Water Management

Azolla requires some amount of standing water for its normal growth being an aquatic fern and a water level that keeps the plants in free floating state is adequate. A minimum water depth of 5-10 cm is recommended for its cultivation (Singh 1989). At lower water depths that allow the roots to touch the surface soil, fragmentation and dispersal of plants is adversely affected but the nutrient deficiencies are corrected.

4.4 Soil Salinity and pH Range

Azolla can be grown successfully in soil having pH between 4.5 and 8, although the optimum pH for its growth is reported between 5-7 (Peters *et al.*, 1980; Singh, 1979). However, soil pH may not cause much problem for the cultivation of *Azolla* because pH shifts towards neutral range after continuous flooding except soils with low organic matter and high sulphate content (De Datta, 1981). It grows poorly in saline soils, which generally have pH above 7 and are deficient in P, Zn and Cu. At about 1.3 per cent NaCl growth of *Azolla* ceases further and the higher concentrations resulted in the death of *Azolla* (Haller *et al.*, 1974). A salt concentration greater than 4000 ppm NaCl reduces *Azolla* growth by 20 per cent (Holst, 1977). Nandabalan and Kannaiyan (1985) found that growth rate of *A. pinnata* was reduced by 31.9 per cent at 0.2 per cent NaCl. However, the fact that *A.filiculoides* has been used to reclaim the saline soils in China (Shang *et al.*, 1987) suggests the possibility of selecting *Azolla* species/strains suitable for different problematic soils. The doubling time and relative growth rates of *Azolla* are influenced by salinity (Kannaiyan *et al.*, 1982; Kannaiyan, 1990).

4.5 Nutrient Management

Amongst the various essential plant nutrients required for normal growth of *Azolla*, phosphorus is considered to be the most important one. Phosphorous is known to play an important role in the growth and multiplication of *Azolla* (Lumpkin and Plucknett, 1980; Kannaiyan, 1981). However, average P content of the soil, which supports fern growth reported, was 25 ppm. Phosphorus level at 20ppm produced more biomass than that at lower level (Kannaiyan, 1981). *A.pinnata* is reported to grow better than *A.microphylla* and *A.mexicana* in soils having low P content (Kushari and Watanabe, 1991). Subudhi Singh and Watanabe (1980) found that the threshold level of P content in *Azolla* appeared to be 0.2 to 0.3 per cent below which the plant growth was proportional to the P content. Fronds of *Azolla* placed in phosphorus deficient solution decreased growth, became red and developed curled roots (Watanabe and Espinase 1977). However, Satapathy and Singh (1980) reported that *A. pinnata* remained green for several months without phosphorous application. The phosphorous deficiency symptoms of *Azolla* decreases, if the water level is low and roots touched the soil (Lumpkin 1977). The importance of phosphorous nutrition was shown to be for nitrogen fixation ability of *Azolla* (Kulasooriya *et al.*, 1980). Potassium deficiency may not cause any problem for *Azolla* cultivation because it is highly efficient in absorption of K from the soil and can rather be used as K fertilizer under such situations and draining of water from fields helps in correcting mineral deficiency (Singh and Singh, 1999).

4.6 Insect Pests and Diseases

The insects from family pyralalidae and chironomidae and snails are reported as major insects in humid tropics and china, which damage *Azolla* (Mochida *et al.*, 1987; Liu and Zheng, 1989). Species of *Chironomus, Pyralis* and *Nymphula* are the important insects of *Azolla* in India (Singh, 1989). *A. pinnata* is highly susceptible to insects and its cultivation during summer is difficult. An application of carbofuran (furadan) @ 2.5-3.0 kg/ha) controls the insects and increases the growth and nitrogen fixing potential of *Azolla* (Singh and Singh 1986). Its effectiveness is further enhanced along with P fertilizer (Singh and Singh 1988). Some products from Neem are also reported to be effective against *Azolla* insects (Reynaud and Franche, 1986). At CRRI (Cuttack) a strain of *A.caroliniana* having good tolerance to the major insects' pests and diseases has been identified (Singh *et al.*, 1991). Fungal infections occur when the culture reached maximum growth (mat is left unharvested for a longer period particularly in hot weather). Common casual organisms are *Rhizoctonia solani fusarivm* sp., *Rhizopus* sp. Common fungicides are able to control this infestation *e.g. bevistin.*

4.7 *Azolla* Application in Rice Fields

The growth of *Azolla* as a dual crop is significantly influenced by density of rice, which can be manipulated by changing the method of planting. Using the double narrow method of rice planting as against the conventional planting produces higher Azolla biomass and rice yield. In double narrow method pair of narrow rows is spaced at 40cm with 7cm gap between 2 narrow rows and hills are maintained 15cm apart. The fern (*Azolla*) is grown in the broader space between 2 pairs of narrow rows. The benefits of this method are particularly higher when *Azolla* inoculation is delayed. *Azolla* can also be successfully grown with randomly planted or direct seeded rice but its biomass is relatively lower than that under line planting. Rectangular planting is better than square planting and the fern growth is higher when rice rows are planted east-west direction.

5. Production of *Azolla* Inoculum

Azolla must not only be protected from death or injury caused by extremes of cold and heat during parts of the year but it must also be propagated and multiplied in sufficient quantities at strategic times of the year for wide scale monocropping in the field or for direct inoculation as an intercrop with the rice. *Azolla* can be maintained in ponds, irrigation channels or slowly moving streams (Lumpkin, 1987a, b). The inoculum density required to ensure good establishment of *Azolla* is high (2-5 t ha^{-1}) to prevent *Azolla* from being over grown by algae or weeds (Watanabe, 1982). The only way to achieve this practically is to multiply the initial inoculums within small areas of fields where it is to be used (Van Hone 1989). A period of two to three weeks' growth is required to accumulate sufficient biomass. In southern China, *Azolla* is most commonly grown in the winter before transplanting of rice whilst in north it is either grown before rice, inter cropped with rice or both (Lumpkin and Plucknett 1982).

The amount of inoculum significantly affects the growth rate of *Azolla*. The rate of inoculation is usually higher for multiplication ponds than for dual cropping. The inoculation of 0.1-0.3 kg/sq.m is recommended for multiplication plots and 0.05-0.1 kgsq.m. (0.5-1.0 t/ha) for dual cropping. It is recommended to inoculate the fern after 1 week of transplanting. In case of *Azolla* raised as a green manure it is inoculated in the main field 15-20 days before transplanting rice.

5.1 Green Manure by *Azolla*

As green manure, *Azolla* is collected directly from ditches/ ponds or nurseries and then applied in the rice fields. For the multiplication and maintenance of *Azolla* the following points should be taken into consideration.

1. The field should be well prepared and levelled properly.
2. Water depth of 2 inches may be maintained
3. Fern *Azolla* is inoculated @ 1 ton/ha this can be reduced or enhanced depending upon the growth conditions.
4. SSP is applied in split doses (25-50 kg/ha). Dose of SSP can be reduced depending on the soil P status. If cattle dung/slurry is used in the field and SSP is applied to the field for rice, there is no need to apply additional SSP for *Azolla*.
5. Furadon @ 2-3 kg/ha can be applied for pest/insect control.

After 2-3 weeks a thick mat is formed which can be incorporated into the soil and then rice can be transplanted. This amounts for 10-20 ton *Azolla* contributing 20-40 kg N/ha.

5.2 *Azolla* as Dual Crop

By this method, *Azolla* is grown along with rice. In dual cropping systems, benefits of *Azolla* can be maximized by proper management of field. The fields are ploughed, leveled and divided into small plots; water level/depth of 5-10 cm is maintained during multiplication by raising bunds.

Fresh *Azolla* is inoculated @ 0.5-1.0 ton/ha to be applied in the field after 7-10 days of transplantation.

Application of basal dose of nitrogen through chemical-N fertilizer or green manure legume is necessary for better production.

The recommended dose of P in inoculum production plots is 4-8 kg P_2O_5/ha at weekly intervals. The highest *Azolla* biomass in dual cropping is obtained at @ 10-15 kg P_2O_5/ha applied in 3 equal splits at 7 days interval.

Incorporated *Azolla* decomposes within 8-10 days and releases 67 per cent of its nitrogen within 20-30 days and can contribute an average of 30 kg N/ha.

Farmers have evolved simple *Azolla* production methods for large-scale multiplication in the fields for easy adoption. Production can be carried out in Nursery plot, Ponds/ditches/canal and in Fabricated tanks of concrete/polythene sheet lined pits or simple earthen ditches

The field selected for multiplication of *Azolla* must be thoroughly prepared and leveled uniformly. The nicely prepared field is divided into 10-25 m^2 plots by raising the bunds to ensure the maintenance of 5-10 cm standing water and facilitate the harvesting of plant material. Fresh *Azolla* is then inoculated at the rate of 50-200 g/m^2 and SSP at 2.5-5.0 g/m^2 is applied every week to ensure rapid multiplication. Furadon at 0.25-0.30 g/m^2 is applied to protect it from insects. After the formation of a thick mat, two third *Azolla* is harvested with the help of bamboo stick and the remaining one third is left for further multiplication. It is possible to undertake production round the year in areas with moderate climate, though the production is poor during summer months. About 1.0-1.5 kg fresh *Azolla*/m^2 can be harvested every week and about 333 tons of fresh biomass can be produced annually from one-hectare area (Singh 1981). *Azolla* can also be produced in bigger plots (100 m^2) and accordingly quantities of all inputs like inoculum (0.5–1 ton/ha), cow dung (1.0–1.5 ton/ha) or cattle slurry (2000-3000 l/ha) or SSP (4-8 kg P_2O_5/ha/week), furadon (2.5-3 kg/ha) can be adjusted. When the production is carried out in ponds/canals, no such addition of fertilizer, insecticide is necessary. Within 2-3 weeks a thick mat is formed and 2/3 rd may be collected and rest is left to grow. 100 kg fresh *Azolla* from 100m^2/week may be obtained.

Azolla nurseries are designed to meet the specific needs of particular environment with regard to both *Azolla* and available construction materials and are thus, difficult to categorize. However, nurseries can be differentiated into two general groups mainly a) Maintenance nurseries and b) Multiplication nurseries.

5.2.1 *Azolla* Maintenance in Nurseries

These are small in size and number and require very intensive, careful management. Special structures such as glasshouse or plastic houses may be required to achieve needed temperature control. Shade, a special water supply and control systems, special plant nutrition measures, intensive pest and disease control and considerable hand labor may all be required to maintain the fern during the extreme conditions such as hot or cold weather. The problem of maintenance is more acute when *Azolla* strains are grown under natural conditions as most of the strains are sensitive to extreme climatic conditions, particularly very high to low temperatures and low relative humidity. Therefore, it is advised to maintain the germplasm in replicates to avoid the risk of losing any valuable material. The N-free IRRI medium is most widely used for growing *Azolla* (Watanabe *et al.*, 1977). The nutrient medium and soil must be changed frequently to ensure that the cultures always remain in active growing conditions. In pots phosphorous can be applied as single super phosphate (SSP) at regular intervals. The pots can be covered with wire net frames to avoid strains getting mixed up or being washed out due to heavy rains. This also protects the plants from strong heat during summer season.

5.2.2 *Azolla* Multiplication in Nurseries

These on the other hand are designed to increase rapidly the supply of *Azolla* inoculum available for agricultural production system (Kong Lin *et al.*, 1964). Because of the large quantity of inoculum required and the difficulty of transporting material in a paddy field/area, most multiplication must take place in or near production fields and must necessarily be managed on a larger scale and with considerably less input per unit area then the maintenance nurseries. It should be stressed that no one should attempt to exploit *Azolla* commercially unless effective procedures have been worked out for A*zolla* production locally.

During winter in higher latitude, subtropical regions or in the temperate zone, it is often necessary to protect *Azolla* from damage due to low temperature or freezing in order to resume a sufficient supply of the fern. The 'winter' period for *Azolla* generally occurs between the time in the autumn when mean temperature reaches or drops below 10°C and approaches or reaches 10°C in the spring and dramatically affects *Azolla*.

5.2.3 Special Structures

Most methods require the construction of some type of protective structures. When night temperature begins to fall below 10°C, *Azolla* is moved in to nursery beds inside these structures and is maintained at a density of 0.75 to 1.0 kg/m^2. Water maintained 6-7 cm deep in the beds is changed once a week. Before the introduction of *Azolla,* the beds may be fertilized with a mixture of 1-2 kg/m^2 of Farm Yard Manure (FYM). *Azolla* inoculum provided must be free from pests. SSP, Ammonium Sulphate and straw ash are applied to accelerate the growth of *Azolla* in nursery. The various structures suggested are:

(i) Arch Shaped Plastic Greenhouse

Long *Azolla* nursery beds are constructed in the field and completely covered with plastic sheets supported by ached bamboo strips. The nursery beds measure 10m long, 1m wide and 13-20 cm deep and beds are placed 1 m apart and positioned in an east and west direction so that the whole bed is exposed to the south to receive maximum winter sunlight. The beds have an arch like frame of 0.5 to 0.6m height made of bamboo strips. The ends of the bamboo strips are struck into the mud on opposite sides of the beds to form the arch. Plastic sheets are stretched over the frame and mud is used to anchor the ends of sheet in the shallow ditch (which was dug around the outside of the frame). Water in the ditches is kept above the lower edges of the sheet so that a seal is formed against cold air and wind. On very cold nights the frames are covered with straw mats to keep the *Azolla* from freezing.

(ii) Plastic or Glass Roofed Greenhouse

Long *Azolla* nursery beds are constructed in the field. Each bed has a mud wall on the north side and a roof composed of plastic sheets or glass window frames. A mud wall about 33cm high is constructed on the north side of the bed. The ends of the bed have sloping mud walls that decline from the top of the north wall to near ground level on the south side. Bamboo strips are positioned across the bed against the top of the north wall Plastic sheets or glass frames are used to form a sloping roof. Rice straw mats are placed over the roof on cold nights.

6. Uses of *Azolla*

Azolla has been reported to be a potential fertilizer due to: (a) High N-fixing ability (b) High biomass due to rapid growth (c) Scavenger of K (d) Solubilise Zn, Fe, Mn (e) Improves soil physico-chemical properties (f) Increases utilization efficiency of urea

(g) *Azolla* mat controls weed growth (h) Can produce compost with a supply of plant available N and (i) Narrows down C:N ratio for better decomposition. In addition *Azolla* can also extract P from eutrophicated water (average P content of soil should be 25 ppm to support growth, clean sewage water, can absorb heavy metals from polluted water, can be a source of biogas, can be used for feeding live stocks, poultry, birds and fish and can be as carrier for microbial cultures (biofertilizers)

6.1 Effect of *Azolla* on Soil Fertility

Azolla application improves the soil fertility by increase in total nitrogen, organic carbon and available phosphorous in the soil (Singh and Singh 1990). Van Hove (1989) found that *Azolla* improves soil structure when incorporated because of its high productivity. The N from *Azolla* is released slowly in comparison to fertilizer N and its availability to the rice crop is about 70 per cent of ammonium sulphate N. However, *Azolla* releases its N faster than BGA (Saha *et al.*, 1982). The total biomass, N content and acetylene reduction activity of *Azolla* provide the estimate of its potential for agricultural use. The environmental condition, nutrient availability, rate and time of inoculation greatly influence the growth of *Azolla* (Kannaiyan, 1993; Singh, 1988, 89; Singh and Singh, 1995).

Under ideal optimal growth conditions it has a potential of fixing more than 10 kg N ha^{-1} day^{-1} (Shi and Hall, 1988). At CRRI (Cuttack) *A. pinnata* fixed 75 mg Ng^{-1}dry wt day^{-1} and produced a biomass of 347 t fresh wt ha^{-1} in a year and it contained 868 kg N as much as in the 1900 kg urea (Singh, 1988). About 30-100 kg N ha^{-1} was shown to be fixed in a month. One crop of *Azolla* provides 20-40 kg N ha^{-1} to the rice in about 20-25 days.

Azolla green manuring produces 10-20 t/ha of green matter containing 20-40 kg N, whereas it's dual cropping provides 20-30 kg N/ha. The supply of N can be further increased by raising more than one crop of *Azolla*, first as a green manure and then as a dual crop alongwith rice (Watanabe, 1987). *Azolla* provides N to rice after its decomposition just like any other organic matter. Incorporated *Azolla* takes about 8-10 days to decompose and releases about 67 per cent of its N with in 35 days (Singh *et al.*, 1981). *Azolla nilotica* showed a significant positive effect on total nitrogen over time in rice crop (Wagner, 1997). The fern is also known to provide potassium upon decomposition to rice (Van Hove, 1989). Ram *et al.* (1994) found that the incorporation of 6,12,18 and 24 t ha^{-1} of fresh *Azolla* into the soil significantly increased its water holding capacity, organic carbon, ammonium nitrogen, nitrate nitrogen, available phosphorous, potassium, calcium and magnesium, while it decreased pH and bulk density.

6.2 Influence of *Azolla* on Rice Grain Yield

In multiplication trials conducted under the All India Coordinated Rice Improvement Project, incorporation of 6 tons *Azolla*/ha was more effective than dual cropping (Pillai, 1982). The increase in grain yield due to *Azolla* green manuring ranged from 0.5-2.0 t/ha whereas dual cropping produced an additional yield of 0.5-1.5 t/ha over the control. The rice yield produced by *Azolla* green manuring followed by dual cropping was equivalent to that from 60 kg N/ha fertilizer application. *Azolla* green manuring performs better with short duration rice varieties whereas dual crop plant is more suited for medium or long duration varieties. Grain yield also varies with *Azolla* strains used (Singh *et al.*, 1991; Singh and Singh 1990). For many centuries, *Azolla* has been used to successfully increase rice yield in Vietnam and Southern China (Fogg *et al.*, 1973; Watanabe and Liu, 1992). Many experiments have demonstrated the effectiveness of *Azolla* as a biofertilizer on rice though the extent of benefit varies according to the climate, the method of application, the species of *Azolla* used and many other factors. The use of *Azolla* increased the rice yields by 112 per cent over unfertilized controls when applied as a monocrop during the fallow season, by 23 per cent when applied as intercrop with rice and by 216 per cent when applied both as a monocrop and an intercrop (Peters, 1978). Tung and Shen (1985) found that *Azolla* grown with rice probably suppresses the growth of rice at early stages. However, at maturity although rice grown with *Azolla* did not have greater height or tiller number, straw and grain yield were higher than the controls. Sisworo *et al.* (1990) found that *Azolla* was equally effective as urea at the rate of 30 Kg N per ha at transplanting and at maximum tillering. Kondo *et al.* (1989) obtained a 27 per cent increase in grain yield with application of *Azolla*. In China the application of *Azolla* as a monocrop and as an intercrop resulted in 30.6 per cent increase in grain yield (Lay *et al.*, 1989). However, Wagner (1997) obtained an increase of 19-103 per cent in grain yield with the application of *A. nilotica*.

6.3 The Application of *Azolla–Anabaena* Symbiont on Crops other than Rice

Azolla can be used as green manure for other aquatic crops other than rice are water Bamboo (*Zizania aquatica*), arrowhead (*Sagittarie sagithfolia*) and Taro (*Colocasia esculenta*). In China, *Azolla* is reported to be used as a green manure for taro (*Colocasia esculenta*) (Anonymous 1982). Incorporation of *Azolla mexicana* into the mud and subsequent intercropping with taro in the Cook Island resulted in 54.6 per cent greater yield than the control, while taro intercropped with *Azolla* in slowly flowing water gave yields that were 87.3 per cent greater than control. (Teckle Haimanot, 1995)

Azolla is also beneficial to wheat when applied in a rotating rice-wheat cropping system (Kothe and Mitra, 1990). *Azolla* applied as a monocrop between the rice and wheat and applied as intercrop with rice has a significant beneficial effect on subsequent wheat crop. Mahapatra and Sharma (1989) found that the application of *Azolla* with *Sesbania* had beneficial residual effects on subsequent wheat crops, raising grain yield by 56-69 per cent .

6.4 Utilization of *Azolla* as Fodder

Azolla is used as a food supplement (fresh or dried) for a variety of animals including pigs in China, rabbits, chickens and ducks in Senegal and Ivory Coast and fish in China (Van Hove, 1989). Sculthorpe (1967) reports that *Azolla* is harvested in large quantities from water bodies in parts of tropical Africa, India and Southeast Asia and utilized as fodder for cattle and pigs. Ali and Leeson (1995), experimenting with broilers found that the use of *Azolla* feed resulted in growth and body weight values similar to those resulting from the use of maize-soybean meal.

In a study on lactating cows, Nik-Khah and Motaghi- Talab (1992) found that *Azolla* could be used as a feed ingredient constituting upto 35 per cent with milk yields and fat percentages being maintained at the same level as with conventional meals. There are number of advantages in producing *Azolla* as a fodder crop such as high nutrient content, can be grown on natural water surfaces, can be fed fresh to animals, no grinding chopping or other processing are necessary, can provide a source of available feed during much of the year and helps to balance feed supplies, grows quickly and yield is high. On a fresh wt. basis *Azolla* contains about 1.3 per cent crude protein while on dried wt. basis it contains between 13-24 per cent and more. Protein and fat content of *Azolla* are higher, while cellulose is much lower than any other forage, accounting for its high quality as a feed. Although lower than rice been in lysine and tryptophan, it does contain all essential amino acids (Tran and Dao, 1973). One hectare of *Azolla* can produce 1-2 tons and fresh fodder/ha/day equivalent to10-30 kg protein/day (Buckingham *et al.*, 1978).

Das *et al.* (1994) found that digested *Azolla pinnata* slurry remaining after biogas production was suitable as a fish pond fertilizer, significantly increasing phytoplankton populations in comparison with other digested raw cow dung. *Azolla* is used widely for swines in China where it is considered equal to high quality sweet potato tops (Chao Mung–Liang and Shi Jin–Shou, 1974; Li fang, 1980). In Vietnam, Tran and Dao (1973) consider that *Azolla* can replace about 50 per cent of the rice bran required in some nations. Different views have been expressed about the acceptance of *Azolla* as a fodder for cattle. Tran and Dao (1973) state that it is used in Vietnam as a substitute green feed for cattle and fed at a rate of 7-10 kg of fresh *Azolla*/day. Ducks and chicken can be raised on a diet including fresh *Azolla*. It has long been recognized as a food for wild fowl in the U.S. and domestic ducks in China. Domestic fowls in Vietnam are fed on fresh *Azolla* @ 0.1-0.3 kg/day (Vo and Tran, 1970). In India Subudhi and Singh (1976) concluded that fresh *Azolla* could replace about 20 per cent of commercial feed in the diet of young chickens and to replace their much commercial feed would require about 9 kg of fresh *Azolla* each day for 100 chickens. *Azolla* is preferred forage for many herbivorous fish. The grass carp (*Ctenopharyngodon idella*), Israeli carp (*Cyrinus carpis*) and *Tilapia* prefer *Azolla* as a source of food to most other aquatic needs. (Lasher, 1967; Edward, 1974; Zon, 1974; Duthu and Kilgen, 1975; Varghese *et al.*, 1976). El- Sayed (1992) found that supplementing the diet of both fingerling and adult Nile Tilapia- *Oreochromis niloticus* L with *Azolla pinnata* had a negative effect on growth of the fish. A rice-*Azolla* fish culture system has proven to be quite successful in Fujian, China (Watanabe and Liu 1992). Pits and ditches are dug in the rice field with double narrow row planting of rice. This allows space for fishing without reducing rice yield. Several strains of *Azolla* are inoculated and three or more kinds of fish (herbivorous and omnivorous) are mixed in the proper ratios. A *supplement* of fish feed is sometimes necessary. This system has been shown to result in 27 per cent recovery of *Azolla* nitrogen by fish and 23 per cent recovery by rice, with 35 per cent incorporation into soil and floodwater (39 per cent loss).

6.5 Weed Suppression

A thick light proof *Azolla* can effectively suppress weed development particularly young seedlings (Kannaiyan *et al.*, 1984). However, *Azolla* cannot suppress weeds that have emerged above the mat. It should be kept in mind that a rice paddy inoculated with an *Azolla* intercrop at an excessive rate can prematurely develop a thick mat. A thick mat develops shortly after transplanting will adversely effect the tillering of young rice seedlings. Krock *et al.* (1991) found that an *Azolla* cover significantly reduced the total amount of weeds particularly the predominant weed *Monochoria vaginalis,* though grasses and hedges could not always be controlled. *Azolla* cover reduced the light intensityby about 90 per cent, reducing photosynthesis in the floodwater and thus, reducing oxygen concentration by more than 50 per cent . Besides reducing the light intensity, *Azolla* cover is known to alter the light quality also.

6.6 Ornamental

Some species of *Azolla* are used as ornamental plants in ponds or aquatic gardens. *Azolla caroliniana* is desirable to both horticulturists and plant physiologists because of its brilliant crimson pigment it produces under stress.

6.7 Mosquito control

Azolla mats were thought to prevent mosquitoes from laying eggs and prevent larvae from coming up in the air (Lumpkin and Plucknett, 1980). In a survey of pools, ponds, wells, rice fields and drains, Ansari and Sharma (1991) found that breeding by *Anopheles* was almost completely suppressed in water bodies that were completely covered with *Azolla*. Breeding of Culex was not completely inhibited but reduced. Rajendran and Reuben (1991) found that immature mosquito populations of *Anopheles subpictus* Grassi, *Culex pseudovishnui* and *C. tritaeniorhynchus* were reduced by a 90 per cent cover of *Azolla microphylla*. Laboratory tests showed that *A pinnata* greatly reduced both oviposition and adult emergence of *Culex* and *Anopheles* but not larval survival.

6.8 Human Consumption

Azolla appears to be fit for human consumption. A few researchers have experimented with the preparation of *Azolla* in soup or in *Azolla*- meatballs as food for man. However, such recepies are yet unpublished (van Hove, 1989). In Tanzania *Azolla* has been reported to be used effectively as a traditional cough medicine (Wagner, 1996).

6.9 Other Uses

Another potential yet undeveloped use of *Azolla* is for the production of hydrogen, a non-polluting, high-energy fuel. When *Azolla- Anabaena* is grown in a nitrogen free atmosphere and/or a water medium containing nitrate, nitrogenase enzyme in the symbiont, instead of fixing nitrogen, evolves hydrogen, using water as the source (Peters, 1975, 76; Newton, 1976). Newton (1976) recorded hydrogen production at the rate of 760 nmol H_2 g^{-1} fresh weight $hour^{-1}$. The rates of hydrogen production can be increased by exposure to a micro aerobic environment, a partial vacuum or argon enriched or carbon dioxide enriched environment or by immobilization of the cells of *Anabaena azollae* isolated from the fern (Hall *et al.*, 1995). Park *et al.* (1991) obtained a production rate of 83 ml H_2 g^{-1} Chl day^{-1}.

Anaerobic fermentation of *Azolla* (or a mixture of *Azolla* and rice straw) results in the production of methane gas which can be utilized as fuel. Moreover, the remaining effluent can be used as a fertilizer because it contains all the nutrients originally incorporated in the plant tissues except for a small percentage of nitrogen lost as ammonia (Van Hove, 1989). Das *et al.* (1994) mixed cow dung and *Azolla pinnata* residues and found that the best ratio was 1:0.4, which gave a gas production 1.4 times that of cow dung alone.

Azolla could also be used for purification of polluted waters. Jain *et al.* (1989) found that *A. pinnata* and *Lemna minor* removed the heavy metals like iron and copper from polluted waters if present in low concentration. Saxena (1995) found that a mixed culture of these two in the ratio of 2:1 was able to sufficiently purify highly polluted effluent from a factory to the extent that it could be used for agricultural purposes.

7. Summary

In this age when mankind is threatened by drastic global environmental changes, the use of sustainable and environmentally appropriate practices in agriculture can greatly contribute to the ecological stability of the planet. The aquatic fern *Azolla* has a high growth rate and symbiotic relationship with nitrogen- fixing endophytic cyanobacteria called *Anabaena azollae*. Such a symbiotic relationship gives this species a competitive advantage over other floating hydrophytes in environments such as rice fields, which are relatively low in available nitrogen. The application of *Azolla* as a biofertilizer on agricultural crops, reducing or replacing. The chemical fertilizers can play a significant role in maintaining or improving the state of the global environment. Besides it is known to suppress weed growth, reduces the loss of applied chemical N fertilizer and checks methane flux from flooded rice fields. This also contains adequate amounts of protein and other nutrients and hence, can be used as a supplemental animal feed. *Azolla* biofertilizer production technology is simple, economical, ecofriendly and well suited for the developing countries of the world.

8. References

Ali M A and Leeson S 1995 The nutritive value of some Asian poultry feed ingredients; *Animal Feed Sci Tech* 55: 227-237.

Anonymous 1982 China and the West: Sharing solutions to agricultural problems. *World Farm* 24: 76-78.

Ansari M A and Sharma U P 1991 Role of *Azolla* in controlling mosquito breeding in Ghaziabad district villages (UP); *Indian J. Malariology* 28: 51-54.

Becking J H 1979 Environmental requirements of *Azolla* in tropical rice production; *Nitrogen and Rice* (pp345-375 International Rice Research Institute, Manila, Philippines).

Buckingham K W, Ela S W, Morris J G and Goldman C R 1978 Nutritive value of nitrogenfixing aquatic fern *Azolla filiculoides. J Agric Food Chem* 26: 1230-1234.

Chao Ming-Liang and Shu Jin- Shou 1974 *Azolla* as a fodder and use of pig dung as rice manure. An investigation of growing red *Azolla* in Xin-Tang Nongye Kexue Changshi 1: 19.

Das D Sikdar K and Chatterjee A K 1994 Potential of *Azolla pinnata* as biogas generator and as fish feed. *Indian J Environ Health* 36: 186-191.

De P K 1939 *The role of blue-green algae in nitrogen fixation in rice fields.* Proc. Roy Soc London, Series B 127:121-139.

De, and Datta S K 1981 *Principles and Practices of Rice Production*, John Willey and Sons, New York.

Dunham D G and Fowler K 1987 Taxonomy and species recognition of *Azolla* Lam; *Azolla Utilisation*, 7-16 International Rice Research Institute, Manilla, Phillipines.

Duthu G S and Kilgen R H 1975 Aquarium studies on the selectivity of 16 aquatic plants as food by fingerling hybrids of the cross between *Ctenopharyngodon idella* and *Cyprinus* carpio. *J Fish Biol* 7: 203-208.

Edward D J 1974 Weed preference and growth of young grass carp in New Zealand *J Mar Freshwater Res.* New Zealand 8: 341-350.

El-Sayed A F M 1992 Effects of substituting fish meal with *Azolla pinnata* in practical diets for fingerling and adult Nile Tilepa, *Orechromis nilotichus* (L); *Aquaculture and Fisheries Mgmt*. 23: 167-173.

Fogg G E, Stewart W D P, Fay P and Walsby A E 1973 *The Blue-Green Algae* Academic Press, London.

Hall D O, Markov S A, Watanabe Y and Rao K K 1995 The potential application of Cyanobacterial photosynthesis for clean technologies; *Photosynthetic Research* 46: 159-167.

Haller W J, Sutton D L and Barlowe W C 1974 Effect of salinity on the growth of several aquatic macrophytes; *Ecology* 55: 891-894.

Hills L V and Gopal B 1967 *Azolla primaeva* and its phylogenetic significance; *Canadian J Bot* 45: 1179-1191.

Holst R W 1977 *Studies Concerning the Growth and Nitrogen Metabolism of Azolla Mexican–Anabaena azollae Symbioses*. Ph.D. Thesis, Southern Illionois University of Carbondale.

Jain S K, Vasudevan P and Jha N K 1989 Removal of some heavy metals from polluted water by aquatic plants. Studies on Duckweed and Water Velvet. *Biol Wastes* 28: 115-126.

Kannaiyan S 1981 *Azolla Biofertilizer for Rice*, INSFFER Training Seminar (1-11 International Rice Research Institute, Los Banos, manila.

Kannaiyan S 1984 Studies on *Azolla* biofertilizers for rice production in Tamil Nadu; *Proc. Int. Workshop to assess the potential of Azolla use in Tropical Asia*. NIFTAL Project, Thailand 1-40.

Kannaiyan S 1990 Biotechnology of Biofertilizer for Rice crop, pp 1-225 Tamil Nadu Agricultural University, Coimbatore, India.

Kannaiyan S 1993 Nitrogen contribution by *Azolla* to rice crop; *Proc. Indian National Science Acad*. B 59: 39-314.

Kannaiyan S 1993 Biofertilizers for rice, Tamil Nadu Agricultural University, Coimbatore.

Kannaiyan S and Nandabalan K 1985 Studies on the influence of neem cake on black rot disease incidence and its effect on the growth and nitrogen fixation in *Azolla*. *International Workshop on Azolla Use*. pp26 Fushon, Fujian Academy of Agric. Sciences, Fuijian, China.

Kannaiyan S, Thangaraju M and Oblisami G 1982 Influence of *Azolla* biofertilizer application as green manure and dual cropping on rice. *Nat Sci Acad News Lett*. 5: 60-62.

Kannaiyan, S. Thangaraju, M and Oblisami G 1982 *Azolla*: a potential biofertilizer for rice production; pp119-126 *In Practical application of Azolla for Rice Production* W Silver and E C Schroder (eds), Martimes Nijhoff Dr. W Junk, Pub. Boston 119-126.

Kannaiyan S, Thangaraju M and Oblisami G 1984 Effect of *Azolla* inoculation in checking the weeds under low land rice. *Sci and Cult* 50: 174-176.

Konar R N and Kapoor R K 1972 Anatomical studies on *Azolla pinnata Phytomorphology* 22: 211-223.

Kondo M, Kobayashi M and Takahashi E 1989 Effect of phosphorous on *Azolla* and its utilization in rice culture in Niger. *Plant and Soil* 120: 165-170.

Kong lin-bo, Su Ziming and Chen Yang Chun 1964 The multiplication of *Azolla* seedlings and the management of *Azolla* seedling fields. *Fuzian Norgye* 2: 13-15.

Kolhe S S and Mittra B N 1990 *Azolla* as an organic source of nitrogen in a rice-wheat cropping system; *Trop Agric* (Trinidad) 67: 267-269.

Krock T, Alkamper J and Watanabe I 1991 *Azolla's* contribution to weed control in rice cultivation. *Plant Res Development* 34: 117-125.

Kulasooriya S A, Hirmburegama W K and de Silva Roy 1980. Effect of light, temperature and phosphorous on the growth and nitrogen fixation in *Azolla pinnata* native to Sri Lanka. *Oecol. Plant* 1: 355-365.

Kushari D P and Watanabe I 1991 Differential responses of *Azolla* to phosphorous deficiency. I. Screening methods in quantity controlled condition; *Soil Sci Plant Nutr* 37: 271-282.

Lasher C W 1967 *Tilapia mossambica* as a fish for aquatic weed control. *Prog. Fish Cult* 29: 48-50.

Lay W L, Huang S N and Wang C T 1989 Effects of *Azolla* application on growth and yield of rice. *Bull. Taichung Distr. Agric. Improvem. Sta., No.* 24: 3-12.

Liu Z Z and Zheng W W 1989 *Azolla in China.* Agriculture Publishing House, Beijing.

Lumpkin T A 1977 *Azolla* in kwangtung Province, People's Republic of China; *Int. Rice Res. Inst. News Letter* 2: 18.

Lumpkin T A 1987 *Symbiotic nitrogen Fixation Technology* (Elkan G H Ed., Marcel Dekker Inc. New York) 55.

Lumpkin T A 1987 Collection, Maintenance and Cultivation of *Azolla*: *Symbiotic nitrogen fixation technology;* 55-93 *In* G H Elkan (Ed), Marcel Dekkar Inc., New York).

Lumpkin T A and Plucknett D L 1980 *Azolla*: Botany, physiology and use as a green manure. *Econ. Bot.* 34: 111-153.

Lumpkin T A and Plucknett D L 1982 *Azolla* as a green manure: Use and management in crop production; *Westview Tropical Agriculture Series, No. 5* Westview Press, Colorado.

Mahapatra B S and Sharma G L 1989 Integrated management of Sesbania, *Azolla* and urea nitrogen in lowland rice under a rice-wheat cropping system. *J Agric. Sci.* (Cambridge) 113: 203-206.

Mandal L N 1984. *J Indian Society of Soil Science*, 32: 572.

Mochida O, Yoshiyasu Y and Dimmano D 1987 *Azolla* Utilisation International Rice Research Institute, Mianila, Phillipines.

Moody K 1981 *Weed Fertilizer interactions in rice; pp 35* Res Paper series no. 64 International Rice Research Institute, Mianila, Phillipines.

Moore A W 1969 *Azolla*: biology and agronomic significance.Bot *Rev* 35: 17-35.

Newton J W 1976 Photoproduction of molecular hydrogen by plant-algal symbiotic system. *Science* 191: 559-561.

Nik-Khah A and Motaghi-Talab M 1992 The use of *Azolla* in lactating cows. *Iranian J Agric. Sci.* 23: 47-56.

Park I H, Rao K K and Hall D O 1991 Photoproduction of hydrogen, hydrogen peroxide and ammonia using immobilized Cyanobacteria. *Intl J Hydrogen Energy* 16: 313-318.

Peters G A 1975 The *Azolla-Anabaena azollae* relationship III Studies on metabolic capacities and a further characterization of the symbiont. *Arch Microbiol* 103: 113-122.

Peters G A 1976 Studies on the *Azolla-Anabaena azollae* symbioses. p. 592-610 *Proceedings of the First International Symposium on Nitrogen Fixation* W E Newton and C J Nyman (eds) Vol. 2 Washington State University Press, Pullman.

Peters G A 1978 Blue-green algae and algal associations. *Bioscience* 28: 580-585.

Peters G A and Mayne B C 1974 The *Azolla-Anabaena* Azollae relationship. VI. Morphological aspects of association. *New Phytol* 80: 583-593.

Peters G A, Ray T B, Mayne B C and Toia Jr. R E 1980 *Azolla-Anabaena* association: Morphological and Physiological studies; *Nitrogen Fixation;* p. 293-309 *In* W E Newton and W H OrmeJohnson (eds) Vol II University Park Press, Baltimore.

Peters G A, Toia Jr R E, Evans W R, Crist D K, Mayne B C and Poole R E 1980 Characterisation and comparisons of five N_2 fixing *Azolla Anabaena* associations. I Optimisation of growth conditions for biomass increase and N content in a controlled environment. *Plant Cell and Environ* 3: 261-269.

Pillai K G 1982 Current status of agrotechnology for rainfed rice culture. *Oryza* 19: 125-140.

Rajendran R A and Reuben R 1991 Evaluation of the water fern *Azolla microphylla* for mosquito population management in the rice land Agro-Ecosystem of South India. *Medical and Vet Entomol* 5: 299-310.

Ram H, Krishna Raja P and Naidu MVS 1994 Effect of *Azolla* on soil properties and yield of mungbean (*Vigna radiata* L); *J Indian Soc. Soil Sci.* 42: 385-387.

Reynaud P A and Franche C 1986 *Azolla pinnata* var. Africana from Molecular Biology to use as green manure. Dekar, Senegal 15.

Saha K C, Panigrahi B C and Singh P K 1982 Blue-green algae, *Azolla* additions on nitrogen and phosphorous availability and redox potential of a flooded soil; *Soil Biol. Biochem.* 14: 23-26.

Satapathy K B and Singh P K 1980 Effect of plant spacing on growth of *Azolla pinnata*, weeds and rice crop; *National Symposium on Biological Nitrogen Fixation in relation to crop production* (TNAU, Coimbatore, Abst.37).

Saxena D K 1995 Purification efficiency of *Lemna* and *Azolla* for WIMCO effluent. *Proc. Natl. Acad. Sci.* India. B 65: 61-65.

Sculpthorpe C D 1967 *The Biology of Aquatic Vascular Plants* Edward Arnold, London.

Shang D H, Wu H, Chem Xp and Cu RS 1987 *Azolla utilization* International Rice Research Institute, Manila.

Shi D J and Hall D O 1988 The *Azolla Anabaena* association: Historical perspective, Symbiosis and Energy metabolism. *Bot Rev* 54: 353-386.

Singh P K 1979 Use of *Azolla* in rice production in India; *Nitrogen and Rice* (International Rice Research Institute, Los Banos, Phillipinnes) 407–418.

Singh P K 1981 Comparative efficiency of *Azolla*, Blue-green algae and other organic manures in relation to N and P availability in a flooded rice soil. *Plant and Soil* 62: 35-44.

Singh P K 1988 *Biofertilizers: Potentialities and Problems* (Sen SP and Palit P Eds. Naya Prakashan Publishers, Calcutta) 109.

Singh P K 1989 Use of *Azolla* in Asian Agriculture. *App. Agric. Res.* 4: 149-161.

Singh P K 2000 Biology of *Azolla* and Blue-green algae. *Biofertilizers, Blue Green Algae and Azolla* (Singh P K, Dhar Wattal Dolly, Pabbi Sunil, Prasanna Radha and Arora Anju, NCCUBGA, IARI, New Delhi-110012) 1-23.

Singh R P and Singh P K 1987 Morphology and nitrogen fixing ability of *Azolla nilotica* Decaisne. *Curr. Sci.* 56: 900-903.

Singh R P and Singh P K 1988 Symbiotic algal nitrogenase activity and heterocyst frequency in seven *Azolla* species after phosphorous fertilization; *Hydrobiologia* 169: 131-318.

Singh A L and Singh P K 1990 Intercropping of *Azolla* biofertilizer with rice at different crop geometry; *Trop. Agric.* (Trinidad) 67: 350-354.

Singh D P and Singh P K 1995 Influence of rate and time of *A. caroliniana* inoculation and its growth and nitrogen fixation and yield of rice (*Oryza sativa*); *Indian J. Agri. Sci.* 65: 10-16.

Singh PK, Singh DP, Manna AB, Singh RP and Bisoi RN 1991 *Biological Nitrogen Fixation associated with rice production* (S K Datta and CS Loger Eds. Oxford and IBH Publication, Calcutta-95).

Sisworo E L, Eskew W, Sisworo H, Rasjid H, Kadarusman H, Solahuddin S and Soepardi G 1990 Studies on the availability of *Azolla* nitrogen and urea nitrogen for rice growth using nitrogen-15; *Plant and Soil* 128: 209-220.

Subudhi BPR and Singh P K 1978 Nutritive value of water fern *Azolla* for chicks; *Poult Sci* 57: 378-380.

Subudhi B P R and Watanabe I 1981 *Soil Sci Plant Nutr.* 27: 237-241.

Swenson H K 1944 The new world species of *Azolla. Ann. Fern. J.* 34: 69-85.

Talley S N, Telley B J and Rains D W 1977 Nitrogen fixation by *Azolla* in rice fields; *Genetic Engineering for Nitrogen Fixation* (Alexander Hallaender ed. Plenum Press, New York) 259-281.

Teckle Haimanot E V D 1995 Comparison of *Azolla mexicana* and N and P fertilization on paddy and taro (*Colocasia esculenta*) yield; *Trop. Agric.* (Trinidad) 72: 70-72.

Tran O T and Dao T T 1973 *Azolla* A green manure, Reitrainse studies, Agric. Prob. 4 38: 119-127.

Tung H F and Watanabe I 1983 Differential response of *Azolla–Anabaena* association to high temperature and minus phosphorous treatments; *New Phytol.* 93: 423-431.

Tung H F and Shen R C 1985 Studies of the *Azolla pinnata-Anabaena azollae* symbioses: concurrent growth of *Azolla* with rice. *Aquatic Bot.* 22: 145-152.

Van Hove C 1989 *Azolla* and its multiple uses with emphasis on Africa, *Food and Agriculture Organization*, Rome.

Varghese T J, Devaraj K V and Shantharam B 1976 Relative growth of the grass carp *Ctenopharygodon idella* VI fed Utricularis and a mixture of *Azolla* and *Lemna; J Inland Fish Soc.* India 8: 206-211.

Vominnkha and Tran Quang Thuyet 1970 *Beo. Hoa Dav (Azolla) Nha Xuat Ban Nang Thou, Hanoi* (Vietnam) 174 pp.

Wagner G M 1997 *Azolla*: a review of its biology and utilization. *Bot Rev.* 63: 1-26.

Watanabe I 1977 *Azolla* utilization in rice culture; *Int. Rice Res. Newslett.* 2: 3-5.

Watanabe I 1978a *Azolla* and its use in lowland rice culture. *Tsuchi Beseibatsu* 20: 1-10.

Watanabe I 1978b Biological nitrogen fixation in rice soils; *Soils and Rice* (The International Rice Research Institute, Los Banos, Philippines) 465-478.

Watanabe I 1982 *Azolla- Anabaena* symbioses, its physiology and use in tropical agriculture (Y R Dommergues and H G Diem Eds., Microbiology of Tropical Soils and Plant Productivity, Dr. W Junk Publishers, London.) 169-185.

Watanabe I 1987 Summer report of the *Azolla* programme of the International Network on soil fertility and fertilizer evaluation for rice; *Azolla Utilisation* (International Rice Res. Instt. Manila, Philippines) 197-201.

Watanabe I and Liu C C 1992 Improving nitrogen fixing systems and integrating them into sustainable rice farming; *Plant Soil* 141: 57-67.

Watanabe I, Espinase C R, Beeja N J and Alimango B V 1977 Utilization of the *Azolla–Anabaena* complex as a nitrogen fertilizer for rice; *Int. Rice Res. Instt. Res. Paper Series no.* 11.

Zon J C, Van J 1974 The grass carp in Holland; *Proc. EWCC Fourth Int. Symp. Aquatic Weeds* 128-133.

Chapter 17

Population Dynamics of the Cyanobacterial Flora in the Foothills of Garhwal Himalayas, Uttaranchal

M.K. Chaudhary and M. Kumar

Contents

ABSTRACT

Cyanobacteria are photosynthetic, oxygen evolving prokaryotic microorganisms. The members of this group are

Glimpses of Cyanobacteria, 2006 271–297
Daya Publishing House, New Delhi

cosmopolitan in distribution. They can survive under most diverse ecological conditions. The capacity to tolerate the adverse ecological conditions makes them a group of pioneers of plant succession. In India, Cyanophyceae have 98 genera and about 833 species. The algal flora of Western Uttar Pradesh (including present Uttaranchal state) remained unexplored till 1965. A little work has been made by Kumar (1970), Khan and Kumari (1972), Bendre and Kumar (1975). The present study was a maiden attempt to explore the Cyanobacterial flora of this region. The work is related to the foothills of Garhwal Himalayas involving five research sites namely Dakpatthar, Haridwar, Rishikesh, Laldhang and Kotdwar. This region is supported by excellent habitats, which offer a rich growth of algal species. Population dynamics of the Cyanobacterial flora in the foothills of Garhwal Himalayas has been worked out during the present studies.

207 taxa (46 heterocystous and 161 non-heterocystous) belonging to 47 genera of 14 families and 5 orders were recorded from this region. Members of Nostocales and particularly family Oscillatoriaceae dominated the area. Majority of species are inhabited as lithophytics. The study of Cyanobacterial population dynamics revealed that this region inhabits 17.2 per cent taxa 'dominant', 20.4 per cent 'very common', 33.1 per cent 'common', 22.2 per cent 'rare' and 6.8 per cent 'very rare'. *Microcystis aeruginosa* v. *major* v. nov., *Eucapsis minuta*, *Crinalium magnum* etc. are very rare in occurrence.

1. Introduction

Cyanobacteria is most primitive group of microorganisms. It constitutes a heterogenous assemblage of organisms ranging from unicellular to branched filamentous forms. Various microbial crusts dominated by Cyanobacteria are found throughout the world on diverse substratum under various climatic conditions (Adhikary, 2000). The members of this group are heterocystous and non-heterocystous, and many of which can fix atmospheric nitrogen and hence, help to maintain the nitrogenous fertility of the soil (De, 1939). Cyanobacteria have a worldwide distribution and the majority of species are cosmopolitan, and often abundant in freshwater. Although, they are predominantly hydrophytes, but forms inhabiting other habitats are not rare. Water is the integral constituent of life and one of the most important natural resources for living beings. Biologically, water makes a highly dynamic ecosystem with its dissolved nutrients and gases, chiefly oxygen. It is postulated that the plant life on our planet started in aquatic environment in the form of unicellular phytoplanktons. During the process of evolution, multicellular structures were developed. Many of these continued to grow in water, while others turned amphibians and terrestrials. Thus, the present domain of life existing on the earth has evolved from water.

In Cyanobacteria, specific species and community composition belong to defined type of water bodies and directly dependent on the biological/chemical nature of water. The Cyanobacteria, which were probably predominant among the aboriginal organisms on the earth, are still quantitatively and qualitatively important in the biosphere. Study of these remarkable organisms is not only of particular intrinsic interest for biology but is also for economic importance. Cyanobacteria possess the ability to synthesize chlorophyll-a. These are also characterized by their ability to form phycobilin, phycocyanin and phycoerythrin pigments. High concentration of these pigments lead to the bluish colour of the organisms as such, these are known as Cyanobacteria or blue-green algae. A more recently discovered group of oxygenic photosynthetic prokaryotes is placed in the order Prochlorales of bacteria. These bacteria share many features with Cyanobacteria but they contain chlorophyll-b as well as chlorophyll-a, lack phycobilin pigments and differ in some other features (Post, 1999; Tomitani *et al.*, 1999).

Blue-green algae are simple aquatic plants that occur naturally in diverse habitats such as rivers, lakes, damp soils, tree trunks, hot springs and snow. They can vary considerably in shape, colour and size, and can be seen under the microscope. Some accumulations may be so large that they are easily seen with the naked eyes. Cyanobacteria are also known as Cyanophytes, blue-green algae and most recently Cyanoprokaryotes. They have a similar external appearance to algae and their requirements for light, nutrients and carbon dioxide are also similar. Some members of blue-green algae can survive in the absence of light and grow using chemicals from its surroundings (Fabbro and Mc Gregor, 2003). Cyanobacteria can survive prolonged periods of drought. A successful revival has been possible even after many years of preservation in dried form. Many forms are able to survive very high temperatures and remain dormant in the soil. The ability of Cyanobacteria to withstand under adverse ecological conditions, capacity to thrive well in hostile environments and entering into a dormant resistant state, have distinguished this group as pioneers of plant succession in both xeroseres and hydroseres, these appear as the first plant community leading to the establishment of higher plant associations.

The process of nitrogen fixation and the occurrence of gas-vesicles are especially important to the success of nuisance species of blue-greens. They are widely distributed over land and water, often in environments where no other vegetation can exist. They were probably, the chief primary producers of organic matter and the first organisms to release elemental oxygen 'O_2' into the primitive atmosphere, which was until then free from O_2. The product of photosynthesis is glycogen. Their fossils have been identified as over three billion years old.

Some members of this group belonging to families Oscillatoriaceae and Nostocaceae have a capacity to change their colour in relation to the wave length (colour) of the incident light. This change in colour is often described as complementary chromatic adaptation or 'Gaidukov phenomenon' *e.g. Oscillatoria sancta, Phormidium tenue* etc. They develop different colours in different wave lengths of light as per their adaptation (Gaidukov, 1902, 1906, 1923). The trichomes of few genera of Cyanophyceae exhibit various types of movements *e.g. Oscillatoria, Trichodesmium, Spirulina, Arthrospira, Microcoleus, Anabaena* etc. Several species of Cyanobacteria grow in abundant quantities so as to colour the entire body of water and such growths are called 'water blooms'. These water blooms are reported to produce toxins and have caused the dangerous hazards in both humans and animals. Blue-green algae have the ability to tolerate a very high temperature upto 85°C (Elenkin, 1914). Kirtikar (1886) was the first to record a thermal alga from India (*Conferva thermalis* f. *Birdwoodi*). Cyanobacteria are also tolerant of low temperature *e.g.* species of *Gloeocapsa* and *Nostoc* are most abundant in fresh-water areas. A number of liverwort, hornwort, moss, fern, cycad and angiosperm species are closely associated with nitrogen fixing Cyanobacteria-typically species of *Nostoc* (Dodds *et al.*, 1995).

The classical taxonomy of the Cyanobacteria is based on the morphological features and their nomenclature is ruled by the Botanical Code, *i.e.* basically two groups–simple filamentous

(Oscillatoriaceae) and heterocystous (Nostocaceae and Stigonemataceae) are seen. The problem of the morphological variability is very common. Drouet (1986) revised the taxonomy profoundly, due to existence of ecophenes *i.e.* organisms sharing the same genotype but expressing distinct morphologies under the influence of environmental factors.

Cyanophyceae have 160 genera and 1500 species in the world. So far 98 genera and about 833 species have been reported from India (Sarma and Khan, 1980). In 1849, Montagne described the first blue-green alga from India, *Calothrix indica* from Assam while, Bharadwaja (1928) was the first Indian to report a new Cyanobacterial species, *Spelaeopogon kashyapi* n. sp. of the family Scytonemataceae. The algal flora of Western Uttar Pradesh (including present Uttaranchal state) remained unexplored till 1965. A little work has been made by Kumar (1970), Khan and Kumari (1972), Bendre and Kumar (1975), Pal (1975) and Kumar and Kumar (2000). Biodiversity of Cyanobacteria of the foothills of Garhwal Himalayan range has been explored well by Chaudhary (2004).

The cosmopolitan distribution of Cyanophyceae is evidence of their ability to withstand and grow in various habitats under a wide variety of environmental factors. Amongst various factors affecting the growth and development of Cyanophyceae *viz.* meteorological and pedological. They directly influence the distribution and abundance of Cyanophyceae in the biosphere. A combination of meteorological and pedological parameters including substrate, light, hydrogen-ion concentration, dessication, temperature etc. have enabled to recognize the population dynamics at different localities during different seasons of the year. Thus, it has been possible to decipher the patterns of dominance with the climatic changes in the foothills of Garhwal Himalayas.

2. Description of Sites

Himalayas represent a remarkable chain of mountains across several states and displaying natural resources. The Shiwalik Himalaya has Tarai and Bhabar regions. The Bhabar belt includes the foot-part of the Shiwalik Hills and usually covers 20-45 km in width, in an undulating topography. Several streams originate from the tract, flowing towards Tarai and adjacent plains. The foothill regions situated in Southwest to East of Garhwal Himalayas (between 29º20′N–30º47′N to 77°70′E-78º80′E) of Uttaranchal (India) have been studied. For the study of population dynamics, the area was divided into five research sites *i.e.* Dakpatthar, Rishikesh, Haridwar, Laldhang and Kotdwar. These are covered under three districts namely, Dehradun, Haridwar and Pauri-Garhwal (Figure 17.1). All sites exhibit a Sub-tropical climate and their elevation ranges between 300-900m aMSL. The whole area lies in between Kumaun and Himachal Himalayas, and is considered as a part of North-West Himalayas by almost all the scholars.

The foot part of Garhwal hills along with the border of UP and UA is situated in the Northern part of the vast and beautiful region of India. This region sustains unique and rich vegetation in various

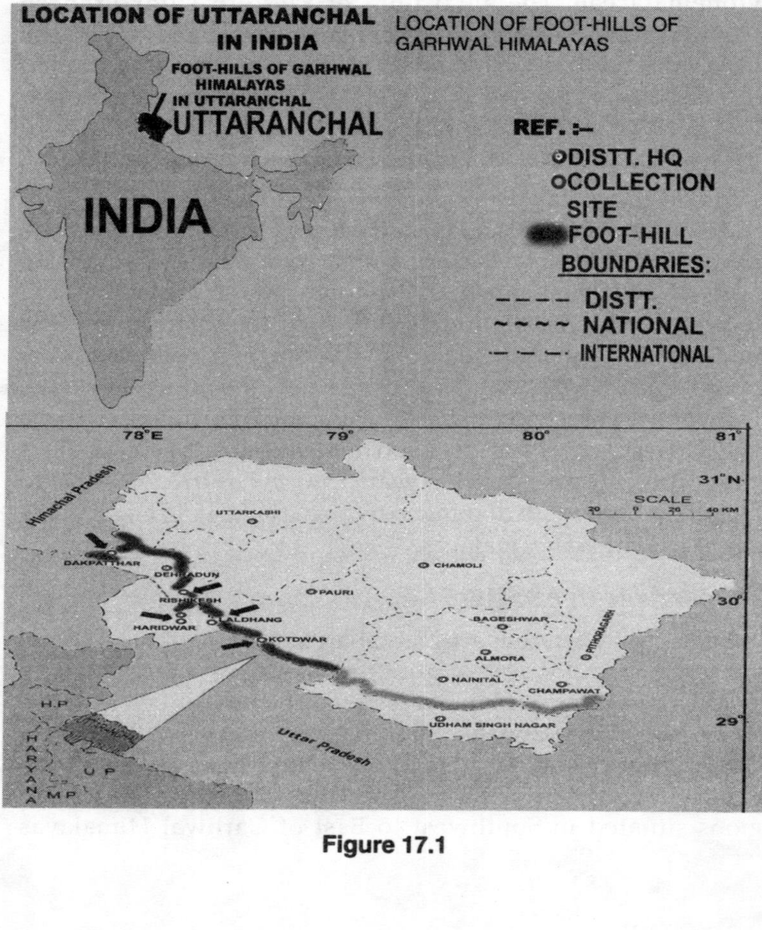

Figure 17.1

habitats ranging from the tracts of Tarai-Bhabar to the high ranges of Lesser Himalayas. It comprises a group of low-lying hills with rivers, rivulets, springs, streams, seepages, ravines, pits, ponds, pools, swamps etc. and stagnant water bodies of different orders- in open water (pelagic) and at the bottom (benthic zones). These excellent habitats offer the growth of various types of vegetation specially the algal flora. This region supports a lot of biodiversity specially in Cyanobacteria because the favourable tropical climate with mineral rich acidic alluvial soils well-irrigated with holy Ganga and Yamuna rivers afford a conclusive environmental condition for the luxurious growth of algae. Now a days, due to increasing population and industrial growth, the environmental pollution led to ecological changes in the Sub-Himalayan range. Currently, these ecological changes respond quickly at micro levels in the environment and inviting rich growth of diverse group of Cyanobacteria. The different studies revealed that this region unfortunately has not received due attention of phycologists or Cyanobacteriologists so far. Therefore, it became important to carry out researches on the diversity and population dynamics of Cyanobacteria of this region.

3. Methodology

Following methodology was adopted for the collection and identification of Cyanobacterial taxa with relation to the population dynamics:

3.1 Collection of Samples

The Cyanobacterial samples for the present study were collected several times in different seasons *i.e.* winter, summer and rainy, of the year from various habitats located in the foothills of Garhwal Himalayas. Cyanobacteria growing on damp soil, alluvial soil, swampy soil, bark of trees, walls of buildings, paddy fields, ponds, ditches, drains, industrial effluents etc. were collected for taxonomic studies from different localities of all research sites of the foothills of Garhwal Himalayan range. These included aquatic, terrestrial, lithophytic, epiphytic Cyanobacterial taxa. The samples were stored into sterile plastic vials, then washed thoroughly with water and preserved in 4 per cent formalin solution.

3.2 Identification of Cyanobacteria

The microscopic slides were prepared from the fixed as well as fresh samples. Preparations were observed under the trinocular research microscope. Camera-lucida drawings of the specimens were sketched in laboratory and these were analyzed on the basis of their taxo-morphological details. Microphotographs were also taken to ensure authenticity of taxa. Identification of the Cyanobacterial taxa was made by consulting the available literature in the field and particularly the monograph of Desikachary (1959).

4. Study of Population Dynamics

Cyanobacteria are unique and ubiquitous prokaryotic organisms in nature. The foothills of Garhwal Himalayas are rich in Cyanobacterial flora spreaded in various types of habitats. A total of 207 taxa of Cyanobacteria belonging to 47 genera of 14 families and 5 orders were recorded from different localities of foothills of Garhwal Himalayas. These were represented by the unicellular and colonial, non-heterocystous and heterocystous filamentous, and heterotrichous forms. Out of total, 207 species, 46 species are heterocystous and remaining 161 species are non-heterocystous. The family of Oscillatoriaceae (85) and Order Nostocales (131) members are dominant in the region (the values put within the parenthesis indicate the number of species). Majority of the species inhabit as lithophytics because they prefer to grow on the rocks probably in the search of a good substratum. These Cyanobacterial species have been recorded for the first time from this region of Uttaranchal state.

For the study of population dynamics, the taxa recorded from different habitats were categorized as dominant, very common, common, rare and very rare on the basis of their presence in various research sites. The sitewise study of population dynamics indicates that the number of dominant taxa were 38 in each site. The highest number (42) of very common taxa was obtained from Laldhang and lowest (25) from Haridwar. The highest number of rare taxa (7) was recorded from Haridwar, while the lowest number was 2 in all rest sites.

Population dynamics of the Cyanobacterial flora in the foothills of Garhwal Himalayas revealed that the highest number of taxa (73) belongs to 'common' group with highest percentage (33.1 per cent) followed by the 'rare' group (49) with 22.2 per cent . The lowest number of taxa (15) was associated with 'very rare' group with 6.8 per cent representation. This region is dominated by 17.2 per cent taxa occurring at all research sites, as such they have been regarded

Table 17.1: Population Dynamics of Cyanophycean Taxa in the Foot–hills of Garhwal Himalayas

Sl.No.	Taxa	Dominant+++++					Very Common++++					Common+++					Rare++					Very Rare+				
		H	D	K	R	L	H	D	K	R	L	H	D	K	R	L	H	D	K	R	L	H	D	K	R	L
1.	*Microcystis viridis*	+	+	+	+	+			–					–					–					–		
2.	*M. bengalensis*	+	+	+	+	+			–					–					–					–		
3.	*M. protocystis*	+	+	+	+	+			–					–					–					–		
4.	*M. aeruginosa*	+	+	+	+	+			–					–					–					–		
	M. aeruginosa v. *major* (v. nov.)			–					–					–					–						+	
	M. aeruginosa v. *minor* (v. nov.)			–					–					–					–				+			
5.	*M.pulverea*	+	+	+	+	+			–					–					–					–		
6.	*M. stagnalis*			–				+	+	+	+			–					–					–		
7.	*M. robusta* v. *minor* (v. nov.)	+	+	+	+	+			–					–					–					–		
8.	*M. lamelliformis*	+	+	+	+	+			–					–					–					–		
9.	*M. orissica*	+	+	+	+	+			–					–					–					–		
10.	*M. elabens*			–				+	+	+	+			–					–					–		
11.	*M. marginata*	+	+	+	+	+			–					–					–					–		
12.	*M. elongata*	+	+	+	+	+			–					–					–					–		

Contd...

Table 17.1–Contd...

Sl.No.	Taxa	Dominant+++++					Very Common++++					Common+++					Rare++					Very Rare+				
		H	D	K	R	L	H	D	K	R	L	H	D	K	R	L	H	D	K	R	L	H	D	K	R	L
13.	*M. flos–aquae*			–				+	+	+	+			–					–					–		
14.	*Chroococcus macrococcus*			–				+	+	+	+			–					–					–		
15.	*C. montanus*	+	+	+	+	+			–					–					–					–		
16.	*C. turgidus* v. *maximus*	+	+	+	+	+			–					–					–					–		
	C. turgidus v. *solitarius*			–					–					+	+	+			–					–		
17.	*C. indicus*			–				+	+	+	+			–					–					–		
18.	*C. tenax*	+	+	+	+	+			–					–					–					–		
19.	*C. minutus*	+	+	+	+	+			–					–					–					–		
20.	*C. hansgirgi*	+	+	+	+	+			–					–					–					–		
21.	*C. pallidus*	+	+	+	+	+			–					–					–					–		
22.	*C. varius*			–				+	+	+	+			–					–					–		
23.	*C. cohaerens*	+	+	+	+	+			–					–					–					–		
24.	*C. minor*			–				+	+	+	+			–					–					–		

Contd...

Table 17.1–Contd...

Sl.No.	Taxa	Dominant+++++					Very Common++++					Common+++					Rare++					Very Rare+				
		H	D	K	R	L	H	D	K	R	L	H	D	K	R	L	H	D	K	R	L	H	D	K	R	L
25.	*Gloeocapsa nigrescens*	+	+	+	+	+			–					–					–					–		
26.	*G. magma*			–					–				+	+	+				–					–		
27.	*G. kuetzingiana*	+	+	+	+	+			–					–					–					–		
28.	*G. decorticans*			–				+	+	+	+			–					–					–		
29.	*G. stegophila v. crassa*			–					–			+		+		+			–					–		
30.	*G. calcarea*	+	+	+	+	+			–					–					–					–		
31.	*G. punctata*			–				+	+	+	+			–					–					–		
32.	*G. atrata*	+	+	+	+	+			–					–					–					–		
33.	*G. crepidinum*			–					–			+		+	+				–					–		
34.	*G. coracina*	+	+	+	+	+			–					–					–					–		
35.	*G. quaternata*			–			+		+	+	+			–					–					–		
36.	*Gloeothece rhodochlamys*			–			+	+	+		+			–					–					–		
37.	*Gl. fusco-lutea*			–					–				+	+	+				–					–		
38.	*Gl. rupestris*			–				+	+	+	+			–					–					–		

Contd...

Table 17.1–Contd...

Sl.No.	Taxa	Dominant+++++					Very Common++++					Common+++					Rare++					Very Rare+				
		H	D	K	R	L	H	D	K	R	L	H	D	K	R	L	H	D	K	R	L	H	D	K	R	L
39.	*Gl. samoensis*	+	+	+	+	+			–					–					–					–		
	Gl. samoensis v. *major*			–					–					–					–							+
40.	*Aphanocapsa pulchra*	+	+	+	+	+			–					–					–					–		
41.	*A. koordersi*	+	+	+	+	+			–					–					–					–		
42.	*A. biformis*	+	+	+	+	+			–					–					–					–		
43.	*A. grevillei*	+	+	+	+	+			–					–					–					–		
44.	*Aphanothece saxicola* v. *major* (v. nov.)			–			+	+		+	+			–					–					–		
45.	*A. naegelii*			–					–				+	+	+				–					–		
46.	*A. bullosa*	+	+	+	+	+			–					–					–					–		
47.	*A. microscopica*			–					–					–			+				+			–		
48.	*A. stagnina*			–					–					–					–						+	
49.	*Synechococcus aeruginosus*			–			+	+	+		+			–					–					–		
50.	*Synechocystis aquatilis*	+	+	+	+	+			–					–					–					–		

Contd...

Table 17.1–Contd...

Sl.No.	Taxa	Dominant+++++					Very Common++++					Common+++					Rare++					Very Rare+				
		H	D	K	R	L	H	D	K	R	L	H	D	K	R	L	H	D	K	R	L	H	D	K	R	L
51.	*S. pevalekii*			–			+		+	+	+			–					–					–		
52.	*Rhabdoderma gorskii*			–					–					–			+				+			–		
53.	*Coelosphaerium kuetzingianum*			–					–			+			+	+			–					–		
54.	*Gomphosphaeria aponina*			–					–					–			+		+					–		
55.	*Merismopedia convoluta*			–					–			+	+			+			–					–		
56.	*M. minima*			–			+	+		+	+			–					–					–		
57.	*M. tenuissima*			–				+	+	+	+			–					–					–		
58.	*M. punctata*			–					–					–			+				+			–		
59.	*M. elegans*			–					–				+	+	+				–					–		
60.	*Eucapsis minuta*			–					–					–					–			+				
61.	*Dactylococcopsis fascicularis*			–					–					–					–			+				
62.	*Entophysalis granulosa*			–					–					–			+				+			–		

Contd...

Table 17.1–Contd...

Sl.No.	Taxa	Dominant+++++					Very Common++++					Common+++					Rare++					Very Rare+				
		H	D	K	R	L	H	D	K	R	L	H	D	K	R	L	H	D	K	R	L	H	D	K	R	L
63.	*Chlorogloea microcystoides*			–					–					–			+	+						–		
64.	*C. fritschii*			–					–				+	+	+				–					–		
65.	*Chroococcidiopsis indica*			–					–			+	+	+					–					–		
66.	*Dermocarpa leibleiniae*			–					–					–			+		+					–		
67.	*D. olivacea* v. *gigantean*			–					–					–			+				+			–		
68.	*D. flahaultii*			–					–			+	+		+				–					–		
69.	*Myxosarcina spectabilis*			–					–					–			+				+			–		
70.	*M. burmensis* v. *major* (v. nov.)			–					–					–				+		+				–		
71.	*Hyella caespitosa*			–					–				+	+	+				–					–		
72.	*Xenococcus acervatus*			–					–					–			+		+					–		
73.	*Arthrospira jenneri*			–					–					–				+			+			–		

Contd...

Table 17.1–Contd...

Sl.No.	Taxa	Dominant+++++					Very Common++++					Common+++					Rare++					Very Rare+				
		H	*D*	*K*	*R*	*L*	*H*	*D*	*K*	*R*	*L*	*H*	*D*	*K*	*R*	*L*	*H*	*D*	*K*	*R*	*L*	*H*	*D*	*K*	*R*	*L*
74.	*A. tenuis*			–					–					–			+				+			–		
75.	*A. platensis* v. *non-constricta*			–					–					–					–							+
76.	*Spirulina subsalsa*			–					–				+		+	+			–					–		
77.	*S. laxissima* f. *major* f. nov.			–					–					–					–			+				
78.	*S. major*			–					–					–				+		+				–		
79.	*S. princeps*			–					–					–			+			+				–		
80.	*S. gigantean*			–					–					–				+		+				–		
81.	*Oscillatoria annae*			–					–			+	+			+			–					–		
82.	*O. margaritifera*			–					–			+			+	+			–					–		
83.	*O. pernorata*			–					–				+	+	+				–					–		
84.	*O. vizagapatensis*			–					–					–			+				+			–		
85.	*O. limosa*	+	+	+	+	+			–					–					–					–		
86.	*O. obscura*			–					–				+		+	+			–					–		
87.	*O. subbrevis*			–			+	+	+		+			–					–					–		
	O. subbrevis f. *major*			–					–					–			+				+			–		

Contd...

Table 17.1–Contd...

Sl.No.	Taxa	Dominant+++++					Very Common++++					Common+++					Rare++					Very Rare+				
		H	D	K	R	L	H	D	K	R	L	H	D	K	R	L	H	D	K	R	L	H	D	K	R	L
88.	*O. curviceps*			–					–			+	+			+			–					–		
	O. curviceps v. *angusta*			–				+	+	+	+			–					–					–		
89.	*O. princeps*			–			+	+	+		+			–					–					–		
	O.princeps v. *pseudolimosa*			–					–					–			+		+					–		
90.	*O. proboscidea*			–					–				+	+	+				–					–		
91.	*O. laete–virens*	+	+	+	+	+			–					–					–					–		
92.	*O. subtilissima*			–					–			+		+		+			–					–		
93.	*O. chlorina*			–			+	+		+	+			–					–					–		
94.	*O. terebriformis*			–					–					+	+	+			–					–		
95.	*O. boryana*			–				+	+	+	+			–					–					–		
96.	*O. chalybea* v. *insularis*			–					–				+		+	+			–					–		
97.	*O. foreaui*			–					–					–				+	+					–		
98.	*O. proteus*	+	+	+	+	+			–					–					–					–		
99.	*O. jasorvensis*			–			+	+		+	+			–					–					–		
100.	*O. mougeotii*			–					–				+	+		+			–					–		

Contd...

Table 17.1–Contd...

Sl.No.	Taxa	Dominant+++++					Very Common++++					Common+++					Rare++					Very Rare+				
		H	*D*	*K*	*R*	*L*	*H*	*D*	*K*	*R*	*L*	*H*	*D*	*K*	*R*	*L*	*H*	*D*	*K*	*R*	*L*	*H*	*D*	*K*	*R*	*L*
101.	*O. tenuis*			–					–			+		+		+			–					–		
102.	*O. raoi*			–			+	+		+	+			–					–					–		
103.	*O. simplicissima*			–					–			+	+			+			–					–		
104.	*O. irrigua*			–					–				+	+	+				–					–		
105.	*O. quadripunctulata*			–			+	+		+	+			–					–					–		
106.	*O. pseudogeminata*			–					–			+		+		+			–					–		
	O. pseudogeminata v. *unigranulata*			–					–					–				+			+			–		
107.	*O. geitleriana*			–					–			+	+			+			–					–		
108.	*O. amoena*			–					–			+	+			+			–					–		
109.	*O. okeni*			–					–				+		+	+			–					–		
110.	*O. salina* f. *major* f. nov.			–					–					–				+			+			–		
111.	*O. claricentrosa*			–					–					–			+				+			–		
112.	*O. formosa*			–					–			+	+	+					–					–		
	O. formosa f. *loktakensis*			–					–					–					–			+				

Contd...

Table 17.1–Contd...

Sl.No.	Taxa	Dominant+++++					Very Common++++					Common+++					Rare++					Very Rare+				
		H	D	K	R	L	H	D	K	R	L	H	D	K	R	L	H	D	K	R	L	H	D	K	R	L
113.	*O. cortiana*			–					–					–					+	+				–		
114.	*O. rubescens*			–					–				+		+	+			–					–		
115.	*O. calcuttensis*			–					–					–			+				+			–		
116.	*O. willei*			–			+	+	+		+			–					–					–		
117.	*Trichodesmium lacustre*			–					–					–					–					+		
118.	*T. erythraeum*			–					–					–				+		+				–		
119.	*Crinalium magnum*			–					–					–					–				+			
120.	*Phormidium fragile*			–					–				+	+		+			–					–		
121.	*P. truncicola*			–					–					–				+	+					–		
122.	*P. rubroterricola*			–					–					–				+	+					–		
123.	*P. purpurascens*			–					–			+		+		+			–					–		
124.	*P. mucosum*			–					–					–			+	+						–		
	P. mucosum v. *arvense*			–					–					–					–			+				
125.	*P. ambiguum*			–					–					–						+	+			–		
126.	*P. pachydermaticum*			–					–			+	+			+			–					–		

Contd...

Table 17.1–Contd...

Sl.No.	Taxa	Dominant+++++					Very Common++++					Common+++					Rare++					Very Rare+				
		H	D	K	R	L	H	D	K	R	L	H	D	K	R	L	H	D	K	R	L	H	D	K	R	L
127.	*P. calcicola*			–					–				+		+	+			–					–		
128.	*P. corium*			–					–			+	+			+			–					–		
	P. corium v. *capitatum*			–					–			+	+		+				–					–		
129.	*P. subfuscum*			–					–					–						+	+			–		
130.	*P. favosum*			–					–			+	+		+				–					–		
131.	*P. autumnale*			–					–					–			+		+					–		
132.	*P. tenue*			–			+	+	+		+			–					–					–		
133.	*P. molle*			–					–					–					–			+				
134.	*Lyngbya mesotricha*			–					–					–			+				+			–		
135.	*L. kuetzingii*			–			+	+	+		+			–					–					–		
136.	*L. spiralis*			–					–					–						+	+			–		
137.	*L. spirulinoides*			–					–			+			+	+			–					–		
138.	*L. gracilis*			–					–			+	+			+			–					–		
139.	*L. borgerti*			–					–					+	+	+			–					–		
140.	*L. birgei*			–					–					–			+		+					–		
141.	*L. shackletoni*			–					–			+			+	+			–					–		

Contd...

Table 17.1–Contd...

Sl.No.	Taxa	*Dominant+++++*					*Very Common++++*					*Common+++*					*Rare++*					*Very Rare+*				
		H	*D*	*K*	*R*	*L*	*H*	*D*	*K*	*R*	*L*	*H*	*D*	*K*	*R*	*L*	*H*	*D*	*K*	*R*	*L*	*H*	*D*	*K*	*R*	*L*
142.	*L. hieronymusii*			–					–			+	+			+			–					–		
143.	*L. cryptovaginata*			–					–					–			+				+			–		
144.	*L. rubida*	+	+	+	+	+			–					–					–					–		
145.	*L. dendrobia*			–					–					–				+	+					–		
146.	*L. corticicola*			–					–			+		+		+			–					–		
147.	*L. aestuarii* v. *constricta*			–			+	+	+		+			–					–					–		
148.	*L. porphyrosiphonis*			–					–			+			+	+			–					–		
149.	*L. allorgei*	+	+	+	+	+			–					–					–					–		
150.	*L. putealis*			–			+	+	+	+				–					–					–		
151.	*L. martensiana*			–			+		+	+	+			–					–					–		
152.	*L. corbierei*			–			+		+	+	+			–					–					–		
153.	*L. major*	+	+	+	+	+			–					–					–					–		
154.	*Symploca muscorum*			–					–					–					+	+				–		
155.	*Microcoleus vaginatus*	+	+	+	+	+			–					–					–					–		

Contd...

Table 17.1–Contd...

Sl.No.	Taxa	Dominant+++++					Very Common++++					Common+++					Rare++					Very Rare+				
		H	D	K	R	L	H	D	K	R	L	H	D	K	R	L	H	D	K	R	L	H	D	K	R	L
156.	M. paludosus			–			+		+	+	+			–					–					–		
157.	M. subtorulosus			–					–					–					+		+			–		
158.	Anabaenopsis circularis	+	+	+	+	+			–					–					–					–		
159.	Cylindrospermum majus			–					–				+	+	+				–					–		
160.	C. sphaerica			–					–				+	+	+				–					–		
161.	C. indicum			–					–					–					+	+				–		
162.	C. stagnale			–					–					+	+	+			–					–		
163.	Nostoc punctiforme			–					–				+	+	+				–					–		
164.	N. linckia			–			+	+	+	+				–					–					–		
165.	N. ellipsosporum v. violacea			–					–					–					–					+		
166.	N. calcicola	+	+	+	+	+			–					–					–					–		
167.	N. sphaericum			–					–					+	+	+			–					–		
168.	Anabaena sphaerica			–					–				+		+	+			–					–		
	A. sphaerica v. tenuis			–					–					–			+				+			–		
169.	A. fertilissima			–					–				+	+		+			–					–		

Contd...

Table 17.1–Contd...

Sl.No.	Taxa	Dominant+++++					Very Common++++					Common+++					Rare++					Very Rare+				
		H	D	K	R	L	H	D	K	R	L	H	D	K	R	L	H	D	K	R	L	H	D	K	R	L
170.	*A. doliolum*			–					–				+	+	+				–					–		
171.	*A. flos–aquae*			–					–			+		+		+			–					–		
172.	*A. oscillarioides*			–					–					+	+	+			–					–		
173.	*A. volzii*			–					–				+	+	+				–					–		
174.	*Aulosira prolifica*			–				+	+	+	+			–					–					–		
175.	*A. aenigmatica*			–			+		+	+	+			–					–					–		
176.	*A. laxa*			–			+	+	+	+				–					–					–		
177.	*A. pseudoramosa*			–					–			+		+	+				–					–		
178.	*A. fertilissima*			–				+	+	+	+			–					–					–		
	A. fertilissima v. *tenuis*			–					–					+	+	+			–					–		
179.	*A. fritschii*			–				+	+	+	+			–					–					–		
180.	*Plectonema tomasinianum*			–					–				+	+	+				–					–		
181.	*P. hansgirgi*			–					–					–					+	+				–		
182.	*Pseudoscytonema malayense*			–					–				+	+	+				–					–		
183.	*Scytonema fritschii*			–			+	+	+		+			–					–					–		

Contd...

Table 17.1–Contd...

Sl.No.	Taxa	Dominant+++++					Very Common++++					Common+++					Rare++					Very Rare+				
		H	D	K	R	L	H	D	K	R	L	H	D	K	R	L	H	D	K	R	L	H	D	K	R	L
184.	*S. millei*			–					–					+	+	+			–					–		
185.	*S. ocellatum*			–					–					+	+	+			–					–		
186.	*S. julianum*			–					–					+	+	+			–					–		
187.	*S. mirabile*			–					–			+	+		+				–					–		
188.	*S. malaviyaensis*			–				+	+	+	+			–					–					–		
189.	*Petalonema alatum* v. *indicum*			–					–					–						+	+			–		
190.	*Microchaete grisea* v. *brevis*			–				+	+	+	+			–					–					–		
191.	*M. elongata*			–					–					–						+	+			–		
192.	*Homoeothrix africana*			–					–					–					–			+				
193.	*Calothrix contarenii*			–					–				+	+	+				–					–		
194.	*C. parietina*			–				+	+	+	+			–					–					–		
195.	*C. fusca*			–					–			+		+	+				–					–		
196.	*C. scytonemicola*			–					–					–						+	+			–		
197.	*C. braunii*			–					–				+	+	+				–					–		
198.	*Rivularia dura*			–					–					–						+	+			–		
199.	*Gloeotrichia pilgeri*			–					–					+	+	+			–					–		

Contd...

Table 17.1–Contd...

Sl.No.	Taxa	Dominant+++++					Very Common++++					Common+++					Rare++					Very Rare+				
		H	*D*	*K*	*R*	*L*	*H*	*D*	*K*	*R*	*L*	*H*	*D*	*K*	*R*	*L*	*H*	*D*	*K*	*R*	*L*	*H*	*D*	*K*	*R*	*L*
200.	*G. indica*			–			+		+	+	+			–					–					–		
201.	*G. intermedia* v. *kanwaensis*			–					–					–			+				+			–		
202.	*G. pisum*			–					–					–						+	+			–		
203.	*G. raciborskii* v. *bombayense*			–					–					+	+	+			–					–		
204.	*Nostochopsis radians*			–				+	+	+	+			–					–					–		
205.	*N. lobatus*			–					–				+	+		+			–					–		
206.	*Mastigocladopsis jogensis*			–					–				+	+		+			–					–		
207.	*Fischerella muscicola*			–					–					–				+		+				–		
	Sitewise Population Dynamics →																									
	Haridwar			38					25					31					24					7		
	Dakpatthar			38					37					46					14					2		
	Kotdwar			38					39					47					15					2		
	Rishikesh			38					36					49					18					2		
	Laldhang			38					42					46					27					2		
Population Dynamics of the Foothills of Garhwal Himalays	No. of Taxa			38					45					73					49					15		
	% of Taxa			17.2					20.4					33.1					22.2					6.8		

Notes: Sites: H: Haridwar; R: Rishikesh; D: Dakpatthar; L: Laldhang; K: Kotdwar; +: Observed; –: Not Observed.

as 'dominant' taxa. The common examples of this group are *Microcystis viridis, M. bengalensis, M. elongata, Chroococcus montanus, C. cohaerens, Oscillatoria limosa, O. proteus, Nostoc calcicola* etc. 20.4 per cent taxa are grouped as 'very common' as they belong to any four sites of the research area. These include *Microcystis stagnalis, Chroococcus macrococcus, Gloeocapsa decorticans, Calothrix parietina, Nostochopsis radians* etc. The 'common' taxa (33.1 per cent) are present at any three research sites. These are *Chroococcus turgidus* v. *solitarius, Gloeocapsa magma, Spirulina subsalsa, Oscillatoria annae,* Phormidium *fragile, Nostoc punctiforme, Nostochopsis lobatus, Mastigocladopsis jogensis* etc.

The 'rare' (22.2 per cent) and 'very rare' (6.8 per cent) taxa on the other hand have been recorded from any two and one-research sites, respectively. The rare group contains *Aphanothece microscopica, Merismopedia punctata, Gloeotrichia pisum, Fischerella muscicola* etc. while, *Microcystis aeruginosa* v. *major* v. nov., *Eucapsis minuta, Crinalium magnum, Homoeothrix africana* etc. are very rare in occurrence in this region (Table 17.1).

5. Conclusions

The studies on the foothills of Garhwal Himalayas in Uttaranchal represent 207 taxa which reveals the richness of the Cyanobacterial population in the state, as such, it is essential to protect 'very rare' taxa, and otherwise they will extinct from the environment. Further investigations are required to explore all possibility of being new taxonomic specimens with interesting features. This will strengthen the genetic pool not only by preserving the taxa under threat but also to bring the new species in the light of the day.

6. Acknowledgements

The authors are thankful to the UGC, New Delhi for providing financial assistance in the form of a Major Research Project entitled, "Biodiversity of Cyanophyceae of the Sub-Himalayan Belt".

7. References

Adhikary S P 2000 Epilithic cyanobacteria on the exposed rocks and walls of temples and monuments of India. *Ind J Microbiol* 40: 67-81.

Bendre A M and Kumar S 1975 Cyanophyceae of Meerut. *Phykos* 14: 1-7.

Bharadwaja Y 1928 *Spelaeopogon kashyapi* n. sp., a new member of the Scytonemataceae. *Ann Bot* (Lond) 42: 69-74.

Chaudhary M K 2004 *An Ecological and Taxonomic Investigation of the Cyanophycean Flora of the Foothills of Garhwal Himalayas of Uttaranchal (India).* Ph D Thesis, MJP Rohilkhand University, Bareilly (UP).

De P K 1939 *The role of the blue-green algae in nitrogen fixation in the rice fields.* Proc Roy Soc London 127: 121-139.

Desikachary T V 1959 *Cyanophyta*. ICAR, New Delhi.

Dodds W K, Gudder D A and Mollenhauer D 1995 The ecology of *Nostoc. J Phycol* 31: 2-18.

Drouet F 1986 *Revision of the Classification of the Oscillatoriaceae.* Monograph Acad. Natur Sci Philadelphia. Fulton Press. Lancaster, Pennsylvania.

Elenkin A A 1914 Über die Thermophilen Algen-formation en. *Bull Jardin Imp Bot Pierre la Grand* 14: 62-110.

Fabbro L and Mc Gregor G 2003 *Blue-Green Algae:* General information NRM Facts Water Series-W3, Queensland Govt Natural Resources and Mines, Australia.

Gaidukov N 1902 Über Einfluss farbigen Lichtes auf die Färbung der Oscillarien n.s. Abh Preuss Akad Wiss, Berlin 1-36.

Gaidukov N 1906 Die Komplimentären Chromatischen Adaptation bei *Porphyra* und *Phormidium. Ber dtsch bot Ges* 24: 1-5.

Gaidukov N 1923 Zur Frage nach der komplimentären chromatischen Adaptation. *Ber dtsch bot Ges* 41: 356-361.

Khan M and Kumari S 1972 Some additions to the algal flora of Dehradun III. Cyanophyta. *J Ranchi Univ.* 8: 298-300.

Kirtikar K R 1886 A new species of alga, *Conferva thermalis Birdwoodii. J Bombay Nat Hist Soc* 1: 135-138.

Kumar H 1970 Cyanophyceae of Sardhana. *Phykos* 9: 79-85.

Kumar M and Kumar H 2000. Cyanobacterial flora of Laldhang. Research paper presented at Nat Symp on Microbes and Human Welfare held at Meerut. Oct 14-16.

Montagne C 1849 Sixiene centurie de Plantes Cellulaires Nouvelles tant di indigenes qu exotiques. Dec 7 A X *Ann Sci Nat Bot* 12: 285-320.

Pal S 1975 A check-list of algae from Ghaziabad. *Phykos* 13: 67-76.

Post A F 1999 The Prochlorophytes–an algal enigma. Biology of chlorophyll a/b containing photosynthetic prokaryotis. pp 115-125 In: Seckback J (ed) *Enigmatic Microorganisms and Life in Extreme Environment.* Kluwer Acad Pub Dordrecht.

Sarma Y S R K and Khan M 1980 *Algal Taxonomy in India. Bot Reco and Monograph-2,* Today and Tomorrow Printer and Publication, New Delhi.

Tomitani A, Okadak Miyashita H, Mattohijs H C P, Ohno T and Tanaka A 1999 Chlorophyll b and phycobilins in the common ancestor of Cyanobacteria and chloroplasts. *Nature* 400: 159-162.

Chapter 18
Cyanobacteria: The Future Food for Man

Iqbal Habib and U.K. Chatturvedi

Contents

ABSTRACT

Cyanobacteria or Blue-green algae are highly successful group of plants, which enjoys a wide distribution in all kinds of habitats. A large number of their inhabit in freshwater ponds and often colour the entire body of water, thus causing water blooms. However, recently a large number of Blue-green algae

Glimpses of Cyanobacteria, 2006 298–304
Daya Publishing House, New Delhi

recognized as a good source of food for human consumption and also of marine and freshwater life. They are a rich source of proteins, vitamins and minerals.

1. Introduction

Cyanobacteria or Blue-green algae besides having economic importance as food also have multifaceted potentialities which includes pharmaceutical as well as therapeutic values. Now a days they are mainly being utilized as a good source of proteins, vitamins and minerals. Researches have been conducted on the multifaceted utility of alga *Spirulina,* at CFTRI, Mysore. Besides its value, *Spriulina* is being utilized as biofertilizer. It is rich in micronutrients and contains nitrogen, phosphorous, calcium and magnesium etc. Cyanobacteria or Blue-green algae do not look like much in the way of life, but they are remarkable organism. The microscopic threads of algae weave together and form carpet on the earth. Each layer of this carpet or mat, depending on its exposure to light and oxygen, performs a unique function in what is a tiny, highly regulated layered ecosystem. These mats of blue-green algae virtually do not need nutrients but are helpful in purifying water in effluent holding ponds and in waste treatment plants in U.S.A. They have been found even to breakdown complex organic molecules, such as petroleum hydrocarbons and pesticides. These algal mats also lead to devastating problems such as algal blooms and oxygen depletion in ponds and rivers.

Blue–green algae grow during the first three billion years of the earth's history during the Pre Cambrian period. It was during that era that these algae absorbed carbon dioxide from the earth's atmosphere and released oxygen, thus helping this earth planet hospitable to other organisms. Cultivation of algae for the production of useful resources is not a new concept, but the most recent demand in the utilization of algal biomass has been sparked by the shortage of food supply and rising face of population. For the past decade several laboratories all over the world exploring the potentiality of algae, particularly micro-algae for commercial application in providing cheap nutritious food. It is now well known that these photosynthetic microorganisms can as a rich source of proteins, vit-A and iron. Mass cultivation of *Spirulina*, a blue-green alga may help in overcoming the shortage of food to some extent. Species of *Spirulina* have been eaten for many years by inhabitants of the northern shores of lake Chad in Africa and Aztec Indians in Mexico. They are the most important source of SCP (Single cell protein) as they can grow profusely in the open ponds and use CO_2 and sunlight. Work is in progress at few centers in India and at several places in different countries of the world such as Japan, Taiwan, Mexico, Israel, Thailand and U.S.A. using *Spirulina* as a source of SCP because of its high nutritive value. At the present time about 2000 metric tonnes of algal proteins and vitamins are being produced as algal health food. They are grown in cemented tanks and then dried or sun dried sold as power or pills in the market.

At the University of California, Barkeley, Oswald in 1970's developed a paddle wheel device for mass production of *Spirulina* and *Dunaliella*. Pond sizes are about 0.4-0.5 hactares for commercial productions. The device is less capital investing, cost-effective and has operating advantages in comparison to the conventional devices such as using airlift or centrifugal pumps etc. Out-door culture for the productions of blue-green algae have also been tried satisfactory. During favourable climates, peak production of 35-40 gm^{-2} per day can be achieved. It is possible to get an average productivity of 2.5 gm^{-2} per day or 75 tonnes ha^{-2} per year but it would required the use of such micro-algae which are more productive and the conditions are favourable. However, the biotechnological production of micro-algae for food needs extensive trials before to achieve a breakthrough. In 16th century, blue-green algae such as *Spirulina* was used as a major source of microbial protein (Clement, 1968). At the First International Conference on microbial protein held in 1967 at Massachusetts Institute of Technology, Cambridge (U.S.A.), the word 'microbial protein' was renamed as 'single cell protein' (SCP) (Scrimshaw, 1968) although this term SCP does not fit to an organism which is multicellular. In India, central Food Technological Research Institute (CFTRI) Mysore, Shri A.M.M. Murugappa Chettiar Research Centre (MCRC) Chennai and National Botanical Research Institute (NBRI), Lucknow. Experiments have also been conducted to obtain SCP from the Sewage, which of course can be utilized as animal feed.

Bernal Diaz del Castillo in 1821 described the biscuits with the name *'tecuilatl'* made from *Spirulina maxima*. Biscuits were made from the dried mats of the alga which was collected from the lake texcoco near Mexico. In1964, J.Leonard, a Belgian botanist which on an expedition, noticed blue-green coloured biscuits sold in several village market. Biscuits prepared from the algae *Spirulina* platensis were marked as *'diha'* and consumed by the local people. New several pilot plant for the production of *Spirulina* powder as a potential source of protein have been established in Japan, U.S.A., India and in some European countries. The first pilot plant named *Sosa texcoco* was commissioned in Mexico in 1973 which produced about 150 tonnes of *Spirulina* powder per annum. Later on, this production increased to about 1000 tonnes in 1982. This powder is exported to U.S.A. where capsules, biscuits and lozenges are prepared by adding Vit.-A and C. Schools in Mexico and U.S.A. use *Spirulina* powder to make biscuits and confectionery which have very high protein content (Sasson 1984).Recently, in India, Dabur Research Foundation has developed *Spirulina* tablets which are sold as *Sunova Spirulina.* As per the advertisement of this foundation. WHO has also hailed *Spirulina* as 'One of the greatest super foods on earth' and is consumed by millions of people in over 40 countries worldwide. In their recent advertisement in a newspaper they have claimed that *Spirulina* is the richest source of protein in the world (6 times more than eggs and 20 times more than milk). It is the richest source of Beta-carotene (Pro-vit. A) in the world, richest source of natural iron and the only source of vit. B_{12} in the world.

2. Food Value

In India in the last two decades, the per capita availability of pulses which are the major source of protein have declined from 75 gm in 1959 to 44 gm/day in 1979. Therefore the cultivation of *Spirulina* could be the right solution in providing cheap nutritious food as it is rich in protein. Vit. A and Iron. Chemical analysis shows that it contains about 65 per cent of food quality protein (Table 18.1), which have 85 per cent digestibility and reasonable balance of essential acids. It is also one of the best source of Beta-carotene as precursor of Vit.-A. A freshly harvested and spray dried *Spirulina* has been found to contain about 3-3.5 mg of carotenes/gm of algal powder. It is the only plant source of Vit. $B_{12.}$ The nutritional profile of *Spirulina* can be compared with that of milk and egg. In one estimate, it is found that the consumption of 2 gm of dry *Spirulina* powder by a person can meet 114 per cent of Vit.-A and 200 per cent of Vit.- B_{12} requirement. For the control of Cholesterol level in human body, the linolinic acid present in *Spirulina* will help in overcoming this problem. This acid is found effective in relieving human pains of tropical pancreatitis and rheumatoid arthritis. Recently, it has received an international attention for its application as can anti-*cancer* substance (Schwartz *et al.*, 1986) and also as a post radiation therapy due to its anti-oxidant property.

Mass cultivation of algae has already been started in several countries such as Japan, Germany, Mexico etc. but inspite of the advancement in biotechnology the commercialization of *Spirulina* has been very slow in India. In the last few years there has been hectic efforts for a large scale production of *Spirulina* products in this country especially at the two main centres-on at Shri A.M.M.Murugappa Chettiar research Center (MCRC), Chennai and the other at CFTRI, Mysore. Chennai Center is the biggest food grade *Spirulina* farm in India. Its annual capacity is about 75 tonnes and the products are marked in India and in other countries as health food, baby food as multi-vitamin tablets.

Table 18.1: The Composition of *Spirulina Fusiformis* Power as Analysed by Michaelson Lab. Inc., California, USA (Constituents are in per 100 of power)

A. MAJOR CONSTITUENTS (IN %)		
Total Protein	–	64.6
Carbohydrates	–	16.1
Fat	–	6.7
Fiber content	–	9.3
Bile pigments and Chlorophyll	–	6
Caloric value	–	136000
Ash content	–	3
B. VITAMINS		
Beta–carotene	–	320,000 IU
Biotin	–	0.22 mg
Cyanobalamin (B_{12})	–	65.75 mg
Folic acid	–	17.6 mg
Riboflavin	–	1.78 mg
Thiamin	–	1.18 mg
Tocopherol	–	0.773 IU
C. MINERALS (mg.)		
Calcium	–	6.58
Phosphorous	–	977
Iron	–	44.7
Sodium	–	796
Potassium	–	1.28

Contd...

Table 18.1–Contd...

D. ESSENTIAL AMINO–ACID (IN %)		
Lysin	–	2.99
Cystine	–	0.474
Methionine	–	1.38
Phenylalanine	–	2.87
Threonine	–	3.04

3. Mass Cultivation of *Spirulina*

For the production of *Spirulina platensis* and other species, generally rectangular or circular cemented tanks are constructed. The size of the tanks may vary according to the convenience and yield required. The total volume in the tank should be approx. 100 liters of volume containing the inoculum and the culture medium. The ratio of the inoculum and the medium should be around 1:4 for the optimum growth; temperature should be maintained between 25°-30°C with low light intensity as *Spirulina* exposed to high light intensity gets lysed easily. *Spirulina* multiplies well at higher pH ranging from 8 to 10.5 but in the initial stages the culture should be maintained at pH 8.5. To get a better yield, it is necessary that the culture must be stirred regularly with the rotators. Regular monitoring of the culture is important to avoid any contamination, though the chances of contamination are very few as the alga growth at a high alkaline pH of 8-11. Harvesting of the alga can be started from the 8th day when large filaments of the alga are seen floating on the surface of water forming thick mats. Harvesting of *Spirulina* is easier than *Chlorella* and *Scenedesmus* as the filaments of *Spirulina* float on water surface due to the presence of gas vacuoles. Fine mesh of steel or nylon or even cotton cloth filters of 30mesh/inch can therefore, harvest it. After harvesting the algal biomass is kept for sun drying which is most suitable and at the same time economical also. The yield is about 8-12 gm. *Spirulina* powder/m^2/day on dry weight basis has been obtained in India, which is equal to 20 tonnes/ha/annum. Dried *Spirulina* powder is packed in aluminium bags and sent for marketing.

Researches conducted by UNDIO Programme in Mexico in 1980 and the test carried out at the National Institute of Nutrition, Hyderabad in 1988 and 1990 under ICMR, New Delhi, have shown no adverse effects on the laboratory animals and human beings. However, there are certain problems which warrant the use of SCP products as human foods such as the presence of large amount of nucleic acids which may lead to the formation of kidney stone and gout, if consumed in high quantity; the possibility for the presence of toxic substances and poor digestibility (Litchfield, 1979). Inspite of these disadvantages, it is hoped that *Spirulina* will become an important supplement to the diet of humans and also of marine and fresh water life.

4. Acknowledgements

The authors are thankful to Prof. B.D. Joshi, Principal, D.D.U. Government Post Graduate College, Sitapur for providing necessary laboratory facilities and to Dr.U.C.Pandey, Reader in Botany, Bareilly College, Bareilly for his valuable suggestions.

5. References

Bothe H 1982 *The Biology of Cyanobacteria*. Blackwell Scientific Publication, Oxford.

Clement G 1968 *Single Cell Protein*. MIT Press, Cambridge.

Davis P 1974 Single Cell Protein. Academic Press,New York.

Kurz W G W and Constabel F 1979 *Microbiol Technology*. Vol. 1. Academic Press, London.

Litchfidd J H 1979 *Microbial Technology* Academic Press, New York.

Richmond A 1986 *Handbook of Micro-Algal Mass Culture*. CRD Press.

Sasson A 1984 *Biotechnology: Challenges And Promises*. UNESCO, Paris.

Scrimsraw N S 1968 *Single Cell Proteins*. MIT Press, Cambridge.

Sherwood M 1974 Single Cell Proteins comes of age. New Scientist 64: 654-639.

Venkataraman L V and E W Becker 1985 *Biotechnology and Utilization of Algae-the Indian Experiences*. CFTRI, Mysore.

Chapter 19
Biotechnological Potentials of Cyanobacteria

Shalini, B.V. Singh and R.K. Gupta

Contents

© R.K. Gupta, M. Kumar & G.S. Paliwal (*Editors*)
Glimpses of Cyanobacteria, 2006 305–320
Daya Publishing House, New Delhi

ABSTRACT

Cyanobacteria(also known as blue green algae) are the largest and most diverse group of photosynthetic prokaryotes.Their habitats vary from fresh and marine water to terrestrial environments.These are oxygen evolving and are also known to fix atmospheric nitrogen. Cyanobacteria also have a unique potential to contribute to productivity in a variety of agricultural and ecological situations. Many cyanobacteria are also capable of using atmospheric dinitrogen (N_2) as the source of nitrogen, and this is commonly termed nitrogen fixation and it may enable us to manipulate the N_2-fixation(nif) and associated genes, and produce strains which fix N_2,evolve H_2 or release ammonia with great efficiency. However, cyanobacteria have the appeal of being a raw, unprocessed food, rich in carotenoids, chlorophyll, phycocynin, and many other bioactive components. *Spirulina* is a filamentous cyanobacteria rich in micronutrients, in beta-carotene, in iron, in B_{12} vitamin and in essential fatty acids. Also *Spirulina* increases the production of gamma interferon,which can possibly make the viruses inactive and play a regulating role related to the human tumoral cells, P-glycoprotein's over-expression. Cyanobacteria also contain vitamins, polyunsaturated fatty acids, and many kinds of complex chemicals. Also Nitrate removal from groundwater may be accomplished by bacterially mediated denitrification. Since nitrate may be taken up effectively by photosynthetic microorganisms, such as cyanobacteria, which require mostly fixed nitrogen, inorganic carbon, and light for growth, the use of photosynthetic organisms would minimize the need of chemicals and fossil fuels for nitrate removal, thus leading to an efficient resource recovery and recycling.

1. Introduction

Microorganisms represent an enormous and largely unexploited source for biotechnological applications. Microbial diversity can be regarded on one hand as a problem, for example due to large variety of microorganisms causing (new emerging) diseases and on the other hand as a solution due to rich biotechnological potential including disease control. Prokaryotes developed the basic types of metabolism and a wide range of activities, which allowed them to colonize all ecosystems and create the biosphere. The abundance of prokaryotes in the open ocean, in soil, and oceanic and terrestrial subsurfaces is 1.2×10^{29}, 2.6×10^{29}, 3.5×10^{30} and $0.25\text{-}2.5 \times 10^{30}$, respectively (Whitman *et al.*1998). Prokaryotes represent likely the largest living surface on earth and this enormous interface between the living and the abiotic world is one of the reasons for their significance in transforming organic matter. It has been estimated that there are ca 7.5×10^{29} viruses in the world ocean (Weinbauer and Herndi 2002). Microorganisms are also associated with plants and animals, not only in the form of pathogens, but also in more beneficial ways such as gut bacteria in ruminants, which transform grass into digestible biomass, or as symbionts in the rhizosphere, which fix molecular nitrogen. Such associations are now under scrutiny to assess their significance for the survival and performance of plants and animals. This kind of interactions between micro-organisms and hosts represents a classical ecological relationship. The knowledge on the diversity of micro-organisms is just the 'tip of the iceberg' is now a commonplace. More refined isolation techniques are required, since this is necessary for many biotechnological applications. Cyanobacteria, *i.e.* photosynthesizing prokaryotes,have never been isolated, until they have been seen by their autofluorescence in epifluorescence microscopes. Once they were detected, it did not take long to isolate them and isolation is indeed easy. This example fuels the hope that we can isolate more prokaryotes once we know what we have to look for knowledge from insitu approaches may thus help here to develop new isolation techniques.

2. Nitrogen Fixation by Cyanobacteria

Cyanobacteria–also called blue-green algae–evolved very early in the history of life, and share some of the characteristics of gliding bacteria on one hand and those of higher plants on the other (Singh *et al.*, 2000). Cyanobacteria can both photosynthesize and fix nitrogen, and these abilities, together with great adaptability to various soil types, make them ubiquitous. Cyanobacteria also have a unique potential to contribute to productivity in a variety of agricultural and ecological situations. Cyanobacteria have been reported from a wide range of soils, thriving both on and below the surface. They are often also types characteristic features of other of sub-aerial environment and many intermittently wet ones such as rice fields. Most paddy soils have a natural population of cyanobacteria which provides a potential source of nitrogen fixation at no cost. Ammonia can be taken up by cyanobacteria through passive diffusion or as ammonium (NH_4^+) by a specific uptake system. The amino acids argenine, asparagine and glutamine have also been reported to serve as nitrogen sources. Nitrate and nitrite are important sources, which later reduce into ammonia. Many cyanobacteria are also capable of using atmospheric dinitrogen (N_2) as the source of nitrogen, and this is what is most commonly termed as nitrogen fixation. Like in many other biological systems, nitrogen fixation in cyanobacteria is brought about by a high molecular weight, oxygen labile, metalloprotein enzyme known as nitogenase. Nitrogenase reduces molecular nitrogen to ammonia in presence of hydrogen. Due to this important characteristic of nitrogen fixation, the utility of cyanobacteria in agriculture to enhance production is beyond doubt. Many studies have been undertaken on the use of dried cyanobacteria to inoculate soils as a means of aiding fertility, and the effect of adding cyanobacteria to soil on rice yield was first studied in the 1950s in Japan. The term 'algalization' is now applied to the use of a defined mixture of cyanobacterial species to inoculate soil, and research on algalization is going on in all major rice producing countries. The average of the results from all these studies have shown an increase in grain yield of 15-20 per cent in field experiments. It has been suggested that the cyanobacteria introduced as a result of algalization can establish themselves permanently if inoculation is done consecutively for 3-4 cropping seasons.

The algae are the simplest members of the plant kingdom, and the blue-green algae are the simplest of the Algae. They have a considerable and increasing economic importance; causing both beneficial and harmful effects on human life. Blue-greens are not true algae. They have no nucleus, the structures that encloses the DNA, and no chloroplast, the structure that encloses the photosynthetic membranes, the structures that are evident in photosynthetic true algae. In fact blue-greens are more akin to bacteria which have similar biochemical and structural characteristics. The process of nitrogen fixation and the occurrence of gas vesicles are especially important to the success of nuisance species of blue-greens. The majority of blue-greens are aerobic photoautotrophs: their life processes require only oxygen, light and inorganic substances. A species of *Oscillatoria* that is found in mud at the bottom of the Thames, are able to live anaerobically. They can live in extremes of temperatures -60°C to 85°C, and a few species are halophilic or salt tolerant (as high as 27 per cent, for comparison, concentration of salt in seawater is 3 per cent). Blue-greens can grow in full sunlight and in almost complete darkness, they are often the first plants to colonize bare areas of rock and soil, as an example subsequent to cataclysmic volcanic explosion. Finally at the onset of adverse environmental conditions, some blue-greens can develop a modified cell, called an akinite. Akinites contain large reserves of carbohydrates, and owing to their density and lack of gas vesicles, eventually settle to the lake bottom. They can tolerate adverse conditions such as the complete drying of a pond or the cold winter temperatures, and, as a consequence, akinites serve as "seeds" for the growth of juvenile filaments when favourable conditions return. Heterocysts and akinites are unique to the blue-greens. The main factors which appear to determine the development of planktonic population are light, temperature, pH, nutrient concentrations and the presence of organic solutes. Heterocystous blue-greens possess the unique ability to sin.antaneously evolve O_2 in photosynthesis (in vegetative cells) and H_2 by nitrogenase catalysed electron transfer to H^+ -ions (in heterocysts), in the absence of N_2 or other substrates of nitrogenase. This is the basis for the attempts of several workers to exploit the potential through the development of a "biophotolytic system" for solar energy conversion, even though to date the thermodynamic efficiency has been disappointingly low. Nevertheless, the utilization of blue-greens in food production and in solar energy conversion may hold immense potential for the future, and could be exploited for mans's economy. Progress in the study of the genetics of blue-greens may enable us to manipulate the N_2-fixation (nif) and associated genes, and produce strains which fix N_2, evolve H_2 or release ammonia with great efficiency.

3. Cyanobacteria as Food

Among cyanobacteria, both edible and toxic species, adapted to almost any of the most extreme habitats on Earth, including deep-sea vents, hot springs, and Antarctica's ice. Edible blue-green algae, including *Nostoc, Spirulina*, and *Aphanizomenon* species have been used for food for thousands of years (Sobel, 1995). Habitat with sufficient algae growth include the Pacific Ocean near Japan and Hawaii, and large freshwater lakes, including Lake Chad in Africa, Klamath Lake in North America, Lake Texcoco in Mexico, and Lake

Titikaka in South America. African and American natives recognized the value of including blue-green algae in their diet and stored dried algae for year-round use and trade. The nutrient content depends on the location and environment in which the algae was grown as altitude, temperature, and sun exposure can greatly effect lipid and pigment composition, *Spirulina* is an algae species grown at sea or in man-made ponds, and the mineral profile will differ from fresh-water algae such as *Aphanizomenon* (Jensen,2000). The nutrient profile is subject to much variation between habitats and harvest procedures which influences the content of vitamins and antitoxidants delivered in the final product. Certain features are common to all blue-green algae, including a high content of bioavailable amino acids and minerals, including zinc, selenium, and magnesium. Industrial standards still vary greatly in term of documenting product composition to the consumer. However, the blue-green algae have the appeal of being a raw, unprocessed food, rich in carotenoids, chlorophyll, phycocynin, and many other bioactive components. *Spirulina* naturally grown in some alkaline lakes of the intertropical area and has been consumed since centuries by several populations. It has 60-70 per cent protein on dry weight basis and is also rich micronutrients, beta-carotene (from which the vitamin A is formed), iron, B_{12} vitamin, and the essential fatty acids. Taking into account its characteristics, *Spirulina* culture could be a solution to improve human health and nutrition in particular in the South countries and also a way to develop industrial crops. Since many years NGOs (Non Govermental Organizations) used *Spirulina* to fight against malnutrition on all the continents. On the other hand, *Spirulina* cultures were not used by nutritionists of national and international organizations of the United Nations like WHO (World Health Organization) and the UNICEF. Nowadays, large Chinese, Indian, Chilean and American companies have already developed industrial crops of this organism. *Spirulina* is not toxic, indeed, it does not have the genus which ensure the synthesis of toxins. On the other hand, many other cyanobacteria are toxic but the very alkaline medium in which *Spirulina* growths does not enable them to develop. Cases of intoxication by *Spirulina* have also been reported.

4. *Spirulina* in Pharmacopoeia

Spirulina is already part of the pharmacopoeia of many countries. Several oral presentations and posters develop the results of research relating to beneficial effect of *Spirulina* components. Thus, the presentation of Belay (USA) reports that *Spirulina* seems to favourably modulate the immune system by increasing macrophage activation, cells T activity and naturally destroying cells activity. It has also been shown that *Spirulina* increases the production of gamma interferon, which can possibly make the viruses inactive, that *Spirulina* aqueous extracts had an antiviral activity particularly against the immune deficient human virus (HIV-1) and the simplex of herpes virus (HSV-1). Polyanionic sulphated polysaccharides of *Spirulina* are also potential candidates for the anti-HIV microbicides.

Gaydou (France) emphasizes the biological interest relating to phycocynin and allocyanin because of their antioxidant properties. As regards the content of vitamins, a dozen of them have been quantified. This whole of results is in favour of use of *S. platensis* like supplement with protein food, essential fatty acids and in vitamins. Many patents were deposited in this field.

Jian-Hong LI (China), polysaccharides coming from *Spirulina* are very effective components. Among its many functions he quotes the improving of immunity, direct affection of the hematopoietic not specialized cells, inhibition of the development of the tumours and the cells of Hela.

Thomas (India) presented studies related to the action of *Spirulina* on the hyperlipidemy and hepatoxicity induced by ingestion of alcohols or drugs anti-tuber culosis and its anti-stress effects. On the local market (India), more than 50 pharmaceutical companies ensure the promotion of *Spirulina* like diet complement.

Costa (Brazil) studied the addition of *Spirulina* in rabbit's food. He observes a reduction in the rate of serum cholestrol. Moreover, he supposes that *S.platensis* has a regulating role related to the human tumoral cells' P-glycoprotein's over-expression.

The acceptation of *Spirulina* by the human can be increased while using it mixed with food(instantaneous soups, pastes,serves chocolate,cold, drinks and cakes). Added to an amount of 1g per portion, it does not change the taste (J Costa). However 1g of *Spirulina* is insufficient to supply the entire iron requirement. Nutritionists have the possibility to evaluate the costs and the effects of the nutritional programme based on *Spirulina* by taking into account local specificities. Our objective is to point out their attention onto this organism as a potential candidate to be used to fight against malnutrition.

5. Cyanobacteria as Disease Control

Spirulina is used for Anemia, hepatitis, gastritis and other inflammations, chronic skin outbreaks, immune strengthen, hypoglycaemia, obesity, overeating, malnourishment, poor skin tone, chronic skin outbreaks and improve mental function. *Spirulina* has a large amount of the blue pigment phycocyanin in biliprotein which has been shown to inhibit cancer-colony formation. *Spirulina* is also nurturing, tonifying and helps to overcome deficiencies all the while cleansing the body. Most of *Spirulina's* protein is made up of biliprotein which benefits those who indulge excessively in animal proteins and refined foods. This micro algae is best suited for athletes who do vigorous exercise giving rise to inflammation. It will also help to keep their sinews and joints flexible. When combined with cereal grass, *spirulina* is a great protection against most inflammation for endurance sports where extreme body heat is generated. *Spirulina* is also the best for weight reduction when compated to all the other micro algae's and wheat grass's.

Orally, blue-green algae is used as a source of dietary protein, B-vitamins, and iron. It is also used orally for weight loss, oral leukoplakia, and obstetric and gynecological disorders. Blue-green algae is also used orally for attention deficit-hyperactivity disorder (ADHD), premenstrual syndrome, allergic rhinitis, diabetes, stimulating the immune system, stress, fatigue, anxiety, depression, improving memory, increasing energy and metabolism, lowering cholesterol, decreasing cardiovascular disease, wound healing, and for promoting digestion and bowel health.

Generally, *Chlorella* is comparable to that of *Spirulina* but contains less protein, much less beta-carotene but more than twice the nucleic acids and chlorophyll. The heavy cell wall of chlorella binds with heavy metals, pesticides and carcinogens and carries them out of the body. *Chlorella* is the safest algae to use for deficiency patterns when compared to all other algae's. Orally, chlorella is used as a food supplement and source of nutrients, including protein, nucleic acids, fiber, vitamins, and minerals. Chlorella is also used orally for cancer prevention, stimulating the immune system, increasing white blood cell counts (*e.g.*, in people with HIV infection or cancer), preventing colds, to protect the body from the effects of radiation (*e.g.*, during cancer therapy), to protect the body from toxic metals such as lead and mercury, and to slow aging. It is also used orally to increase beneficial flora in the gastrointestinal tract in order to improve digestion, and to help treat ulcers, colitis, Crohn's disease, and diverticulosis. *Chlorella* is also promoted for the prevention of stress-induced ulcers; treatment of constipation, bad breath, and hypertension; as an antioxidant; to reduce serum cholesterol; to increase energy; to detoxify the body; and as a source of magnesium to promote mental health, relieve premenstrual syndrome (PMS), and reduce asthma attacks. It is also used orally for fibromyalgia.

Topically, *Chlorella* is used for treating ulcers, postirradiation dermatitis, vulval leukoplakias, and trichomoniasis. Recently becoming popular, this micro-algae has been found nutritionally less potent in chlorophyll and protein when compared to other micro algae but has larger doses of beta-carotene (vitamin A).

6. Cyanobacteria as Ecological Indicators

It is believed that eutrophication in lakes and seas is caused by pollution. Recent research shows that 65 per cent of the lakes in our country are eutrophic. Wherever conditions of temperature, light, and nutrition are appropriate, increased growth of blue-green algae or dinoflagelletes may occur. This phenomenon is called water bloom or red tide. Some blue-green algae and dinoflagelletes may produce toxins, hepatotoxic cyclic peptides such as *microcystins* and *nodularins.*

7. Cyanobacteria as an Important Source of Nutrients and a Treasure Chest of Human Medicine

Many antibiotics are acquired from actinomyces and fungi, for example, penicillin, streptomycin, chloromycetin, terramycin (oxytetracyclin), cycloheximide, amphotericin, nystatin, blasticidin and kasugamycin. There are hundreds of antifungal antibiotics, most of which are produced by a few species of Actinomycetes of the Streptomyces genus, and fungi of the Penicillium and Aspergillus genera and their close relatives. After these antibiotics were discovered, they served as effective medicines in controlling diseases in humans, domestic animals and cultivated plants.

Microalgae contain more than ten vitamins, five polyunsaturated fatty acids, and many kinds of complex chemicals.

Some of the vitamins that are produced by algae are of particular commercial interest, for example, vitamin B_{12}, vitamin E(tocopherol), nicotinic acid, vitamin C (ascorbic acid), d-Ca-pantothenate, vitamin A, thiamine·HCl(B1), pyridoxine(B6), riboflavin(B2), folic acid and biotin. Vitamins and polyunsaturated fatty acids are a source of nutrition and medicines for humans. It has been reported that under certain conditions up to 85 per cent of the dry weight of microalgae is lipids, and 30.229 per cent of the total oil and fat content is Ψ-linolenic acid.

Pharmaceutical and biologically active compounds have already received considerable attention from researchers in the areas of pharmacy, antibiotics and biologically active compounds. Both cell extracts and growth media extracts of many unicellular algae (*e.g. Chlorella vulgaris, Chlamydomonas* pyrenoides) in vitro are effective against both Gram-positive and Gram-negative bacteria. Studies in this area began at the Roche Research Institute of Marine Pharmacology. A wide range of in vitro active anti-fungal activities have also been reported from extracts of green algae, diatoms and dinoflagellates. Similarly, many microalgae have been screened to date and most of them contain and/or excrete pharmacologically active compounds. For example, the dinoflagellates *Gymnodinium* sp. and *Gonyaulax* sp. produce alkylguanidine compounds that affect the central nervous system, and extracts from *Dunaliella tertiolecta* and *Rivularia* firma show a range of pharmaceutical activities. In *R. firma* the activity is due to brominated bi-indoles.

The red unicellular alga *Porphyridium cruentum and Gracilaria lichnoides* have been found to produce lipid arachidonic acid, which is conventionally obtained from animal tissue such as the adrenal gland and liver. It is both an essential component of the humans diet and the precursor of a number of C20 compounds like prostaglandins, prostacyclins and thromboxanes. A kind of immobilized microsome prostaglandin is synthesized from alga-derived arachidonid acid. Optimal growing conditions (*e.g.* 320C, 8000lux) give a maximum arachidonic acid yield of (0.46mg·l-1·h-1) in actively growing cultures. It is also produced in high concentration by the freshwater chryophytes *Ochromonas danica, O. malbamensis* and *Poteriochromanas stipil*. Microalgae such as *Ochromonas* sp., *Prymnesium parvum Alexandrium catenella, Dinophysis aenta,* and *Gymnodinium breve* and a number of blue-green algae produce toxins, but they have potential pharmaceutical application. Some blue-green algae, especially *Lyngbya spiralis*, have also shown to have anti-parasitic activity against *Trichomonas* sp.

Organisms which thrive in extreme alkaline environments offer us the opportunity to appreciate the range of adaptive possibilities that evolution can bring to bear on fundamental biological processes and they constitute unique models for investigations on how biomolecules are stabilised when subjected to extreme conditions (Grant, 1993). Alkaliphilic micro-organisms offer a multitude of actual or potential applications in various fields of biotechnology. Micro-organisms are highly efficient in their ability to produce many kinds of bioactive compounds. A large number of antibiotics have been shown to be produced by various types of bacteria, such as actinomycetes. Screening bacteria from alkaline habitats or those grown under extreme cultural conditions remains a profitable area for investigation. Some new antibiotics were produced by certain bacteria when an alkaline medium with high alkalinity (pH 9 to 10.5) was used(Horikoshi,1991). It is estimated that protein production for human nutrition must be increased in coming years. One of the ways to augment protein production is not only to increase protein production through conventional sources but also to produce proteins by microbial biomass. The production of microbial biomass as a source of protein is still conceivable on both the industrial scale and on a lowskill level employing mass developments in natural ponds. *Spirulina* has long been a staple food in Africa and many countries(Jones,1994).

Spirulina is today receiving growing attention for its unique nutritional characteristics. It is now being sold mainly in health food stores as powder, granules or flakes and as tablets or capsules. Commercial production of *Spirulina* today takes place in Mexico, Taiwan, Thailand, the USA, China, Japan, Australia and Israel(Grant,1990). A special value of *Spirulina* as a plant source is related to the absence of cellulose in its cell wall. Cell protein is thus readily digested and assimilated by the human body, a feature which is important for people suffering from intestinal malabsorption or older people under strict diets with difficulty digesting complex protein. There are some advantages in the use of *Spirulina* spp. as conventional source of protein. *Spirulina* powder has the highest protein content (60-70 per cent) of any natural food, far more than fish flesh (15-20 per cent), soybeans (35 per cent), dried milk (35 per cent), peanuts (25 per cent), fresh eggs (12 per cent) or grains (8-14 per cent). The spectrum of amino acid of this organism is similar to that of other micro-organisms (*i.e.* somewhat deficient in methionine, cysteine and lysine). Thus, it appears that the high protein content together with its aminoacid composition make *Spirulina* as a source of nonconventional protein of considerable interest. Due to uricacid production in the course of purine metabolism and impending side effects such as renal stones and gout, a major concern lies with the total nucleic acid content (4.6-6 per cent of dry weight), which is lower than that of bacteria and yeast. In addition to proteins, the composition of *Spirulina* powder and its vitamin content reflect another important benefit as a human foodstuff as it is a rich natural source of vitamins. Another crucial nutritional feature of *Spirulina* concerns its essential fatty acid composition. The presence of high concentrations of g-linoleic acid (4 per cent of dry weight) synthesised in *Spirulina* is of considerable importance for human nutrition as it is one of the best sources of g-linoleic acid after human milk and vegetable oils. Fatty acids perform important physiological tasks as precursors of the thromboxanes and leukotrienes, which have a number of essential functions in the body and are involved in the immune system's defence mechanisms and inflammatory reactions.

The organism has one very useful characteristic: *Spirulina* grows naturally in alkaline lakes containing Na_2CO_3, $NaHCO_3$ or other minerals and a source of fixed nitrogen and it is readily cultured in hypersaline alkaline environments which prevent the invasion and growth of other contaminant organisms and the lake flora tends to be monospecific, permitting the recovery of *Spirulina* by simple filtration (de la Nove *et al.*, 1992 and Rhee and Gotham, 1981). The yields are impressive. A pond devoted totally to the growth of *Spirulina* spp. can produce 125 times as much protein as the same amount of area devoted to corn, 70 times as much protein as fish farming and 600 times as much as cattle ranching. The value of the final product does not give justification or impetus to expensive research and development or expensive processing. It is of urgency for developing countries to assess the availability of high quality nutritional material: The economics of *Spirulina* both on small as well as big farms make it a potentially profitable option.

8. Nitrate Removal from Groundwater

Nitrate (NO_3^-) concentrations in groundwater have increased globally (Kapoor and Viraraghavan, 1997). Wastewater, fertilizers, and livestock farming are major sources of nitrate in groundwater supplies (Hem, 1992). Groundwater in many locations is used as a supply for drinking water, and high nitrate concentrations present a potential risk to public health, particular to infants (Gangolli,1994). In the United States, the Environmental Protection Agency has set a maximum contaminant level (MCL) for nitrate in drinking water of 0.71 mM (10 mg of NO_3^--N liter^{-1}) (Pontius, 1993.). Similar MCLs for nitrate have been established in Canada and Europe. As a result, many public and private groundwater supply wells have been shut off as drinking water sources because their nitrate levels exceed the MCL (Jeanfils *et al.*, 1993). As water demands in many urban and agricultural areas increase, technology for treating nitrate-contaminated groundwater is becoming increasingly urgent.

Nitrate removal from groundwater may be accomplished by bacterially mediated denitrification, or chemically and physically based technologies (Jeanfils, Canisius, and Burlion. 1993). These treatment processes, however, require input of external energy sources (*e.g.*, electricity or organic carbon) and/or chemical additives and generate concentrated waste streams that must be disposed. Therefore, they are often problematic and expensive (Grant,1990 and Hoffmann, 1998). Since nitrate may be taken up effectively by photosynthetic microorganisms, such as cyanobacteria, which require mostly fixed nitrogen, inorganic carbon, and light for growth, the use of photosynthetic organisms would minimize the need of chemicals and fossil fuels for nitrate removal, thus leading to an efficient resource recovery and recycling. However, limited information on engineering photosynthetic system for treating nitrate-contaminated drinking water supplies is available. With cyanobacteria, nitrate is taken up by a common high-affinity transport system involving the NrtABCD permease (an ABC-type transporter) and to a lesser extent enters the cells by diffusion(Lomas, 1996, and Luque, 1994). Once inside the cell, nitrate is reduced to nitrite by nitrate reductase, and nitrite is further reduced to ammonium by nitrite reductase (Herrero, and Flores. 1997). Ammonium is then incorporated into carbon skeletons mainly through the operation of the glutamine synthetase-glutamate synthase cycle (Flores and Herrero, 1994). Fixed nitrogen storage components, such as C-phycocyanin (Boussiba, and Richmond, 1980) and/or cyanophycin (Lawry and Simon,1982), may be formed and accumulated in the cells under certain physiological conditions.

Regulation of nitrate assimilation has been found to be controlled by several regulatory products that affect expression of the *nrtABCD*, *narB* (encoding nitrate reductase), and *nirA* (encoding nitrite reductase) genes that together form the *nirA-nrtABCD-narB* operon in several cyanobacteria (Herrero, 1997 and Suzuki, 1993). Nitrate uptake can be influenced by availability of phosphate ion ($PO_4^{3}-$), which has an important role in cellular energetics as part of ATP (adenosine triphosphate) and which influences the activity of many enzymes required for cell metabolism, including the nitrate reduction process(Ahlgren, 1988).Also, nitrate uptake can be affected by spectral quality and light intensity (Azuara, 1983 and Rhee, 1981)., temperature (Lomas, 1996 and Talbot,1993), and cell density and nitrate level (Lau, 1995.), as well as physiological acclimation of the culture (Lavoie,1985).

Microalgal treatment of wastewater has been investigated for over 4 decades as an environmentally sound alternative to remove nutrients and heavy metals from wastewater sources. However, very few investigations have considered applying this technology to the treatment of drinking water (Hoffmann,1998). The chemical composition (*e.g.*, the pH, dissolved inorganic carbon, nutrient levels, and metals) of groundwater differs from that of wastewater, and the feasibility of growing microalgae (a term used for cyanobacteria and unicellular algae) in groundwater has yet to be evaluated.

9. References

Ahlgren G 1988 Phosphorus as growth-regulating factor relative to other environmental factors in cultured algae. *Hydrobiologia* 170:191-210.

Azuara M P and Aparicio P J 1983 In vivo blue-light activation of *Chlamydomonas reinhardtii* nitrate reductase. *Plant Physiol* 71:286-290.

Boussiba S and Richmond A 1980. C-phycocyanin as a storage in the blue-green alga *Spirulina platensis. Arch Microbiol* 125:143-147.

de la Noüe J, G. Laliberté, and D. Proulx. 1992. Algae and waste water. J *Appl Phycol*. 4:247-254.

Flores, E and A Herrero. 1994. Assimilatory nitrogen metabolism and its regulation; pp487-517 *In D. A. Bryant* (Ed) The molecular biology of cyanobacteria Kluwer, Dordrecht, The Netherlands.

Gangolli S D, P A van den Brandt, V J Feron, C Janzowsky, J H Koeman, G J A Speijers, B Spiegelhalder, R Walker, and J S Wishnok. 1994. Nitrate, nitrite, and N-nitroso compounds. *Eur J Pharmacol Environ Toxicol Pharmacol* Sect 292:1-38.

Grant W D 1993. Hpypersaline Environments; pp 13-17 *In Trends in Microbial ecology*, R Guerrero and C Pedros-Allio (Eds).

Grant, WD, Mwatha, WE, Jones, BE1990 Alkaliphiles: ecology, diversity and applications. *FEMS Microbiol Rev*75: 255-270.

Hem J D 1992 Study and interpretation of the chemical characteristics of nature water. *USGS water supply paper 2254*, 3rd ed US Government Printing Office, Washington, DC.

Herrero A and E Flores 1997 Nitrate metabolism,, *Cyanobacterial Nitrogen Metabolism and Environmental Biotechnology; pp1-33 In* A K Rai (Ed) Narosa Publishing House, New Delhi, India.

Hoffmann J P 1998 Wastewater treatment with suspended and nonsuspended algae. *J Phycol* 34:757-763.

Horikoshi K 1991 General view of alkaliphiles and thermophiles; pp3-31 *In Suoerbugs: Micro-organisms in Extreme Environments*, KHorikoshi and WD Grant (eds) Springer Verlag Berlin.

Jeanfils J, Canisius M F and Burlion N 1993 Effect of nitrate concentrations on growth and nitrate uptake by free-living and immobilized *Chlorella vulgaris* cells. *J Appl Phycol* 5:369-374.

Jensen G S, Ginsberg DI, *et al.*, 2000 Consumption of *Aphanizomenon flos aquae* has rapid effects on the circulation and function of immune cells in human. *JANA* 2:50-58.

Jones B E, Grant W D, Collins N C, Mwatha, WE Alkaliphiles.1994 Diversity and Identification. *In Bacterial Diversity and Systematics*,(Ed FGPriest),pp 195-230 Plenum Press New York.

Kapoor A and Viraraghavan T 1997 Nitrate removal from drinking water—review. J Environ Eng 123:371-380.

Lau P S, Tam N F Y, and Wong Y S 1995 Effect of algal density on nutrient removal from primary settled wastewater. *Environ Pollut* 8:59-66.

Lavoie A and Noüe J de la 1985 Hyperconcentrated cultures of *Scenedesmus obliquus: A new approach for wastewater biological tertiary treatment? Water Res* 19:1437-1442.

Lawry N H and R D Simon 1982 The normal and induced occurrence of cyanophycin inclusion-bodies in several blue-green algae. *J Phycol* 18:391-399.

Lomas M, Glibert P M, and Berg G M 1996 Characterization of nitrogen uptake by natural population of *Aureococcus anophagefferens* (Chrysophyceae) as a function of incubation duration, substrate concentration, light and temperature. J Phycol 32:907-916.

Luque I, E Flores and A Herrero 1994 Nitrate and nitrite transport in the cyanobacterium *Synechococcus* sp PCC 7942 are mediated by the same permease. *Biochim Biophys Acta* 1184:296-298.

Pontius, F W 1993 Nitrate and cancer: is there a link. *J Am Water Works Assoc* 85:12-14.

Rhee G Y, and Gotham I J 1981 The effect of environmental factors on phytoplankton growth: light and the interactions of light with nitrate limitation. *Limnol Oceanogr* 26:649-659.

Singh P K, Dhar D W, Pabbi S, Prasanna R and Arora A 2000 *Biofertilizers: Blue Green Algae and Azolla*, IARI, Venus Printers and Publishers, New Delhi.

Sobel D 1995 Rethinking medicine: improving health outcomes with cost-effective psychosocial interventions. *Psychosomatic Medicine* 57:458-469.

Suzuki I, Sugiyama, T and Omata T 1993 Primary structure and transcriptional regulation of the gene for nitrite reductase from the cyanobacterium *Synechococcus* sp PCC 7942. *Plant Cell Physiol* 34:1311-1320.

Talbot P and de la Noüe, J 1993 Tertiary treatment of wastewater with *Phormidium bohneri* (Schmidle) under various light and temperature conditions. *Water Res* 27:153-159.

Whitman W, Coleman D and Wiebe, W. Prokaryotes1998 The unseen majority *Proc Nati Acad Sci* 95:6578-6583.

Weinbauer, M and Herndl, G Viruses and heterotrophic microplankton 2002 *Encyclopedia of life Support systems* In Press.

Chapter 20

Spirulina: A Blue Green Revolution

S. Singh, A. Ganguly and A. Tiwari

Algal Biotechnology Laboratory, Department of Biological Sciences, Rani Durgavati University, Jabalpur – 482 001 (M.P.), India

Contents

Glimpses of Cyanobacteria, 2006 321–349
Daya Publishing House, New Delhi

ABSTRACT

Spirulina is the only single ,natural source providing the highest amount of protein (71%) ever known to man . The potential biotechnological applications of *Spirulina* in the field of food, feed, medicines, cosmetics, biochemical and in and in the environmental deplollution are well established and documented. It is commercially important because of its tremendous nutritional values, therapeutic qualities, high protein content, low fat, high vitamin B12 as well as essential fatty acids like g-linolenic acid.The phytonutrients (pigments and polysaccharides) of *Spirulina* makes this alga useful in cosmetic ,nutraceutical, pharmaceutical and aquaculture industries. It is largely marketed in different forms of recipes and products. *Spirulina* has been grown in different modes of cultivation system under varying environmental conditions outdoors. But still much more is required to be done as far as its cultivation, popularization and commercialization is concerned in India. Strain improvement, identification of efficient cultivation system and optimization of mass cultivation in different climatic zones of India needs special attention. Apart from richest source of nutrients for malnourished population, training and popularization of *Spirulina* cultivation specially among weaker section of society could be proved an additional income generating system. Present article summarizes potentials and prospects of *Spirulina* and specially efforts of our group in the development of biotechnology of *Spirulina* cultivation in Madhya Pradesh (India).

1. Introduction

In nature myriad of microorganisms are present in habitats ranging from hot deserts to cold climates. Since time immemorial mankind have been striving for searching different types of food material, ranging from microorganisms to higher plants and from lower animals to higher animals, to fulfil their basic needs. Since the beginning of the civilization mankind has been relying on various plants and their products but recently man has been trying to utilize non-conventional food from microorganisms including micro-algae and their products.

Inspite of the vast potential resources, very few micro-algal strains especially blue-green algal strains have been commercially exploited. The studies have mainly been focused on the strains contributing immensely to the nitrogen budget of various aquatic ecosystems. In addition, to the nitrogen fixing ability certain species of cyanobacteria are also proved to be a very good source of food, feed, drugs, colours and industrially important compounds etc.

History of consumption of cyanobacteria dates back to 1521 when Bernal Diaz del Castillo reported that *Spirulina maxima* was harvested from lake Texcoco (Ciferri, 1983). According to his report *Spirulina* was dried and sold for human consumption in Tenochtitlan (Mexico City). It is still being used as food by the Kanembu tribe in the lake Chad area of Africa and known as *dihe*. *Spirulina* is a spiral shaped, non-heterocystous cyanobacterium. The spiral nature of the filaments contributes to its name *Spirulina*. It can grow to great length and trichome diameter ranges from 6–12 μm (Tomaselli, 1997) and are easy to harvest. It grows naturally in warm and highly alkaline lakes in sub-tropical areas. *Spirulina* has been regarded as a good candidate for various biotechnological applications and their potential in the conversion of light energy into renewable forms of useful chemicals for food, feed, pharmaceuticals and other industries has often been claimed, assessed and well-documented. These organisms synthesize compounds like amino acids, proteins, sugars, complex polysaccharides, vitamins and growth promoting substances for their own metabolism. It is the only since natural source providing the highest amount of protein (71 per cent) ever known to man. The protein content in *Spirulina* is three times that of Soyabean, five times that of meat and the protein quality is among the best (Figure 20.1). The protein yield per unit area per year is the highest as compared to other protein yielding crops. Its immense commercial importance is due to its tremendous nutritional values. (Table 20.1), therapeutic qualities, low fat, high vitamin (B_{12}), provitamin as well as essential fatty acid like γ-linolenic acid (Vonshak and Richmond, 1978). *Spirulina* also is useful in aquaculture particularly as feed for tropical fishes and special feed for shrimp farming. In recent years due to the increasing interest in exploitation of microorganisms for various industrial applications (Sutherland, 1998), the attention towards the polysaccharide producing cyanobacteria greatly increased in particular towards the strain that produce abundant capsule slime and also release large amount of polysaccharide. The polysaccharides are traditionally used as viscosity agents for relation, thickening (Sutherland, 1996) and emulsifying. New

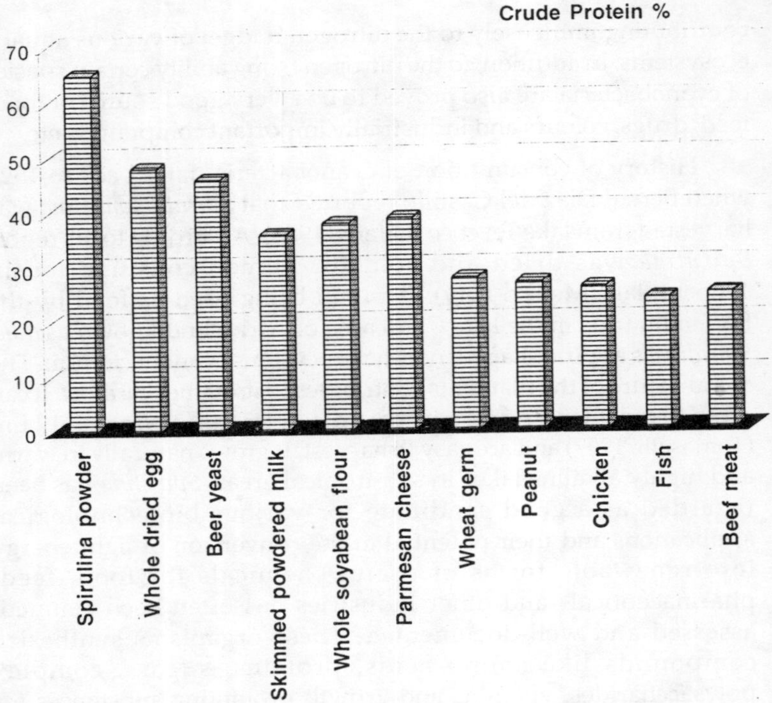

Figure 20.1: Crude Protein Content in Different Protein Source

bioactivities are currently being developed for a variety of commercial uses as in pharmaceuticals as antiviral (Hayashi and Hayashi, 1996; Hayashi *et al.*, 1996) and anti inflammatory agents and for testing the cytotoxicity of leukemic cells through apoptosis, in cosmetic (Sutherland, 1994), health foods and ion absorption applications, bioadhesive oral drug delivery system (Garesh, 1998). The polysaccharides have promising potentials in their utilization such as wastewater treatment, clarification of water in reservoirs and sedimentation of colloids in chemical and food industries. The sulphated polysaccharide from *S. platensis* possess inhibitory properties against various types of viruses such as Herpes simplex virus type–1, Human Cytomegalovirus, Meseals virus, Mumps virus, Influenza virus and HIV–1. This novel sulphated polysaccharide named as Calcium Spirulan has been reported by Hayashi *et al.* (1996). *Spirulina* is also of special interest for entomologist and

Sericulturists due to its enhancing effect in the yield of mulberry consuming silkworm and growth of various insects used in researches. *Spirulina* are extracted for the production of phycocyanin known commercially as "lime-blue" used as a blue colorant for food and cosmetics like pimple lotion, facial mask, hair oil etc. *Spirulina* is largely being produced worldwide and marketed in different forms of recipes and products. A large number of formulations are available for the preparation of food using *Spirulina* powder like soup, noodles, candies, appetizers, vegetable pate, low calorie bread and bread spread (Henrikson, 1994). One of potential application that has gained increasing attention from medical scientists during recent years, is its utilization in pharmaceuticals and nutraceuticals. *Spirulina* is also used as an agent for combating water pollution.

Table 20.1: Properties of *Spirulina*

	Physical Properties	
1.	Appearance	Fine powder or jelly like
2.	Colour	Dark blue green
3.	Odor and taste	Mild like seaweed
4.	Bulk and density	0.35–60 kg lt^{-1}
5.	Particle size	64 mesh through
	General Properties	
1.	Protein	65–70 per cent
2.	Carbohydrate	15–25 per cent
3.	Fats	06–08 per cent
4.	Mine	07–13 per cent
5.	Moisture	05–07 per cent
6.	Fibers	08–10 per cent

Owing to the varied potentials of *Spirulina* its biomass is produced commercially in Japan, China, USA, Mexico, Africa, and several other countries, (*e.g. Chlorella* and *Spirulina* sp.) primarily as health food. This article summarizes and discusses the progress made so far in the field of *Spirulina* cultivation and highlights various areas of commercial applications.

Outdoor cultivation of *Spirulina* is greatly influenced by major factors (*viz.*, Light, pH, temperature, nutrients and osmoticum of the medium) affecting growth. Due to high temperature, rate of evaporation changes osmoticum of the growth medium (Vonshak, 1987) which could be detrimental to the very survival of the culture. Development of high salinity tolerant strains could sustain such growth constraints. For the first time high salinity (Li^+–R, Na^+–R) tolerant strains of *Spirulina platensis* have been successfully isolated and characterized (Singh *et al.*, 2000; Chaubey, 2000). Furthermore, nitrate nitrogen regulation of growth, photosynthesis, macromolecular contents and nitrate reduction have been investigated in these genetically modified strains (Deomurari, 2001) since efforts in this direction is very much required for commercial exploitation of *Spirulina* sp.

2. Biomass Production

Biomass defined as a total organic matter synthesized by living organisms in a unit area in a unit time. Since the size of economically important microorganisms is very small therefore, their biomass production is slow and expensive. For the commercial exploitation of the microorganism it is necessary to enhance its biomass production in a unit area. During mass cultivation, *Spirulina* cultures are exposed to high temperature, high irradiance and salinity. Work on development of high salt tolerant strains has been carried out in our lab (Patel Rekha, 2003). Mass production of *Spirulina* requires an inexpensive yet reliable enclosure for growing the cultures. The design of these enclosures represents a compromise between the cost of the investments in relation to the expected returns and the desire to establish conditions for highest possible output rates. *Spirulina* grows luxuriantly in all kinds of fresh water habitat. In last two decades much progress has been made for the mass cultivation of *Spirulina*. The commercial production of *Spirulina* is much easier as compared to other microorganisms. It requires less energy for growth and development (Venkataraman, 1989). In addition this organism has a unique capacity to grow at high temperature, high alkaline habitat (Ciferri, 1983; Richmond, 1988). Large scale and laboratory production of *Spirulina* under controlled and natural conditions have been demonstrated in France, (Clement, 1975); Japan, (Clement, 1978); Taiwan, (Soong, 1980); Germany, (Soeder, 1978); Israel (Hu *et al.*, 1997); Malaysia (Rafiqul *et al.*, .2005) and India (Venkataraman and Nigam, 1978)

2.1 Culture Media

Spirulina has been known to grow profusely in the highly alkaline water bodies. So high pH is an essential feature required for the growth of *Spirulina* as has been reported by Zarrouk (1966). Being non-heterocystous forms it requires the presence of nitrate source in the growth medium. Various media generally used for cultivation of are Zarrouk's medium (1966), C.F.T.R.I. medium (Venkataraman, 1983)& Rao's medium (Rao,1979). Studies on comparative growth on different media have also been performed in our lab to identify best suitable medium (Singh *et al.*, 2004) Previously the cultivation of *Spirulina* was limited to fresh water using expensive inorganic nutrients. Now a days different anthropogenic wastage are supplemented with suitable nitrogen and bicarbonate sources for the mass cultivation of *Spirulina* (Mc Gary, 1971; Korasic *et al.*, 1979; Oron *et al.*, 1974; Chung *et al.*, 1978; Faucher *et al.*, 1979; Tel-or *et al.*, 1980).

Dharampal *et al.* (1992) had suggested N : P : K as an alternative cheap source of nitrate in place of $NaNO^3$. The outdoor mass cultivation of *Spirulina* is possible because it requires high bicarbonate and high pH for its growth. These conditions act as a barrier against the contamination of other microorganisms.

2.2 Temperature and Light

Temperature of the culture has a direct effect on the growth of *Spirulina*. It is a mesophillic alga and requires temperature of 35–37°C for optimal growth. The impact of temperature has been studied by Vonshak *et al.* (1983). Mass cultivation of photoautotrophs as compared to that of heterotrophs is hampered due to the profound effect of light. The effect of light and dark cycles are considered as the most important factors limiting the outdoor productivity of algal biomass (Richmond, 1990). The effect of both these factors on photosynthetic activity has been extensively studied by Laws *et al.* (1983), Kok (1953), Terry (1986), Qiang *et al.* (1998), and Bolychevtseva (2003).

2.3 Reactors

Spirulina production in large scale has increased many folds all over the world. The leading producers being China with a turn over of 800 tonnes, followed by USA 500 tonnes, India and

Bangladesh 450 tonnes per year as per the data in the year 2002. *Spirulina* has been cultivated in various mode of cultivation system outdoors or in laboratory conditions using clean water or waste water medium (Singh *et al.*, 1997). In mass cultivation of *Spirulina*, it is difficult to obtain high productivity in open ponds because of high fluctuations in temperature and light intensity throughout the day and year. For maximum exploitation of photosynthetic potentials of the cells it is desired to efficiently utilize solar irradiance available outdoors by designing and developing suitable photobioreactor (Richmond *et al.*, 1991).

Many types of photobioreactor including closed photobioreactor have been used for the cultivation of *Spirulina* (Torzillo *et al.*, 1986) to avoid contamination by external sources. In our laboratory Glass tubular photo-bioreactor have been designed and used for the cultivation of microalgae *viz.*, *Nostoc muscorum, Anabaena variabitis* and *Spirulina* etc. (Thakur, 2000; Ravi Kiran, 2001; Verma, 2002) under laboratory conditions. Simultaneously successful efforts have been made to grow *Spirulina* outdoors in earthen pots, glass tanks and plastic bottles as well. In open ponds it is difficult to obtain high productivity because the temperature and light intensity vary throughout the day and year. Further open ponds require a large surface area and chances of contamination be there (Table 20.2). For efficient utilization of solar irradiation by the microalgae it is necessary to develop and use suitable photobioreactors. The most common closed, photobioreactors, which has been used for the cultivation of microalgae, are given below:

1. Vertical or Horizontal tubular
2. Helical or Serpentine (Hoshimo *et al.*, 1992 and Carlozzi and Torzillo, 1996)
3. Inclined or horizontal thin panel (Hu Qiang *et al.*, 1997).

The sufficient light supply is a critical factor in the designing of photobioreactor.

Extensive investigations have been conducted in Israel on the possibility of large scale cultivation of *Spirulina platensis*, utilizing brackish water unsuitable for human consumption or agricultural use. In Negev desert the cultures was grown in shallow (0.2 m) black plastic lined channels 50–300 m long (Ciferri, 1983). In the developing country like India, cultivation of *Spirulina* on large scale in open ponds and solar heat drying is most suitable for village level industries (Chaudhary *et al.*, 1983; Becker and Venkataraman, 1984; Fox, 1988).

Table 20.2: Advantages and Disadvantages of Open and Closed Systems

Parameters	*Open System (Cuvette, container, stirred vessel, natural water, raceway pond and inclined surface devices)*	*Closed System [Plastic sleeves, Fermenter like tank, tubular (PBR) and Laminar (PBR)]*
Contamination risk	Extremely high	Low
Space required	High	Low
Water loss	Extremely high	Almost none
CO_2 loss	High	Almost none
Biomass quality	None susceptible	Susceptible
Variable as to cultivable species	Cultivation possibilities are restricted to few species	High, nearly all microalgal species may be cultivated
Flexibility of production parameters	Change of production between the possible varieties nearly impossible	Change of production without any problem
Process control	Dependent on exterior condition	Manually controlled
Standardization	Not possible	Possible
Weather dependence	Absolute	Insignificant
Threshold production period	6–8 weeks	2–4 weeks
Biomass concentration during production	Low, approx. 0.1–0.2 g L^{-1}	High, approx. 2–8 g L^{-1}
Efficiency of treatment process	Low, time consuming, large volume flow due to low concentration	High, faster
Reproducibility of production parameters	Dependant on exterior condition	Possible within certain tolerance

The use of algae for the production of commercially important commodities has great potential (Chaumont, 1993 and Venkataraman *et al.*, 1995). Initially biological systems like bacteria, fungi, plants and animals have been used extensively for bioprocessing. Although the concept of utilizing algae as a biochemical reactor is not a new one but the progress in this respect is relatively very slow. The reasons for the slow progress of algal technologies could be due to poor performance of axenic cultures, light dependent growth and lack of culture techniques.

3. Health Food

The growing demand of food material forever increasing population has forced the scientists to search for some alternative non-conventional resources of food. Till now men has been exploiting only plant resources of terrestrial ecosystem for their food requirements. Recently fresh water and marine plant resources have been exploited to fulfil our dietary requirement. In this connection *Spirulina* is proved to be a very promising microorganism.It has a very high protein contents, low fat and high vitamins content and is also rich in essential fatty acids like Ψ-Iinolenic acid (Vonshak and Richmond, 1978).

The United Nations World Food Conference in 1974 declared that *Spirulina* was "The best food for tomorrow". *Spirulina* has the approval of the food and drugs authority of the United States for being sold as a natural food. In Japan and in the United States, *Spirulina* tablets are often consumed by business executives to combat the increasing stress. Today it is a part of diet for space travelers and U.S. Olympic team for quick energy synthesis. German's used it for treating obesity and is given to nursing mothers in many South East Asian Countries. The accumulated knowledge on the ecology, physiology, biochemistry and biotechnology has made *Spirulina* as health food, yet more work is required in terms of strain selection and better understanding of the factors which govern growth and light utilization in dense cultures to improve this growing agro-industries. Various food and feed additive of *Spirulina* are given in Table 20.3.

4. Therapeutics

Soil Actinomycetes and bacteria have been the source for majority of antibiotics and therapeutic compounds. In addition researchers are searching for some antibiotic therapeutic compounds from a group of microbe that has been overlooked in the past, the cyanobacteria. Thus cyanobacteria, especially *Spirulina* has emerged as one of the most promising agent for synthesizing potentially new therapeutic compounds (Table 20.4). The water soluble pigments and the water soluble polysaccharide known as "Calcium Spirulan" have been reported to show immense therapeutic potentials. Efforts for optimizing this polysaccharide under different light regimes have been made in our laboratory (Thadani, 2002)

Table 20.3: Some Reported Applications and Benefits of *Spirulina* as Food and Feed

Sl.No.	*Formulation*	*Consumer*	*Benefits*	*References*
1.	Novel Food Pastalina: Whole Wheat Noddle/*Spirulina* Protein Powder 10% Alga/*Soya*/ Egg), Pills Powder Feed	Human	Quick energy synthesizer, reduces stress	Hill *et al.*, 1980; Kaempfel *et al.*, 1995; Delay *et al.*, 1996
2.	Powder Feed	Fish, Prawn and Shrimp	Enhancement of colour and fulfill the nutrient deficiency and egg production	Ciferri, 1983; Saxena, 1983; Henrikson Mgbenka *et al.*, 1997, Gomeskaran, Nandeesha *et al.*, Olvera-Novoa *et al.*, Langdon *et al.*, 1999; Nandeeesh, 2001
3.	Powder Feed	Poultry	Reduces early mortality and increases the characteristic color of the broiled chicken	Lyamin *et al.*, 1997; *et al.*, 2000, Al-B *et al.*, 2001
4.	Feed additives (Pellet form)	Weanling pigs	Improved disease resistance	Grinstead *et al.*, 2000
5.	Feed additives (spray dried, pellet form, sun dried powder and moist *Spirulina* biomass)	Pets and animals	Improved feather quality, coat quality, appetite, healing ability and fertility	Berestov *et al.*, 1999

In India, the major nutritional deficiency diseases identified have been the protein deficiency, malnutrition, iron deficiency, anemia and endemic goiter. Microalgae have the longest history as

Table 20.4: Some Reported Therapeutic Applications of *Spirulina*

Sl.No.	*Health Disorders*	*Spirulina Formulation*	*Effects*	*References*
1.	Acute anemia and pernicious anemia	Powder	Improve the haemoglobin level	Becker, 1984; Seshadri and Seshagiri, 1986; Richmond 1992; Belay, 1997
2.	Obesity	Tablets	Significant reduction in body weight	Becker *et al.*, 1986
3.	High Cholesterol	Tablets	Lowering the cholesterol level and improving the lipid profile	Noriaki *et al.* 1988, Henrikson. 1994 Ramamoorthy, 1996; Fong *et al.*, 2000
4.	Pancreatitis, Hepatitis, Cirrhosis	Tablets	Curative effect	Venkataraman and Becker, 1988
5.	Immune function	Tablets and Polysaccharides of *Spirulina*	Enhances immune response particularly primary immune response by stimulating macrophage function	Baojaing *et al.*, 1994; Pascaud, 1993; Hayashi *et al.*, 1994 Henrikson, 1994; Schiffrir *et al.*, 1997; Saeki *et al.*, 2000, Hirahashi *et al.*, 2002
6.	Cancer	Tablets	Chemopreventive	Schwartz *et al.*, 1988; Schwartz, 1990; Lisheng *et al.*, 1991; Fedkovic *et al.*, 1993; Babu 1995; Mathew *et al.*, 1995; Mishima *et al.*, 1998; Lee *et al.*, 2000

Contd...

Table 20.4–Contd...

Sl.No.	*Health Disorders*	*Spirulina Formulation*	*Effects*	*References*
7.	Viral infection	Water extract Calcium Spirulan	Showed activity against: Herpes Simplex HIV, HSV, Human Cytomegalo virus, Influenza A virus, Mumps virus and Measels virus	Hayashi *et al.*, 1993; Hayashi *et al.*, 1996b; Lee *et al.*, 2001
8.	Bacterial infection	Cell extract, polysaccharide	Showed activity against: *Bacillus subtillis, Streptococcus aureus, Saccharomyces cerevisiae* and *Candida albicans, E. coli, Salmonella* sp., *Shigella* sp., *Klebsiella* sp., *Vibrio* sp., *Bacillus* sp. and *Streptococcus* sp.	Balloni *et al.*, 1980; Gerwick *et al.*, 1987; Singh *et al.*, 2005
9.	Diabetes	Water soluble and insoluble fraction of *Spirulina*	Decrease in serum glucose level in both fasting and glucose loading case	Hosoyamada *et al.*, 1991; de Carie *et al.*, 1995; Taka *et al.*, 1996
10.	Heavy metal and Drugs induced toxicity	Water soluble fraction of *Spirulina* extract	Significant decrease in blood urea nitrogen (BUN) and serum creatinine levels	Richmond, 1984; Fukino *et al.*, 1990; Henrikson, 1994 Yamane *et al.*, 1998; Shastri *et al.*, 1999; Upasani, *et al.*, 2001
11.	Prolong radiation exposure	Crude ethanol precipitate (CEP) of *Spirulina platensis*	Significant reduction in radiation induced micronucleus frequencies and the possibilities of radical scavenging mechanism	Schwratz *et al.*, 1987; Qishen *et al.*, 1989; Zhang, *et al.*, 2001

food supplement, since the nutritional properties have been proved to be equal to other conventional plant proteins. The nutritionally rich, non-toxic alga holds great promise for its therapeutic utilization. It is well established that nutrient deficiency is associated with consistent changes in immune response such as number of T-cells, lymphocyte response to antigens, phagocyte function, IGA antibody response, compliment activity and production of cytokines. *Spirulina* provides such a nutritional role in modulating immune system function favorably (Belay, 2002). It particularly affects the primary response by stimulating macrophage functions, phagocytosis and IL-1 production in mice (Hayashi *et al.*, 1994). Polysaccharides of *Spirulina* can improve the body immune functions (Baojaing *et al.*, 1994). Phycocyanin of *Spirulina* plays a major role in renal failure causes by mercury and pharmaceutical drugs. It has been shown that *Spirulina* can inhibit HIV–1 replication in human derived T-cells and in human peripheral blood mononuclear cells (Ayehunie *et al.*, 1996). Various clinical and biochemical uses of *Spirulina* are summarized in Table 20.4. The antibacterial properties of water soluble polyasaccharide of *Spirulina platensis* on different pathogenic forms were studied in our lab and was found to be positive (Singh *et al.*, 2005).

5. Spirulina in Cosmetics Industry

Recent years has seen a shift in the need for production of natural pigments for use as food colours and in cosmetics industry because of the proven carcinogenecity of coal tar based dyes. This commercial interest has led to the exploitation of microalgae particularly cyanobacteria which are technologically amenable for production of pigments and other byproducts. Chlorophyll *'a'* is present in all algae and in higher plants. Although chlorophylls are not of serious commercial interest nevertheless, chlorophyll *'a'* extract of *Spirulina* having iron–oxides and higher alcohols is patented as strong deodorant (Hasting, 1968).

Lipid soluble carotenoid are composed of isoprene units and can be divided into tow groups *viz.*, carotene containing hydrocarbons and xanthophylls containing epoxy, hydroxy, ketonic, carboxylic, glucosidic derivatives of carotenoids. Cantaxanthin of *Spirulina* is used for colouring of fancy gold fish (Matsuno *et al.*, 1979).

The only known source of phycobillins are algae belonging to Rhodophyceae, Cryptophyceae, and Cyanophyceae (Goodwin, 1974). It can be broadly classified into two groups one blue-coloured phycocyanin and other is red coloured phycoerythrin. It occurs as C-phycocyanin and C-phycoerythrin in various species of *Spirulina*. These two pigments are used in eye shadows, eye liner, lipistics etc. High concentration of nutrients *viz.*, amino acids, vitamins, minerals, trace elements, fatty acids, γ-linolenic acid, chlorophyll and polysaccharides available as balanced natural formula in this micro alga provides an exceptional caring power. The high concentration of chlorophyll represents, more life force than of most other plants. Formulated with proven and safe cosmetics ingredients Spiruskin, German based products meet the individual need of all skin types to provide a natural beauty and balance. We have also formulated in our laboratory one-face cream based on *Spirulina* which is under trial. A product summary of Spiruskin and its application are given in Table 20.5. Although certain cosmetic formulation have been demonstrated at C.F.T.R.I. (Mysore), India yet it could not be much exploited commercially in India by Indian Companies despite its tremendous potentials and large market in the country. This needs greater attention at various levels in the society.

6. *Spirulina* in Bioremediation

Cyanobacteria are known to be the primary producers in various aquatic ecosystems and could be used as biological agents in environmental depollution *i.e.* to help in reducing inorganic and organic load in the ecosystem. Such great potential has already been exploited and documented.

6.1 Inorganic Pollutants

Spirulina can use nitrogen and phosphorus in its various metabolic activities thereby decreasing their levels below the accepted standards in polluted water bodies (Gabbisu and Hall, 1993). In this cyanobacterium management of the photosynthesis and respiration interplay, either to increase photosynthesis respiration or to separate these two processes may affect their ability to utilize exogenous mineral resources. The dibasic salts of calcium and magnesium mainly cause the hardness of the fresh water. The growth of *Spirulina* is inhibited by higher concentration of these two salts therefore, *Spirulina* has inbuilt mechanism to precipitate these salts. These unique characteristic features of the *Spirulina* could be used to reclaim hardness of water. *Spirulina* has been shown to substantially reduce eutrophic nutrients (*i.e.* NO_3^- and PO_4^-) from industrial wastewater (Shrivastav, 1996).

Table 20.5: List of Commercially Available Body Care Products Developed from *Spirulina platensis* by Sanatur GmbH, D-78224 Stingen, Germany

Sl.No.	*Types of Product*	*Use/Impact of Product*
1.	Beauty Cream Mask	Hydroactive Fore-very skin type
2.	Clarifying and Firming mask	For every skin type
3.	Cleansing Milk	Gentle and through for every skin type
4.	2 Phase Eye Makeup Remover	Gentle and soft for every skin type
5.	Soft Plankton Peeling	For clear skin complexion
6.	Toning Lotion	Refreshers and gives a radiant glow for every skin type can be used as an after shave lotion for men
7.	Active Moisturizing Gel	For every skin type, including sensitive skin
8.	Eye Zone Repair Gel	For every skin type, including sensitive skin
9.	Throat and Decolleate Cream	Hydro Lipid complex for a beautiful decolleate contains UVA and UVB protecting factors
10.	Body Firming Cream	Tightens, supports against cellulite
11.	Body Cream	Moisturizes and smoothes supports the natural skin resistance
12.	Vitalizing Leg Gel	Refreshes and tones. Revives heavy legs.
13.	Sensitive Hair and Body Shampoo	For healthy skin and beautiful hair. Mild and gentle.
14.	Nutritive Cream	For dry skin. Reduces wrinkles upto 33.2 per cent when applied daily.
15.	Balancing Cream	For oily and mixed skin. Reduces wrinkles upto 31.2 per cent when applied daily. Ideal for every skin type when using makeup.
16.	Anti Pigment Cream	Whitening complex for face and body with UVA and UVB light protectors. Anti aging care for hands.

6.2 Organic Pollutants

Water pollution is a big problem of great impact mostly due to intensive rearing of farm animals, municipal sewage and industrial effluents. Disposal of wastewater from Swine farms can create serious ecological problems. High concentrations of ammonia, total nitrogen and phosphorus cause eutrophication of recipient water with harmful effects on the whole ecosystem. The biological recycling of various domestic and industrial organic wastes is possible only through some aquatic autotrophic microorganisms (Garett, 1981; Oswald, 1981; Robinson and Toerien, 1982; Mitchell and Richmond, 1988). Domestic waste materials were treated mainly to limit pollution problem and to minimize other hazards (Pelczar *et al.*, 1977). The treated effluents and sludge can be used for irrigation purposes and as plant fertilizers. *Spirulina* is an aquatic cyanobacterium and has ability to grow in the aquatic habitat rich in organic pollutants (Singh *et al.*, 1997). The limitations for the cultivation of *Spirulina* in raw sewage are the availability of limited amount of nitrogen and bicarbonate sources. This limitation can be overcome by the addition of 1 per cent sodium bicarbonate and 0.1 per cent sodium nitrate in sewage wastes (Pal and Amla, 1992; Shrivastava, 1996). The cyanobacterium *Spirulina* grown on wastes can be used as dietary supplements for animals and for many other purposes like energy sources, fuel, fertilizers, fine chemicals etc. (Chiu *et al.*, 1980).

Removal of heavy metals from soil and water or their remediation from waste streams has been a long-term challenge. Many studies have demonstrated that both prokaryotes and eukaryotes have the ability to remove metals from contaminated water or waste streams (Pavel and Tomas, 2000). *Spirulina* has also the property of phytoremediation capacity and can remove low concentration of cadmium from fresh water (Rangsayator *et al.*, 2002) and has biosorption capacity. *Spirulina* has been reported to be very efficient biosorbent of metals Cr^{3+}, Cd^{2+} and Cu^{2+} (Chojnacka *et al.*, 2005) and efficient in removal of lead (Gong *et al.*, 2005). Similar studies of the effect of the metal ions cadmium and copper on *Spirulina* has been studied in our lab (Singh *et al.*, 2003) to determine their toxicity level.

7. Future Prospects of *Spirulina*

Even today starvation and malnutrition are still widespread in vast portion of the world and that, notwithstanding the success of the various "green revolution", present- day agriculture seems incapable of satisfying the most basic human need, adequate nutrition. By 2020 there will be an immense increase in the population with an unpredictable need for food and unfortunately according to various reports, growing area peaked in 1981 and has fallen more than 5 per cent since in ten years. Yields are down as irrigated area has began to shrink and fertilizers are being curtailed due to diminishing returns. Big losses of fertile land continued to result in deforestation, desertification and soil depletion. Thus the production of microorganisms in general and *Spirulina* in particular to be used as food or feed has evident advantages. The future of cyanobacterial biotechnology depends on expansion of the range of products to be manufactured as well as improvements in bioprocess engineering. So far, few cyanobacterial strains producing biotechnologically important compounds have been documented, characterized and exploited commercially. Screening programs must identify new cyanobacterial strains and their products in near future for human welfare.

Spirulina is one food that has gained a lot of nutritional importance in the past few years. It has a potential to bring about a nutritional revolution in the developing countries owing to its multinutritional and clinical properties. But the mass production of *Spirulina* is still a very expensive process making it unaffordable to the poorer class, which is maximum affected with dietary malnutrition. Hence it is necessary to develop, low cost biotechnology for large- scale production of *Spirulina* using locally available resources. Furthermore, mass awareness programs are required to educate common people along with educated class to popularize the miracle nutritional values of *Spirulina*. In India low cost cultivation of *Spirulina* at village level can generate extra employment and income for marginal laborers and women given an opportunity to them. At present less significant amount of biomass is being produced by few private organizations in India as compared to other countries like China, Japan, USA, Italy, Taiwan and Mexico etc. *Spirulina* biotechnology has great potentials and future in India as well in sector like rural self-employment programs. Apparently at least two ways strategies are required to be adopted; one development of appropriate biotechnology for mass production of *Spirulina*, outdoors suitable for climatic conditions of Indian subcontinent, secondly all out effort at various levels to popularize and training of unemployed younger generation. At *Morarka Foundation* in India to encourage decentralized production at lower cost, many novel methods for *Spirulina* cultivation have been evolved. The process has been standardized for its cultivation at household levels by farmers in rural areas this has enabled to offer *Spirulina* production as an income generating activity.

In addition many innovative applications of *Spirulina* have also been developed for wastewater recycling and for enrichment of microbial cultures used in agriculture. This has further expanded the scope of its consumption at reasonable rates benefiting both producers as well as consumers.

Acknowledgements

We are thankful to the Head, Department of Biological Sciences, R.D. University, Jabalpur for laboratory facilities.

References

Al-Batshan HA, Al-Myfarrej SI, Al-Homaidan AA and Qureshi MA 2001 Enhancement of chicken macrophage phagocytic function and nitrite production by dietary *Spirulina platensis*. *Immunopharmacol Immunotoxicol* 23: 281-289.

Ayehunie S, Belay A, Hu Y, Baba T, Ruprecht R 1996. Inhibition of HIV-1 replication by an aqueous extract of *Spirulina platensis*, In: 7th *IAAA Conference Knysna South Africa*, April 17.

Babu M 1995. Evaluation of chemoprevention of oral cancer with *Spirulina*. *Nutrition and Cancer* 24(2): 197-202.

Baojiang G 1994. Study on effect and mechanisms of polysaccharides of *Spirulina platensis* on body immune function improvement. In: 2nd *Asia Pacific Conference on Algal Biotechnology*, 24pp.

Becker E W and Venkataraman L V 1984. Production and utilization of blue-green alga *Spirulina* in India. *Biomass* 4: 105-125.

Becker E W 1986. Nutritional properties of micro algae potential constrain. In: *Handbook of Microalgal Mass Culture* (Ed. A Richmond) CRC Press Florida USA 339 pp.

Belay A, Kato T, Ota V, Grobbelaar JU 1996. *Spirulina* (*Arthrospira*): Potential application as an animal feed supplement (ed. BMA Karoon, B A Whitton) *Journal of Applied Phycology* 8(4-5): 303-311.

Belay A 2002.The potential application of *Spirulina* (*Arthrospira*) as a nutritional and therapeutic supplement in health management. *The Journal of American Nutraceutical Association* 5(2): 27-48.

Berestov V A and Kudryavtser V B 1999. *Spirulina*: a feed additive *Krolikovodstvo-i-Zverovodstvo* 6: 11-12.

Carie G Z, De Cano MS, De Mule MZ and Steyerthal N, Piantanida M 1995. Effect of *Spirulina platensis* on glucoseuric acid and cholesterol and cholesterol levels in the blood of rodents. *Int J experiment Bot* 57: 93-96.

Carzillo, P and Torzillo, G 1996. Productivity of *Spirulina* in a strongly curved outdoor tubular reactor. *Appl Microbiol Biotechnol* 45: 18-23.

Chaumont, D 1993. Biotechnology of algal biomass production: A review of system for outdoor mass culture. *J Appl Phycol* 5: 593-604.

Chaudhari P, Krishnamoorthi K and Kotangale L 1983. *Indian J Environ Health* 25: 275-281.

Chiu R J, Liu H I, Chen C C, Chi Y C, Shao H, Soong P and Has P L C 1980. The cultivation of *Spirulina platensis* on fermented swine manure. In: *Proc Int Symp Biogas Microalgae and Livestock Waste,* Taipei Taiwan Republic of China.

Chojnacka K, Chojnacki A and Gorecka H 2005. Biosorption of Cr^{3+}, Cd^{2+} and Cu^{2+} ions by blue-green algae *Spirulina* sp: kinetics, equilibrium and the mechanism of the process. *Chemosphere* 59(1): 75- 84.

Choubey A 2000. Isolation and Characterization of high salinity tolerant (Li^{+}–R) strain of *Spirulina platensis* for biotechnological exploitation. *Dissertation Thesis* MSc Biochemistry R D University Jabalpur (MP).

Chung P O, Pond W G, Kingsbury J M, Walker Jr EF and Krook L 1978. Production and nutritive value of *Arthrospira platensis* a spiral blue-green alga grown on swine wastes *J Anim Sci* 47: 319-330.

Clement G 1975. Production et constituants caracteristiques des algues *Spirulina platensis* et *maxima. Ann Nutr Anim* 29: 477-488.

Clement G 1978. Une nouvelle resource: L' Algue *Spirulina. Soc Phycol France Bull* 23: 57-61.

Ciferri O 1983. *Spirulina* the edible microorganism. *Microbial Rev* 47(4): 551.

Deomurari N 2001. Nitrate nitrogen regulation of growth photosynthesis macromolecular contents extracellular nitrite release and nitrate reductase of wild type *Spirulina platensis* and its high salt tolerant (Li^+–R and Na^+–R) mutant strain. *Dissertation Thesis* MSc Biochemistry R D University Jabalpur (MP).

Faucher O, Coupal B and Leduy A 1979. Utilization of seawater urea as a culture medium for *Spirulina maxima. Can J Microbiol* 25:752-759..

Fedkovic Y, Astre C, Pinguet F, Gerber M Y, Chou M and Pujol H 1993.Spiruline et cancer. In: Deumenge F., Durand-Chastel H, Tuolemont A Eds. Algue de vie Musee Oceanographique Bulletin de Institut Oceanographique Monaco. Numero special 12: 117-120.

Fong B, Gheung M and Lee M 2000. Effect of dietary *Spirulina* on plasma cholesterol and triglycerides in mice. In: *Abstracts* 4th *Asia-Pacific Conference on Algal Biotechnology* 150pp.

Fox R D 1988. Nutrient preparation and low cost basin construction for village production of *Spirulina* In: *Algal Biotechnology* (Eds. T Stadler J Mallion MC, Verdus Y Karamanos, H Morvan and D Christiaen) Elsevier Amsterdam 355-364 pp.

Fukino H, Takagi Y and Yamane Y 1990. Effect of *Spirulina* (*S. platensis*) on the renal toxicity induced by inorganic mercury and cisplatin Eisei Kagaku 36: 5.

Gabbisu C and Hall D O 1993. Removal of phosphate by foam-immobilized *Phormidium laminosum* in batch and continuous flow bioreactor *J Chem Tech Biotechnol* 57: 181-189.

Garett M K and Fallowfield H J 1981. Algal biomass from farm waste: A pilot plant study In: *Energy from Biomass* (Ed. W Palz, P Chartier and D O Hall). Applied Science Publishers London 691-696 pp.

Geresh S 1998. Modifications of polysaccharide of algal origin. A presentation in International Meeting on "Polysaccharide Biotechnology": An interdisciplinary Approach in Ben-Gurion University of the Negev Beer- Sheva Israel.

Gerwick W, Reyes Sand Alvarado B 1987. Two malyngamides from the Caribbean cyanobacterium *Lyngbya majuscula Phytochem* 26: 1701-1704.

Goodwin T W 1974. *Sterols in Algal Physiology and Biochemistry* (Ed. W D P Stewart) Oxford: Blackwell 266-280 pp.

Gong R, Ding Y, Liu H, Chen Q and Liu Z 2005. Lead biosorption and desorption by intact and pretreated *Spirulina maxima* biomass *Chemosphere* 58(1): 125-30.

Grinstead G S, Tokach M D, Dritz S S, Goodband R D, Nelssen J L 2000. Effects of *Spirulina platensis* on growth performance of weanling pigs. *Animal Feed Science and Technology* 83: 3-4: 237-247.

Haro V M A and Araujo A D R 1997. Use of live and inert microalgae as feed for *Artemia franciscana* (Anostraca: Artemiidae) *Revista de Biologia Tropical* 45(2): 803-811.

Hasting J W 1968. Bioluminescence *Annual Review Biochemistry* 37: 597.

Hayashi K, Hayashi T and Morita N 1993. An extract from *Spirulina platensis* is a selective inhibitor of Herpes simplex virus type penetration into Hela cells. *Phytother Res* 7: 76-80.

Hayashi K, Hayashi T and Kojima I 1996. A natural sulfated polysaccharide, calcium spirulan, isolated from *Spirulina platensis: In vitro* and ex *vivo* evaluation of anti-herpes simplex virus and anti HIV activities *AIDS Res Hum Retroviruses* 12: 1463-1471.

Hayashi T and Hayashi K 1996. Calcium Spirulan an inhibitor of enveloped virus replication from a blue-green alga *Spirulina Journal of Natural Products* 59: 83-87.

Henrikson R 1994. Microalga *Spirulina* superalimento del futuro Ronore Enterprises 2a ed. Ediciones Urano Barcelona Espana 222pp.

Hill C 1980. The secrets of *Spirulina:* Medical discoveries of Japanese Doctors C A Boulder Creek University of the Trees Press in Co-operation with the Journal of Nutritional Microbiology.

Hirahashi T, Matsumoto M, Hazeki K, Saeki Y, Ui M and Seya T 2002. Activation of human innate immune system by *Spirulina* augmentation of interferon production and NK cytotoxicity by oral administration of hot water of *Spirulina platensis Int Immunopharmacol* 2:423-34.

Hosoyamada Y, Takai T and Kato T 1991. Effects of water soluble and insoluble fractions of *Spirulina* on Serum lipid components and glucose tolerance in rats *J Jpn Soc Nutr Food Sci* 44:273-277.

Hu Qiang, Zhengyu HU Cohen, ZVI and Richmond A 1997. Enhancement of eicosapentaenoic acid (EPA) and g-Iinolenic acid (GLA) production by manipulating algal density of outdoor cultures of *Monodus subterraneus* (Eustigmatophyta) and *Spirulina platensis* (Cyanophyta) *Eur J Phycol.,* 32: 81-86.

Kaempfel U, Berghausen K H, Lieflander M 1995. Are cyanobacteria suitable for feeding human beings? A contribution on the essential amino acids of cyanobacteria. *Deutsche Lebensmittel Rundschau* 91(2): 50-52.

Kok B 1953. Experiments on photosynthesis by *Chlorella* in flashing light In: Burlew JS (ed.) Algal culture from laboratory to pilot plant. Carnegie Institution of Washington Washington DC 63-75 pp.

Kosaric N, Nguyen H T, and Bergougnou M A 1979. Growth of *Spirulina maxima* on cow-manure wastes. *Biotechnol Bioengg* 21: 2169-2173.

Langdon C and Onal E 1999. Replacement of living microalgae with spray dried diets for the marine mussel *Mytilus galloprovincialis Aquaculture* 180(3-4): 283-294.

Laws E A, Satoru T, Hirata J and Pang L 1987. Optimization of microalgae production in shallow outdoor flume. *Biotechnol Bioeng* 32: 140-147.

Lee J B, Srisomporn P, Hayashi K, Tanaka T, Sankawa U and Hayashi T 2001. Effects of structural modification of Calcium *Spirulana* sulfated polysaccharide from *Spirulina platensis* on anti viral activity. *Chem Pharm Bull* 49(1): 108-110.

Lyamin M Y, Solov'-eV A A, Kiseleva S V, Zaytsev S I and Charnova N I 1997. Renewable food energy sources in aquaculture. Mekhanizatsiya-i-Elektrifikatsiya-Sel'skogo-Khozyaistva 6: 25-28.

Mathew B, Sankaranarayanan R, Nair P, Varghese C, Somanathan T, Amma P, Amma N and Nair M 1995. Evaluation of chemoprevention of oral cancer with *Spirulina fusiformis Nutr Cancer* 24:197-202.

Matsuno T, Nagata S, Iwhashim Koike and T, Okada M 1979. Intensification of colour of fancy red corp with Zeaxanthin and myxoxanthophyll major carotenoid constituents of *Spirulina* Nippon Suisan Gukkaishi 45(5): 627.

McGarry M G, 1971. Unicellular protein production using domestic wastewater. *J Agric Sci* 4: 213-223.

Mgbenka B O and Orji R 1997. Use of fresh palm fruit extract as a feed ingredient in the diet of larval African catfish *Clarias gariepinus. Journal of Applied Aquaculture* 7(4): 79-91.

Michell S A and Richmond A 1988. Optimization of growth medium for *Spirulina* based on cattle waste *Biological Waste* 25: 41-50.

Mishima T, Murata J, Toyoshima M, Fujii K, Nakajima M, Hayashi T, Kato T and Saiki I 1988. Inhibition of tumor invasion and metastasis by Calcium Spirulan (Ca-SP) a novel sulfated polysaccharide derived from a blue green algae *Spirulina platensis Clin Exp Metastasis* 16: 541-550.

Nandeesha M C, Gangadhar B, Varghese T J and Keshavanath P 1998. Effect of feeding *Spirulina platensis* on the growth proximate composition and organoleptic quality of common carp *Cyprinus caprio* L *Aquaculture Research* 29(5): 305-312.

Nandeesha M C, Gangadhar B, Manissery J K and Venkataraman L V 2001. Growth performance of two Indian major carps catla (*Catla catla*) and rohu (*Labeo rohita*) fed diets containing different levels of *Spirulina platensis Bioresou Technol* 80:117-20.

Noriaki Nakaya M D, Yasuhiko Homma M D and Yuichira Goto M D 1988. Cholesterol lowering effect of *Spirulina. Nutr Rep Int*. 37: 1329.

Olvera Novoa MA, Dominguez cen L J, Olivera Castillo L and Martinez Palacios C A 1998 Effect of the use of the microalga *Spirulina maxima* as fish meal replacement in diets for *tilapia Oreochromis mossambicus* (Peters) *fry. Aquaculture Research* 29(10): 709-715.

Oron G, Shelef G and Levi A 1974. Growth of *Spirulina maxima* algae in effluents from secondary waste water treatment plants. *Biotechnol Bioengn* 16: 881-896.

Oswald W J 1981. Algae as solar energy converters In: *Energy from Biomass* (Ed. W Palz, P Chartier and D O Hall). Applied Science Publisher London 633-46 pp.

Pascaud M 1993. The essential polyunsaturated fatty acids of *Spirulina* and our immune response. In Doumengouge F, Durand Chastel H, Toulemont A Eds. *Spiruline* algue de vie Musee Oceanographique Bulletin de I' Institut Oceanographique Monaco Numero special 12:49-57.

Pal D and Amla DV 1992. Commercial fertilizer N : P : K an alternative nitrogen source for cultivation of *Spirulina platensis* in sewage. *Cryptogamic Algol* 13(4): 247-255.

Patel R, 2003. Mass cultivation of *Spirulina platensis* using low cost technology for utilization as a animal feed *PhD Thesis* RD University Jabalpur MP India121pp.

Patterson G M L 1996. Biotechnological applications of cyanobacteria *Journal of Scientific and Industrial Research* 55: 669-684.

Pavel K and Tomas R 2000. Bioremediation of heavy metal pollution exploiting constituents metabolites and metabolic pathways of livings A Review CCC 65(8) : 1205-1247.

Pelczar M J, Reid R D and Chan E C S 1977. *Microbiology* Tata McGraw Hill Publishing Company New Delhi.

Qishen P, Baojiang G and Kolman A 1989. Radioprotective effects of extract from *Spirulina platensis* in mouse bone marrow cells studied by using the micronucleus tests *Toxicol Letters* 48:165-169.

Rafiqul I M, Jalal K C A and Alam M J 2005. Environmental factors for optimization of *Spirulina* biomass in laboratory culture *Biotechnology* 4: 119 -22.

Ramamoorthy A and Premakumari S 1996. Effect of supplementation of *Spirulina* on hypercholesterolemic patients *J Food Sci Technol* 33: 124-128.

Rao D L N 1979. Studies on mineral nutrition,mass cultivation and nutritional values of the blue-green algae *Spirulina platensis Geitler PhD Thesis* IARI New Delhi 47pp.

Ravikiran D 2001 Optimization of mass cultivation of *Spirulina platensis* in glass tubular photobioreactors. Dissertation thesis *MSc Microbiology* R D University Jabalpur India..

Richmond A 1984. Microalgae of economic potential. In: Richmond A ed. *Handbook of Microalgal Mass Culture* CRC Press Inc Boca Raton USA 199- 243pp.

Richmond A 1988. *Spirulina* in microalgal biotechnology. MA Borowitzka and LJ Borowitzka (ed) Cambridge University Press Cambridge 85 pp.

Richmond A, Torzillo G, Sacchi A and Materassi R 1991. Effect of temperature on yield and night biomass loss in *Spirulina platensis* grown outdoors in tubular photobioreactors. *Applied Phycology* 3: 103-109.

Richmond A 1992. Mass culture of Cyanobacteria. In Mann N, Carr N Eds. *Photosynthetic Prokaryotes* 2nd ed. Plenum Press New York and London. 181-210 pp.

Robinson R K and Toerien D F 1982. The algae: A future source of protein. In: *Development in Food Proteins* ed. BJF Hudson Applied Science Publishers London and New York 289-335 pp.

Saeki Y, Matsumoto M, Hayashi A, Azuma I, Toyoshima K and Seya, T 2000. Innate immunotherapy for cancer, anti-tumor, effects of BCG-CWS in implanted tumor model summary of a paper presented at the 59th Annual Meeting of Japanese Cancer Association Oct 4-6.

Saxena P N, Ahmad M R, Shyan R, and Amla D V 1983. Cultivation of *Spirulina* in sewage for poultry feed. *Experientia* 39: 1077-1083.

Schwartz J and Shklar G1987. Regression of experimental hamster cancer by beta-carotene and algae extracts. J Oral Maxillofac Surg 45: 510-515.

Schwartz J, Shklar G, Reid S and Trickier D 1988. Prevention of experimental oral cancer by extracts of *Spirulina Dunaliella* algae *Nutr Cancer* 11: 127-134.

Schwartz J, Flynn E and Shklar G 1990. The effect of catrotenoids on the antitumor immune response *in vivo* and *in vitro* with hamster and mouse immune effectors. In: Bendich A, Chandra R, Gerrard K, Cerami A, Takaku F Eds. *Micronutrients and Immune Functions-Cytokines and Metabolism*. New York Academy of Sciences, 92-109 pp.

Seshadri C V and Seshagiri S 1986. *Spirulina* the wonder gift of nature: A compendium. MCRC Publications Madras India.

Shastri D, Kumar M M M and Kumar A 1991. Modulation of lead toxicity by *Spirulina fusiformis Phytother Res* 13: 258-260.

Soeder C J 1978. Economic consideration concerning the autotrophic production of micro algae at the technical scale. *Arch Hydrobiol Beih Ergeb Limnol* 11: 259-273.

Soong P 1980. Production and development of *Chlorella* and *Spirulina* in Taiwan. Algae Biomass G Shelef and C J Soeder ed. 97-113 pp Elsevier/North Holland Biomedical Press Amsterdam.

Shrivastava A 1996. Production of Single cell protein (SCP) through mass cultivation of *Spirulina platensis* in industrial effluent *MSc (Microbiology) Dissertation thesis* R D University Jabalpur India.

Singh S, Thakur I P S and Shrivastava A 1998. Cultivation of high protein rich alga *Spirulina platensis* in gelatin factory effluent of Jabalpur *XIX All India Botanical Conference* 26-28 Dec. Gurukul Kangri University Haridwar India.

Singh S, Thakur I P S and Shrivastava A 1997. Growth adaptation and low cost cultivation of high protein rich alga *Spirulina platensis* in gelatin factory effluent for animal feed. *Symposium on Algal Biology and Biotechnology* 6-8 Feb CAS in Botany Banaras Hindu University Varanasi India.

Singh S, Patel R, Datta P and Choubey A 2000. Isolation and characterization of high salinity tolerant (Li^+R) strain of *Spirulina platensis* for biotechnological exploitation 41st Annual

Conference of Association of Microbiologist of India Birla Institute of Scientific Research Jaipur India Nov 25-27.

Singh S, Ganguly A and Singh S K 2003. Impact of heavy metals (Cu and Cd) on the growth and survival of high protein rich alga *Spirulina platensis*. *Vislesana Research Journal* RD University Jabalpur 8(1): 63-70.

Singh S, Patel Rand Ganguly A 2004. Cultivation of *Spirulina platensis*,.the richest source of protein in different culture media for evaluation. International Symposium on Microbial Diversity: Challenges Opportunities and Relevance in New Millenium Department of Biological Sciences R D University Jabalpur India Nov 19th–21st.

Singh S, Ganguly A and Baisya S 2005. Antibacterial properties of Polysaccharide from *Spirulina platensis*. 46th annual conference of Association of Microbiologists of India Department of Microbiology Osmania University Hyderabad (AP) India Dec 8-10.

Takai Y, Hossayamada Y and Kato T 1991. Effect of water soluble and water insoluble fractions of *Spirulina* over serum lipids and glucose resistance of rats *J Jpa Soc Nutr Foods Sci* 44: 273-277.

Thadani P 2002. Optimisation of polysaccharide prodution of *Spirulina platensis* under different light regime. *MSc (Biochemistry) Dissertation Thesis* R D University Jabalpur India.

Sutherland I W 1994. Structure function relationships in microbiol exopolysaccharides *Biotech Adv* 12:393-448.

Sutherland I W 1996. Extracellular polysaccharides. In Rehm H J, Reed G (Eds) *Biotechnology* 6 VCH Weinheim 615-657.

Sutherland I W 1998. Novel and established applications of microbiol polysaccharides *Tibtech* 16: 41- 46.

Tel-or E, Boussiba S and Richmond A E 1980. Products and chemical from *Spirulina platensis*. In: *Algae Biomass* G Shelef and C J Soeder ed Elsevier North-Holland Biomedical Press Amsterdam 611-618 pp.

Terry K 1985. Photosynthesis in modulated light: quantitative dependence of photosynthetic enhancement on flashing rate

Biotechnol Biengg 28: 988–995.

Tomaselli L 1997. Morphology, ultrastructure and taxonomy of *Arthrospira* (*Spirulina*) *maxima* and *Arthrospira* (*Spirulina*) *platensis*. Vonshak A (ed) *Spirulina platensis* (*Arthrospira*): Physiology cell-biology and biotechnology. Taylor and Francis London 1-16 pp.

Torzillo G, Pushparaj B, Bocci F, Balloni W, Materassi Rand Florenzano G 1986. Production of *Spirulina* biomass in closed photobioreactors Biomass 11: 61-74.

Thakur I P S 2000. Study on immobilization induced responses in diazotrophic cyanobacterium *Nostoc muscorum* for biotechnological applications. *Ph D Thesis Microbiology* RD University Jabalpur India.

Upasani C D, Khera A and Balaraman R 2001. Effect of lead with vitamin EC or *Spirulina* on malondialdehyde conjugated dienes and hydroperoxides in rats *Indian J Exp Biol* 39: 70- 74.

Venkataraman L V, Nigam B P 1978. Mass culturing of fresh water algae for utilization as a protein source *Phykos* 18(1&2): 83-95.

Venkataraman L V 1983. *Spirulina:* A monograph. Published by Central Food Technology Research Institute Mysore India.

Venkataraman L V and Becker E W 1988. Production of food, feed, biochemicals and bioenergy from micro algae. Proc International Biotechnology Symposium Paris 910-922 pp.

Venkataraman L V 1989. *Spirulina* state of art and emerging prospects *Phykos* 28: 231-250.

Venkataraman L V, Bhagyalakshmi N and Ravishanker G A 1995. Commercial production of micro and macro algae-problems and potentials *Indian J Microbiol* 35: 1-19.

Verma S K 2002. Effect of increasing photon flux density on the biomass production of wild type, U^{2+}–R and Na^{+}–R mutant strain of *Spirulina platensis* grown in glass tubular photobioreactor under laboratory conditions *MSc (Microbiology) Dissertation Thesis* R D University Jabalpur India.

Vonshak A and Richmond A 1978. *Spirulina* culture in Israel *Arch Hydrobiol* 11 :274-280.

Vonshak A 1987. Biological limitation in developing the biotechnology for algal mass cultivation Science De L'eau 6: 99-103.

Wei F H, Chen H F and Luo XL 1991. Study on raising oncomelania using freshwater micro-algae. *Chinese Journal of Schistosomiasis Control* 11 (5): 286- 287.

Yamane Y, Fukino H, Icho T, Kato T and Shimamatsu H 1998. Effect of *Spirulina* (*Spirulina platensis*) on the renal toxicity induced by inoganic mercury and para -amino phenol. Summary of abstracts 108th conference of the pharmaceutical society of Japan 58pp.

Zarrouk C 1996. Contribution a 1 etude dune cyanophycee influence de divers facteurs physiques et chimiques sur la photosynthese de Spirulina maxima (Setch et.Gardner) Getlier (PhD thesis).

Zhang Ha, Un A P, Sun Y and Deng, Y M 2001. Chemo- and radio-protective effects of polysaccharide of *Spirulina platensis* on haemopoietic system of mice and dogs. *Acta Pharmacol Sin* 22: 1121-1124.

Authors Index

Subject Index

Subject Index

The Editors

Dr. Rajan Kumar Gupta (b. 1963) Obtained his M.Sc. and Ph.D. degree from Banaras Hindu University and worked on Ecophysiology of Antarctic Cyanobacteria for his Ph.D. degree. Since past fifteen years he has been working on Antarctic cyanobacteria.

Dr. Gupta has a great desire to explore the unexplored region of the earth and unearthing the heroes of the soil (microbes) which has led him to the land of superlatives and was deputed by Govt. of India for his participation as Biological Scientist in Antarctica twice as a result of which he participated in XIth and XIVth Indian Scientific Expeditions to Antarctica during 1991-92 and 1994-95. He has visited several countries like Mauritius, Japan, Nepal Thailand and South Africa for presentation of his work in the field of cyanobacteria. Dr. Gupta has worked on various aspects of cyanobacteria *i.e.* morphology, ecology and nitrogen fixation and published about 30 technical papers in various National and overseas Journals and more then 10 articles in various books. Four students are working under his supervision for their D.Phil degree of HNB Garhwal University. He has worked on Use of Cyanobacteria as Biofertilizer in Antarctica as well as in Foothills of Garhwal Himalaya and is the member of number of organizations in India and abroad. He is the Fellow of the Society for Environment and Ecoplanning and International Botanical Society and Chaired various sessions in the conferences in India and abroad. Presently Dr. Gupta is working as Reader in Department of Botany, Govt. P.G. College, Rishikesh (Dehradun), Uttaranchal.

Dr. Mukesh Kumar (b. 1963) obtained his M. Sc. and M. Phil. degrees in Botany and M. Ed. in Education from Meerut University, Meerut- 250 005 (now C.C.S University, Meerut). He took his doctorate degree on

Polyhouse Technology from H. N. B. Garhwal University, Srinagar- 246 174, Pauri (Uttaranchal). Presently he is working as a Reader at the Department of Botany, Sahu Jain (P.G.) College, Najibabad–246 763, Bijnor (U.P.).

Dr. Kumar is well known for his researches in the fields of Polyhouse Technology and Cyanobacteria. He is a member of several National and International Research Organizations. He has successfully accomplished a couple of Research Projects sanctioned in the fields of his specialization by the UGC, New Delhi. He has published more than 17 technical papers to various journals and chapters in reputed books. Two research scholars have already been awarded their Ph.D degrees and four more are currently working for it under his supervision.

Professor G.S. Paliwal (b. 1938), presently a Senior Consultant at the Regional Centre, National Afforestation and Eco-development Board, located at Agricultural Finance Corporation Ltd., NRO, New Delhi – 110 058, is well-known for his contributions to several facets of Development Botany, Electron Microscopy, Tree Biology and Systematic Botany. He has authored/edited 7 books, five of which have won him national awards and over 160 research papers, besides supervising over 100 doctoral theses. Prior to his present assignment, he served at the Department of Botany, University of Delhi, Delhi – 110 007, as a Lecturer (1963–1977), a Reader (1977–1978), Professor and Head, Department of Botany, H.N.B. Garhwal University, Srinagar – 246 174, Pauri (Uttaranchal; 1978–1998), and in between also as Dean, Faculty of Science. During the span 2000–2002, he was awarded the UGC, New Delhi-sponsored Emeritus Professorship, at the University of Delhi.

Professor Paliwal has had extensive training under the internationally known plant scientists like Prof. P Maheshwari, Prof. B.M. Johri and Prof. S.C. Maheshwari (in India) and at the laboratories of C.R. Metcalfe (Royal Botanical Garden, Kew (England), Prof. H.D. Behnke (University of Heidelberg, Germany), and Cornell University, Ithaca (USA) and has also traveled extensively to England, France, Holland, Switzerland, Canada, Czechoslovakia, Poland, Hungary and Japan.

Presently he is compiling two books for the benefit of the common man as well as the elite citizens on: (A) The Tree Wealth of India and (B) The Tree Wealth of Delhi, a subject which is closest to his heart. He is also carrying out identical studies on the trees of Chandigarh, Agra, Lucknow, Kanpur, Bhopal, and Jaipur.